Signals and Communication Technology

This series is devoted to fundamentals and applications of modern methods of signal processing and cutting-edge communication technologies. The main topics are information and signal theory, acoustical signal processing, image processing and multimedia systems, mobile and wireless communications, and computer and communication networks. Volumes in the series address researchers in academia and industrial R&D departments. The series is application-oriented. The level of presentation of each individual volume, however, depends on the subject and can range from practical to scientific.

Indexing: All books in "Signals and Communication Technology" are indexed by Scopus and zbMATH

For general information about this book series, comments or suggestions, please contact Mary James at mary.james@springer.com or Ramesh Nath Premnath at ramesh.premnath@springer.com.

Sundaravadivazhagan Balasubaramanian
Gnanasankaran Natarajan • Pethuru Raj Chelliah
Editors

Intelligent Robots and Drones for Precision Agriculture

Springer

Editors
Sundaravadivazhagan Balasubaramanian
Department of Information Technology
University of Technology and Applied Sciences-AL Mussanah
Al Mussanah, Oman

Gnanasankaran Natarajan
Department of Computer Science
Thiagarajar College
Madurai, Tamil Nadu, India

Pethuru Raj Chelliah
Edge AI Division, Reliance Jio Platforms Ltd.
Bangalore, Karnataka, India

ISSN 1860-4862 ISSN 1860-4870 (electronic)
Signals and Communication Technology
ISBN 978-3-031-51197-4 ISBN 978-3-031-51195-0 (eBook)
https://doi.org/10.1007/978-3-031-51195-0

This Springer imprint is published by the registered company Springer Nature Switzerland AG
The registered company address is: Gewerbestrasse 11, 6330 Cham, Switzerland

Paper in this product is recyclable.

Preface

The agriculture domain is seeing several innovations, disruptions and transformations with the faster maturity and stability of digital technologies. There are two important phenomena here:

1. Digitization
2. Digitalization

The digitization process is being enabled through the growing power of the Internet of Things (IoT) and edge technologies such as diminutive sensors, stickers, actuators, microcontrollers, single board computers (SBCs), beacons, LED systems, specks, RFID tags, bar codes, micro- and nanoelectronics, etc. These technologies are for transitioning all kinds of physical, mechanical, electrical and electronic systems into digital systems. Digitized systems can find, connect and communicate in the vicinity and with remote ones through a kind of networking. Digital systems through interactions and collaborations generate a massive quantity of multi-structured data. With 5G communication networks being pervasively deployed and managed, digital data being generated in large amounts gets transmitted to centralized data stores (databases, data warehouses, data lakes, etc.), data analytics platforms, etc.

Then comes the digitalization process. The core idea here is how to make sense of digital data collected from different and distributed sources. With the accumulation of IoT sensors and devices, the amount of poly-structured data becomes huge. Therefore, the complex and time-consuming process of transitioning data into information and to knowledge is being supported through a host of path-breaking digitalization technologies such as artificial intelligence (AI)-driven data analytics methods.

AI is the prime factor for producing actionable insights out of data heaps, including business workloads, IT services and devices (equipment, machinery, appliances, instruments, robots, drones, consumer electronics, wares, connected vehicles, etc.). In this book on precision and smart agriculture, we have incorporated well-written chapters explaining the growing contributions of intelligent robots and drones. There are chapters illustrating how cognitive robots empower smart farming. Further on, there are several trendsetting and breakthrough technologies such as 5G, edge computing, cloud-native computing, cybersecurity, blockchain, digital twins and machine and deep learning algorithms for transforming agriculture processes to significantly enhance agriculture yields. This book is to demystify the enabling technologies and tools towards smart agriculture.

Al Mussanah, Oman — Sundaravadivazhagan Balasubaramanian
Madurai, Tamil Nadu, India — Gnanasankaran Natarajan
Bangalore, Karnataka, India — Pethuru Raj Chelliah

Contents

Intelligent Computing with Drones and Robotics for Precision Agriculture

Vijayakumari Kaliannan and Fatema Khalifa Said Al Saidi

1 Introduction

The acronym IoT stands for the Internet of Things. It will connect devices, machineries, and gears to the Internet with the help of technologies like wireless sensor networks, etc. As of now, above 9 billion things are linked to the Internet. In the future, it will cross over 20 billion things. We will also say this as an M2M (machine-to-machine) connectivity. It can be extended to households, smart farming, smart city, and smart agriculture. RFID, sensor-connected devices, and smart networks are examples of IoT enablers.

In accordance with Gartner's research, the Internet of Things (IoT) refers to a system of physical objects equipped with embedded technology to facilitate communication, as well as to sense and interact with either their internal attributes or the surrounding environment [1]. IoT technology is a trending technology of today's world. It brings a new revolution in each field of the common man's lifetime by building the whole thing smart and intellectual. By using IoT in smart farming, farmers can easily find live data about soil moisture and temperature to increase the crop yield and product quality. Smart farming is one of the high-tech, capital-intensive methods for creating clean and maintainable food for the people.

The yield is observed with the support of sensors measuring light, moisture, and temperature. These sensors are integrated into systems, such as IoT-based smart farming. Planters can monitor field conditions from any position. Associating with traditional farming, IoT-based smart farming is extremely excellent and efficient [2].

In terms of ecofriendly issues, IoT-based smart farming can produce huge benefits and efficient water usage. By controlling the environmental conditions, we can yield more crops efficiently. Everyone has a little yard and farmland, but we don't

V. Kaliannan (✉) · F. K. S. Al Saidi
Trinity College for Women, Affiliated to Periyar University, Namakkal, Tamil Nadu, India

S. Balasubaramanian et al. (eds.), *Intelligent Robots and Drones for Precision Agriculture*, Signals and Communication Technology,
https://doi.org/10.1007/978-3-031-51195-0_1

have time and energy to maintain or manage the land, so the authors have introduced the concept of IoT with the smart farming system to easily absorb the soil moisture, water management, yield of crops, and so on [3].

2 Internet of Things

The Internet of Things (IoT) is completed through two arguments: the Internet and things. The word "things" in the IoT refers to several IoT plans with distinctive characters and capacity to perform isolated sensing, actuation, and live observation of a definite data arrangement. IoT gadgets can also interchange facts in real time with other associated nodes and applications, either directly or indirectly, or to combine facts from several nodes, to process them, and to transmit them to multiple attendants [4]. It consists of the following components:

- Input/output interface for sensors.
- Interface for connecting to the Internet.
- Memory and storage interface.
- Audio/video interface.

Wearable devices, smart watches, observing IoT smart homes, IoT smart transportation systems, and IoT smart devices for healthcare are examples of IoT gadgets [5].

2.1 IoT Architecture

Figure 1 indicates three-layered architecture of fundamental structure of IoT.

Application layer – learning content distributed.
Network layer – used to transfer data using online network.
Perception layer – collecting the request information from the network.

2.2 Smart Farming

Smart farm (SF) refers to an integration of information technology (IT) with farm equipment and sensors for crop growing and food production. IoT is being employed in a variety of industries, including smart homes, smart cities, smart healthcare, and smart agriculture. Farmers employ IoT technology to improve agricultural efficiency in areas like irrigation, fertilizer, harvesting information, and climate forecasting by monitoring with sensors and to make better decisions. They will be able to increase production yield and the efficiency of farming [6].

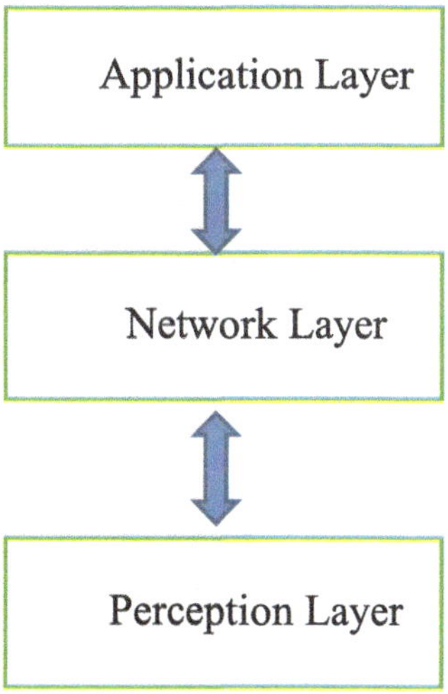

Fig. 1 Internet of Things architecture. (Source: Own image)

"Smart farming" is a different thought that refers to farms to operate the Internet of Things (IoT), robotics, drones, and artificial intelligence (AI) to enhance the product measure and value while reducing the amount of mandatory human labor for production [7].

Nowadays, the major problem in smart farming is shortage of labor; also, some of the laborers do not know how to do work efficiently in a short time. The major reasons for labor scarcity contain higher wages in nearby cities. To tackle these issues, modern agriculture is looking to incorporate robotics and sensors with software technologies [8].

IoT aims at supporting farmers in connecting the supply-demand gap by confirming great yielding, productivity, and eco-friendly preservation. Precision agriculture is a technique of retaining [9].

Figure 2 indicates the characteristic of smart farming using IoT technologies.

3 Sensors

In linked agriculture, IoT applications are used in many types of probes to receive data in actual time. Agriculturalists and agro-managers rely on connected apparatus to inspect earth circumstances and monitor yield and strength of livestock and activate drones and agricultural vehicles, which are typically in remote places with limited broadband connectivity.

Advancements in sensor technology, including the incorporation of low-power, miniaturized, and disposable tracking solutions along with the advent of 5G, empower farmers to explore more opportunities for using sensing technology in various scenarios. This technology will assist them individually or in combination in responding to changing eco-friendly environments, enrolment, controlling, and demand conditions [10].

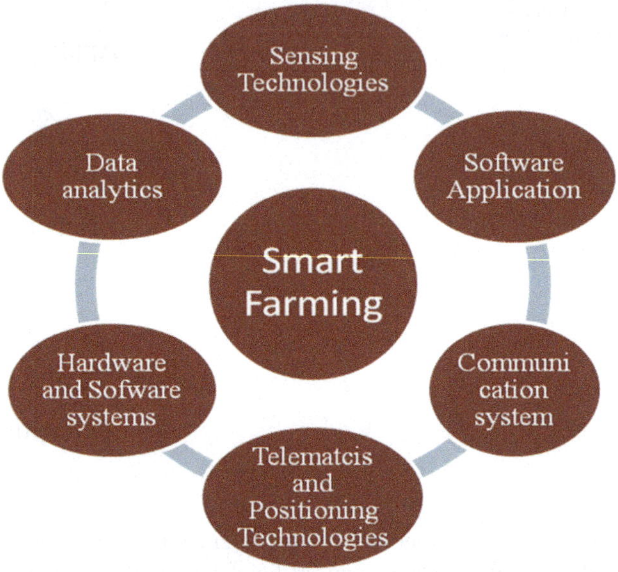

Fig. 2 Smart farming. (Source: Own image)

Gateways are computer hardware devices that allow data to pass from one network to another. They're important in the smart farm environment, since they give faithful access to sensor networks, cameras, and actuators. When it arises for livestock observation, irrigation controls, and perimeter investigation, poor data routing might have negative consequences [11].

Advancements in IoT edge handling have allowed apps to run implanted in devices rather than transfer raw data across gateways. Device administration and decision-making can take place at or near the end points where IoT edge processing generates data. Certain LTE-M and narrowband IoT (NB-IoT) cellular modules allow procedures to connect directly to the mobile IoT network, with inbuilt handling and loading capabilities to support agricultural applications that demand flexibility right inside the module [12].

3.1 Modules

Adding cellular abilities or Wi-Fi networking to devices and other field equipment is best done with modules. Device creators and integrators in the agro-industry must trust the complete use of modules above a "chip-down" approach satisfactory for other functional blocks of the strategy due to radiofrequency engineering difficulties and strict governing acquiescence requirements. For farm IoT solution integrators, this method to connectivity explanations provides many choices and the greatest time-to-market ROI (return on investment) to satisfy the sector's needs [13].

3.2 *Precision Farming*

Gathering data in real time, powerful analytics, and interconnected devices to fine-tune responsiveness to crop and livestock variability. IoT agriculture helps growers to understand the processes of their business better, from soil to animal circumstances, and in observing water control [14], improve your procedures, and put your data to work for long-term growth and results.

3.3 *Smart Irrigation*

The integration of sensor-based water monitoring and management, fostering connectivity, leads to the reduction of water wastage, promotes crop health, and enhances agricultural yields. Farmers can utilize sensor-generated data to optimize irrigation practices, satisfying demand while also preserving natural resources. Over time, these sustainable approaches contribute to increased profitability while simultaneously conserving water [15].

What Are Agricultural Sensors, and How Do They Work?

Agricultural sensors are devices used in a smart agriculture that compromise data that helps farmers to observe and enhance crops by regulating changes in the environment.

It will be controlled by using the application on mobile.

3.4 *Benefits of Agricultural Sensors*

Agricultural sensors have the following advantages:

- They are produced to fulfill rising food demand by enhancing harvests while consuming the fewest assets possible, such as rainwater, enrichers, and germs. This is accomplished through supply preservation and field mapping.
- They are easy to operate.
- They are less cost.
- They can be employed for contamination and overall condition of temperature in addition to farming purposes.

Drawbacks of Agricultural Sensors

Agricultural sensors have the following problems:

- Internet access is required for smart farming and IoT technology. This isn't available in underdeveloped countries like India.
- There is a widespread belief in the industry that users are always not willing to accept the most up-to-date IoT devices with agricultural sensors.
- Essential infrastructure such as smart grids, traffic systems, and cellular towers are not universally available. This obstructs the spread of its use much more.

4 Types of Smart Agricultural Sensors

Figure 3 indicates the drones used as a smart sensor in agriculture for smart farming.

In agriculture, smart sensors give data that benefits agriculturalists to monitor and enhance their yields also keeping up with altering environmental and ecosystem elements.

There are several different types of sensors that are utilized in farming for smart agriculture.

4.1 Optical Sensors

Figure 4 indicates optical sensors in agriculture.

Fig. 3 Smart agricultural sensors. (Source: Retrieved from website https://www.tractorjunction.com/blog/types-of-smart-sensors-in-agriculture-for-farming-in-india/)

Fig. 4 Optical sensors. (Source: Retrieved from website https://www.tractorjunction.com/blog/types-of-smart-sensors-in-agriculture-for-farming-in-india/)

These devices utilize light to assess various characteristics of dust particles across different frequencies. They are typically installed on vehicles or drones, enabling the collection and analysis of data related to soil reflectance and plant color. Visual sensors are capable of identifying mud, organic materials, and soil moisture content.

4.2 Electrochemical Sensors for Soil Nutrient Detection

Figure 5 indicates electrochemical sensors for soil nutrient detection for smart farming.

It facilitates the gathering of soil chemical information. Electrochemical sensors, specifically designed for detecting nutrient levels in soil, serve as material sensors. Soil samples are typically sent to a laboratory for comprehensive analysis. For specialized tests, such as pH measurement, ion-selective electrodes are employed to detect the movement of specific ions, such as nitrate, potassium, or hydrogen.

4.3 Mechanical Soil Sensors for Agriculture

Figure 6 indicates mechanical soil sensors for agriculture.

These instruments work by cutting down the earth and recording the power using heaviness scales or weight cells. When a device penetrates the earth, it records the land forces resulting from the soil's critical movement. The proportion of the strength essential to enter the soil moderate to the fore area of the tool promised with the soil is quantified in a unit of pressure called soil mechanical resistance [16].

Fig. 5 Electrochemical sensors. (Source: Retrieved from website https://www.tractorjunction.com/blog/types-of-smart-sensors-in-agriculture-for-farming-in-india/)

Fig. 6 Mechanical soil sensors. (Source: Retrieved from website https://www.tractorjunction.com/blog/types-of-smart-sensors-in-agriculture-for-farming-in-india/)

4.4 Dielectric Soil Moisture Sensors

Figure 7 indicates dielectric soil moisture sensors in smart agriculture.

It assesses soil moisture content. Humidity sensors, in conjunction with rain gauges, are strategically placed across the farm. This arrangement allows for continuous monitoring of soil moisture levels, particularly during periods of reduced vegetation cover.

4.5 Location Sensors in Agriculture

Figure 8 indicates dielectric location sensors in agriculture.

The variety, space, and tallness of several points within the mandatory area are resolved by these sensors. They depend on GPS satellites to accomplish this.

4.6 Electronic Sensors

Figure 9 indicates the electronic sensors in smart farming.

It's a device that's mounted on tractors and other field equipment to monitor how well they're working. The data was then sent to devices or mail to people via cellular and cable communication networks.

4.7 Airflow Sensors

Figure 10 indicates the airflow sensors in smart farming.

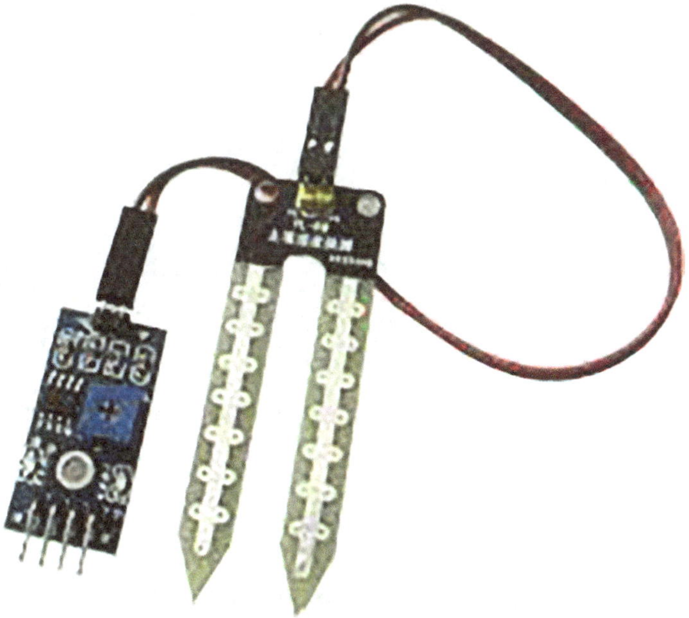

Fig. 7 Dielectric soil moisture sensors. (Source: Retrieved from website https://www.tractorjunction.com/blog/types-of-smart-sensors-in-agriculture-for-farming-in-india/)

Fig. 8 Dielectric location sensors in agriculture. (Source: Retrieved from website https://www.tractorjunction.com/blog/types-of-smart-sensors-in-agriculture-for-farming-in-india/)

Fig. 9 Electronic sensors in agriculture. (Source: Retrieved from website https://www.tractor-junction.com/blog/types-of-smart-sensors-in-agriculture-for-farming-in-india/)

Fig. 10 Airflow sensors in agriculture. (Source: Retrieved from website https://www.tractorjunction.com/blog/types-of-smart-sensors-in-agriculture-for-farming-in-india/)

Its quantities can be taken at certain areas at traveling. The necessary power is determined by the density needed to penetrate a specified depth of soil with a given volume of air. Various soil attributes, including compaction, structure, soil composition, and moisture content, contribute to unique characteristics.

4.8 Agricultural Sensors

Figure 11 indicates several types of agricultural probes that are used in smart farming.

Fig. 11 Agricultural sensors in agriculture. (Source: Retrieved from website https://www.tractor-junction.com/blog/types-of-smart-sensors-in-agriculture-for-farming-in-india/)

Systematically, this probe calculates and monitors the air temperature, soil humidity at various depths, rainfall, leaf humidity, chlorophyll levels, wind speed, droplet temperature, wind direction, relative moisture, solar energy, and atmospheric pressure.

There is a long list of IoT sensors that are utilized in agriculture:

(a) Weather monitoring.
(b) Automation of greenhouse.
(c) Management of crop.
(d) Livestock management and supervision.
(e) Intelligent precision for agriculture using sensors.
(f) Agricultural drones.

5 Small-Scale Farmers

Various sensors in agriculture can assist small-scale agriculturalists with a variety of instruments and gadgets. Some farmers are unable to purchase all of the agricultural sensors required. Farmers simply have to pay a minimal cost, which is the purchase of a smartphone. A smartphone can save them money because it has many apps that work with devices in your field [17].

5.1 Will Farmers Benefit from the Usage of Sensors in Agriculture in the Future?

Agricultural robotics and technology are widely used by farmers due to the lack of workers and the demand for feeding nutrition to an ever-increasing worldwide population.

Machine education envisions robots and sensors to perceive and learn from their environment through smart sensors. They have been using them for more than a year.

Using online cloud facilities and a control panel, a new novel technology permits farmers to remotely observe insect groups in their area and make fast action to preserve their harvests.

5.2 How Do Agricultural System Start with Smart Sensors?

Sensors have a plethora of applications in agriculture. Smart gadgets help farmers to improve your plantation's routine, production, and takings in a variety of ways. Using and operating agricultural sensors, on the other hand, are not for everyone.

There are special considerations regarding agricultural sensors that you should be aware of before using them.

You must select the sensors for your work to make an explanation of agriculture. The accuracy and validity of the data gathered are serious to the achievement of your manufactured goods, and it will be strong-minded by the excellence of your sensors [18].

5.3 The Brain of Your Device

Every smart sensor in agriculture solutions should focus on data analytics. To acquire meaningful visions created on the achieved data, you must have a persuasive capacity of data analytics and put on suitable algorithms.

5.4 Maintenance of Your Device

If the sensors are used at a specific location, they become damaged. Preserving your system is a task that is especially important for sensors in farming. With this instance, you must ensure that your gadget is long-lasting and simple to maintain.

5.5 Reliable Infrastructure for Your Device

In the need of a solid internal infrastructure, make sure that your smart farming software runs well. The internal systems must be entirely secure to prevent failure due to security issues.

5.6 Connectivity of Device

The necessity to transfer data from a large number of agricultural devices and infrastructure is a significant barrier to the smart farming implementation.

We trust that this information on sensors in agriculture is adequate and reliable. This information will undoubtedly assist and guide you in your search for a clever and dependable farm sensor [19].

6 Applications

6.1 Machine Navigation

- Tractors and huge plugging systems can be run robotically from the consolation of home using GPS, just like the children's toy vehicles.
- Those incorporated automated gadgets are surprisingly accurate and self-regulate while distinctive terrains are detected, making labor-in-depth operations less complicated [9].
- Smartphones can comfortably find their movements into their work progress.

Figure 12 indicates how machine navigation is done in smart farming.

6.2 Harvesting Robotics

Selecting plants with agriculture robots alleviates the challenge of worker shortages. These robots can manage the challenges of collecting fruits and vegetables for a day, a week, and a month. These technologies appoint a mixture of photograph and robotic hands to perceive which berries to pluck, thereby regulating the quality. Apples have a primary effort on agricultural robot harvesting in high-functioning expenses.

Figure 13 indicates the harvesting robotics in smart farming.

Fig. 12 Machine navigation. (Source: Retrieved from website https://finance.yahoo.com/news/agricultural-technology-companies-finding-themselves-134026705.html)

Fig. 13 Harvesting robot. (Source: Retrieved from website https://www.ese.upenn.edu/centers-institutes/)

6.3 Remote Sensing

Figure 14 shows how drones monitor the farm or field activities.

6.4 Computer Imaging

Figure 15 indicates the computer imaging sensor in smart farming.

Computer imaging serves as a method for sensing visual information. It is placed across the field of a farmer to reproduce photos of the farm that is digital in nature.

Crop monitoring, weather conditions, and soil quality are some of the applications in smart farming.

Fig. 14 Remote sensing. (Source: Retrieved from website https://www.upl-td.com/pdf/Annual_Report_2020-21.pdf)

Fig. 15 Computer imaging. (Source: Retrieved from website https://www.bloomingbackyard.com/alkaline-soil-plants/)

7 Challenges

This segment discourses some of the main obstacles that must be overcome to establish the Internet of Things. To be widely accepted by the IoT community, answers for these concerns must come from industrial, collective, authorized, economic, and corporate backgrounds.

7.1 Standards and Interoperability

Standards are critical in establishing new technology marketplaces. Interoperability will be more challenging if various industrialists do not utilize similar values, necessitating the deployment of additional gateways to translate from one standard to another. Furthermore, a corporation that controls many aspects of a vertical market may be able to dominate that industry, suffocating competition and erecting obstacles to entry for smaller players and entrepreneurs. Different data standards can also tend to lock customers into a product family [20].

7.2 Security

As the Internet of Things connects more devices, malware gains decentralized access points. Devices that are less expensive and located in physically dangerous areas are vulnerable to interfering.

With policy-driven approaches to security and provisioning, expect to see a variety of strategies and vendors tackling these concerns.

8 Conclusion

IoT technology is being tested by researchers all around the nation to boost farm production in a form that complements the service that is already in. We discussed agricultural network architecture, platform, and topology to help farmers gain access to the IoT backbone and increase crop productivity. This chapter also contains a detailed analysis of present and upcoming technologies in IoT agricultural applications, communication protocols, and other unique technology. This chapter detailed several skills used in the field of IoT farming agriculture. Numerous important characteristics of IoT-based farming agriculture, such as technologies, industries, and national legislation, have been delivered to support diverse stakeholders. The management has activated to maintain IoT in agriculture, and it is estimated that IoT in agriculture revolutionizes traditional farming techniques.

References

1. A.A. Laghari, K. Wu, R.A. Laghari, A. Mureed Ali, Review and State of Art of Internet of Things (IoT). https://link.springer.com/article/10.1007/s11831-021-09622-6
2. Y. Hajjaji, W. Boulila, I. Romdhani, A. Hussain, Big Data and IoT based applications in smart environments: A systematic review. Comput. Sci. Rev. **39** (2021) Elsevier, https://www.sciencedirect.com/science/article/abs/pii/S1574013720304184?via%3Dihub
3. G. Idoje, T. Dagiuklas, M. Iqbal, Survey for smart farming technologies: Challenges and issues. Comput. Electr. Eng. **92**, 107104 (2021) ISSN:0045-7906. https://www.sciencedirect.com/science/article/abs/pii/S0045790621001117?via%3Dihub
4. T. Kassanuk, M. Mustafa, P. Panse, R. Sivanand, K. Phasinam, T. Santosh, An internet of things and cloud based smart irrigation system. Ann. Roman. Soc. Cell Biol., 20010–20016 (2021) https://www.annalsofrscb.ro/index.php/journal/article/view/8847
5. T. Lee, S. Mckeever, J. Courtney, Flying free: A research overview of deep learning in drone navigation autonomy. Drones **5**(2), 52 (2021) https://www.mdpi.com/2504-446X/5/2/52
6. S.L. Ullo, G.R. Sinha, Remote sensing 2072–4292 "Advances in IoT and smart sensors for remote sensing and agriculture applications". (2021). https://doi.org/10.3390/rs13132585. https://www.mdpi.com/2072-4292/13/13/2585
7. M.R.M., M.K. Saiteja, G.J., S.N., N.K.G.N., IOT based crop monitoring system for smart farming, In: *2021 6th International Conference on Communication and Electronics Systems (ICCES)*, (2021), pp. 562–568, https://www.ijraset.com/best-journal/domains-of-iot-devices-in-the-field-of-agriculture
8. H.H. Pajouh, A. Dehghantanha, R.M. Parizi, M. Aledhari, H. Karimipour, A survey on internet of things security: Requirements, challenges, and solutions. Internet Things **14**, 100129., ISSN:2542-6605 (2021). https://doi.org/10.1016/j.iot.2019.100129
9. N.G. Rezk, E.E.-D. Hemdan, A.-F. Attia, A. El-Sayed, M.A. El-Rashidy, An efficient IoT based smart farming system using machine learning algorithms. Multimed. Tools Appl. **80**, 773–797 (2021). https://doi.org/10.1007/s11042-020-09740-6
10. S.V. Gaikwad, A.D. Vibhute, K.V. Kale, S.C. Mehrotra, An innovative IoT based system for precision farming. Comput. Electron. Agric. **187**, 106291., ISSN:0168-1699 (2021). https://doi.org/10.1016/j.compag.2021.106291
11. J. Wang, M.K. Lim, C. Wang, M.-L. Tseng, The evolution of the internet of things (IoT) over the past 20 years. Comput. Ind. Eng. **155** (2021) Elsevier, https://www.sciencedirect.com/science/article/abs/pii/S0360835221000784
12. D. Chaudhary, S. Sharma, S. Khar, R.K. Srivastava, Internet of Things (IoT) and Big Data: Future of farming. Agric. Eng. Today **47** (2023) ISSN:0970-2962
13. P. Geetika, The Internet of Things: IoT applications and security challenges. Innov. IT **2**(1) (2015) ISSN:2395-1192
14. A.M.S. Poornima, Internet of things in agriculture: A review. Agric. Rev. **39**(4) (2018) ISSN:0253-1496
15. M.R. Kumari, M.B. Narayan, P.P. Prava, Application and security in Internet of Things (IOTs). Int. J. Technol. **9**(1) (2019) ISSN:2231-3907
16. K. Raquib, M. Hannan, Internet of Things (IoT) and it's needs. Al-falah Sch. Eng. Technol. **12**(1) (2020) ISSN:0975 9638
17. S.N. Kumar, G. Saileswar, IoT (Internet of Things): Prospects and challenges in India. Invertis J. Manag. **12**(1) (2020) ISSN:0975 6310
18. S. Shakuli, Agribot: An Internet of Things based farmbot. Asian J. Res. Soc. Sci. Hum. **11**(10) (2021) ISSN:2249-7315
19. S. Rishi, Overview of IOT (Internet of things). Academicia Int. Multidiscipl. Res. J. **11**(10) (2021) ISSN:2249-7137
20. O.M. Kumar, Application of Internet of Things in agriculture. Asian J. Multidimens. Res. **10**(10) (2021) ISSN:2278-4853

Smart Farming and Precision Agriculture and Its Need in Today's World

Sreya John and P. J. Arul Leena Rose

1 Introduction

Farming and agriculture have always been an alternative to man to meet his needs and expenses. In many cases, it has been the only source of livelihood. This trend began thousands of years ago and has evolved in such a way that it even helped in maintaining a balance in the economic affairs of many nations. Today, the agricultural sector serves as the backbone of many countries. It plays a major role in the social, cultural, and economic aspects of a nation [1]. The whole food industry of the world is based on the agricultural sector, and it provides employment to many people. Along with these advantages, there are also a few disadvantages of the expanding agricultural sector which cannot be neglected [2]. To meet the needs of a huge population within a short span of time, the experts are expanding the agricultural sector in an unhealthy manner. To increase the yield and productivity, a high number of fertilizers and pesticides are used which are harmful to the ecosystem [3]. This affects the surrounding flora and fauna resulting in their extinction. The clearing of forests and grasslands to transform them into agricultural lands are also affecting the climatic conditions of the concerned region. Another demerit is the overexploitation of natural resources such as land and soil [4]. As the farmland increases in size, a large amount of fresh water is utilized for irrigation purposes. This also affects the amount of groundwater resources and causes scarcity if continued for a longer time. As a result, studies and research were done to tackle these issues and find an alternative to conserve our environment without affecting the quality and quantity of crop production.

S. John · P. J. Arul Leena Rose (✉)
Department of Computer Science, College of Science and Humanities, SRM Institute of Science and Technology, Kattankulathur, Chennai, India

S. Balasubaramanian et al. (eds.), *Intelligent Robots and Drones for Precision Agriculture*, Signals and Communication Technology,
https://doi.org/10.1007/978-3-031-51195-0_2

Agriculture has undergone many stages of development based on the circumstantial necessities of man and has benefited in numerous ways. This advancement in the field of farming and agriculture was not a sudden breakthrough but the result of long-term research and experiments involving various technological aspects. Through the innovations at each stage, they took conventional farming that involved manual labor, time, and energy to another level where each of these was replaced by technology which in turn saved time and resources, improved yield production, and made all other jobs easier. This way, ordinary farming was upgraded using various smart ideas which resulted in smart farming. The introduction of technology in the farming sector was a breakthrough and provided solution to all the problems [5]. Within a short span of time, Artificial Intelligence-based systems overtook the traditional farming methods. These systems were controlled by robots and wireless networks which provided high-quality yield with less impact on the environment. Hence, farming techniques were modified by integrating the advanced technologies with the conventional methods to be known as smart farming. In other words, it is the process of incorporating modern technology-based software and systems such as drones, sensors, etc. to extract information and data to improve the quality and quantity of agricultural products. A similar revolution, precision agriculture made smart farming more realistic. While smart farming concentrated on the wider aspects of the agricultural sector, precision agriculture did wonders in the core area [6]. To be precise, it is all about measurements and observations of individual crops and acting based on the data collected by smart farming devices. Precision livestock farming is another area of precision agriculture that manages and monitors the welfare of farm animals using technology. Devices such as wireless networks, cameras, drones, computers, etc. are used for this purpose, and they monitor the various factors that affect the farm animals. Some of the main benefits of precision livestock farming are improved animal health and growth, high milk production, and early detection of diseases. Like smart farming, precision agriculture and farming are also widely adapted as it is a multidisciplinary approach that involves collaborations with veterinarians, scientists, engineers, climatologists, etc. These both are closely associated processes, and together they provide us with high-quality yield production.

2 The Evolution of Farming: From Conventional to Smart

The history of farming and cultivating crops for food production began centuries ago. It was a significant phase in the evolution of mankind. This type of farming is labor-intensive and requires more effort and dedication from the farmers. Various external factors such as climate, soil condition, temperature, etc. also influenced the traditional agriculture domain significantly. Any type of variations in these factors affected agriculture as a whole, and visible changes were seen in the yield production. This type of farming never guarantees a profitable income. As time went by, the human population began increasing in an uncontrollable manner which resulted in an inflated need for food, land, water, other natural resources, etc. People started clearing more and more land for cultivation as there came a drastic demand for food. This affected nature more severely. Deforestation for crop cultivation and housing resulted in soil erosion, landslides, and many other disasters. It is one of the major reasons behind the extinction of many varieties of plants and animals. Another reason is the increased use of fertilizers and pesticides which are harmful to farm-friendly microbes and animals. They are also responsible for the disintegration of the soil quality. These harmful chemical substances enter the water bodies through running water and deplete the water quality. This causes a threat to the living organisms in the oceans and other water bodies. As the farmlands are expanding day by day, the demand for freshwater resources is also increasing. The process of irrigation requires a lot of water, and the primary source is groundwater. This groundwater reliance has reached an extent where the study reports show overused and mismanaged groundwater resources in many parts of the world. The establishment of industries and urbanization is also a reason behind the deterioration of land and other natural resources.

Climatic changes and their causes on the environment are other aftereffects of traditional farming. Recent studies have shown that agriculture contributes largely to the increased presence of many harmful pollutants such as methane in the atmosphere. Methane is a greenhouse gas, and the excess amount of it can cause a greenhouse effect which results in damages such as ozone layer depletion. This increases the temperature which causes a rise in the sea level. Damage to the ozone layer will allow the harmful ultraviolet rays to reach the Earth's surface which when come in direct contact with the skin results in major skin diseases such as sunburns, cancer, etc. In this way, conventional farming was creating many problems for humankind and its surroundings [7]. As a result, man was forced to find an alternative to these challenges which gave way to the establishment of smart farming. This farming method utilizes technologies to improve crop production as well as the environment. Technology took the agriculture domain to a higher level where it not only made the often-tedious tasks easier but also provided many solutions to various complications of conventional farming. Smart farming and precision agriculture are two such technology-driven farming concepts that focus on managing and preparing the agricultural industry with frameworks to include advanced technologies such as big data, the Internet of Things, and machine learning for tracking, monitoring, analyzing, automating, and executing operations.

3 Intelligent Computing and AI: The Brain Behind Smart Farming

Intelligent computing is an integration of advanced computing technologies and various other fields such as Artificial Intelligence (AI), machine learning (ML), natural language processing (NLP), and deep learning (DL) to mimic human abilities such as learning, decision-making, reasoning, and problem-solving. These systems have the ability to analyze and process large volumes of data to make decisions and thus improve their performance over time based on the experiences and interactions. The introduction of these technologies to the agricultural sector brought many high-level transformations and advances to farming and cultivation [8].

Artificial Intelligence has been a trending topic in computer science for many years. It is a generally used term to describe a computer that uses intelligence to perform a particular task. In other words, it is the ability of a machine to learn from previous experiences to solve a particular problem. This form of learning from experiences and replicating human intelligence is accomplished through various algorithms which are defined specifically for this purpose [9]. As time passed by, many advances happened in this branch of science and resulted in the development of subsections known as machine learning (ML) and deep learning (DL).

Machine learning technologies are one of the widely used mechanizations in recent advances. It is a subset of Artificial Intelligence where the machines are trained instead of explicitly programmed, to behave in the same way as that of the human brain. Machine learning is a revolution as it doesn't have to program each application just like that of the traditional computing technologies. Instead, data which are commonly in the form of images, text files, etc. are collected and used in training the system. Such trained systems can produce good results when tested with data that is similar to that of the trained ones. Some examples of machine learning that we experience in the farming sector are the detection of diseases and pests on the crops [10]. The machine learning algorithms are categorized into three types of learning. They are supervised, unsupervised, and reinforcement learning. Regression and classification algorithms come under the category of supervised learning where at each stage the model is taught to perform the actions using labels and features acquired from the data [11]. There won't be any associated labels to the features that are collected from the data in the case of unsupervised learning. Here, the goal is achieved only based on the features and patterns [12]. Clustering and association rules are the common types of unsupervised learning. Based on the similarity among the patterns, the data will be categorized accordingly. However, decision-making is the main element in reinforcement learning [13]. This is nothing but the act of training the system to behave optimally in a particular environment so that it can provide an accurate result. Even if there are many challenges while adopting machine learning techniques, their usages and numerous benefits to the mankind make them powerful. This gave way to other sectors and industries such as health, farming, automobile, finances, etc. to rely on machine learning technologies for further advances.

Deep learning is another subfield of Artificial Intelligence that uses the neural networks to analyze and process data which is similar to the working of a human brain. It is said to be more advanced and powerful than the machine learning as it uses a large volume of dataset to perform the sophisticated tasks. In agriculture, DL is used for tasks like identifying diseases in crops by analyzing images of leaves. It can automatically detect visual signs of diseases, enabling timely interventions [14]. The unimaginable growth of all these industries in recent years is a clear evidence to depict the effectiveness of advanced technologies. The contributions of each sector have boosted us socially and economically [15].

Smart farming, driven by the integration of Artificial Intelligence (AI) technologies, is revolutionizing traditional agricultural practices. The cutting-edge potential of AI can be seen in a variety of applications, from crop health prediction to real-time data analysis for decision-making [16]. These developments have the ability to transform farming methods and significantly boost production. Real-time data analysis is one of AI's most important contributions to smart farming. Agricultural landscapes are surrounded by sensors and IoT devices, which create an uninterrupted stream of data. AI algorithms are excellent at analyzing these data, obtaining useful insights, and presenting them in a way that is easy to understand for informed decision-making [17]. Farmers can receive instantaneous information regarding pest activity, temperature changes, and soil moisture levels, enabling timely interventions that maximize resource use and crop yield. Additionally, the predictive power of AI offers a significant advancement in the way farmers monitor crop health [18]. By the use of highly sophisticated machine learning algorithms, AI has the ability to foresee the probable disease breakouts by analyzing the historical data along with the real-time data. This helps the farmers to take active measures to find remedies and protect their crops from the diseases at the earliest. Also, it promotes environmentally friendly practices by minimizing the need for excessive pesticide or fertilizer application. The power of AI goes beyond an individual farm and includes an interconnected agricultural environment. It can provide thorough knowledge which has the power to guide the entire supply chain by combining data from multiple sources such as satellites, market trends, weather forecasts, etc. These data-driven strategies help the farmers to make decisions on what and when to plant and harvest the crops which enhance efficiency, cut waste, and improve the profitability of the entire agricultural sector [19].

However, integration of AI in smart farming has its own challenges and drawbacks. Some of these are listed below:

- Accessibility and quality of data: These systems require high-quality data to perform accurately. But a major challenge faced while integrating AI in smart farming is the unavailability and inconsistency of the datasets. Therefore, ensuring the quality and availability of data has to be taken significantly.
- Privacy and security of data: Safeguarding famers' details and crop information is of utmost importance as these are sensitive and valuable data. It is important to provide data privacy and security to the farmers while using technology-induced farming methods.
- Initial investments and cost: Implementation of AI-based equipment is costly and requires expertise to do the installations. The small-scale farmers and the ones that are new to the field may find it difficult to meet such needs. Therefore, proper funding and guidance from the authorities have to be sanctioned and taken care of.
- These commonly faced challenges have to be addressed and rectified as the farmers' proficiency in technology can vary depending on various factors. Other than this, there are other challenges such as unpredictable environmental factors, ethical concerns, etc. which also can occur while adapting to AI-based smart farming.

4 Cultivating Insights: ML and DL in Precision Agriculture

Machine learning and deep learning techniques have demonstrated their prowess in analyzing complex agricultural data. This section will showcase how these algorithms are employed to predict crop yields, detect diseases, and enhance crop management strategies.

Technology has taken the agriculture domain to a higher level where it not only makes the often-tedious tasks easier but also provides many solutions to various complications of conventional farming. Smart farming and precision agriculture are two such technology-driven farming concepts that focus on managing and preparing the agricultural industry with frameworks to include advanced technologies such as deep learning and machine learning for tracking, monitoring, analyzing, automating, and executing operations [20]. There are numerous applications of machine learning and deep learning algorithms in the field of smart farming. Currently, technology-oriented devices are involved from selecting a particular land for farming to processing the cultivated crops into edible food items. In the previous sections, we have seen the key features and basic functionalities of machine learning and deep learning techniques. Now, let us see how these techniques are employed in the agricultural sector to predict crop yields, detect diseases, and enhance crop management strategies.

4.1 Maintaining Crop Health

Constant monitoring of crops to improve their health is a time-consuming process. There are various external and internal factors that influence the health of the crops. Ample amount of rainfall, temperature, sunlight, and proper soil conditions are some of the necessary aspects in plant growth [21]. Even though the plant receives all these requirements, it doesn't guarantee a successful crop production. Internal aspects such as damaged seeds, diseases, pest infestation, etc. can make them barren and unhealthy. The traditional way of tackling this issue is the manual observation of the plants. This process requires a lot of effort and time. And the person who does this should be an expert in all these fields, especially on varieties of diseases that affect plants. Performing these tasks manually will be difficult, and farmers with less knowledge will find it impossible to identify the diseases [22]. In this case, precision agriculture can be used as a solution. Various automated systems are developed under precision agriculture to identify the diseases affecting agricultural crops at an earlier stage. There are also systems based on big data and other technologies to monitor external factors such as weather changes and rainfall. The data from such systems can be analyzed to find out the optimum conditions that are necessary for healthy crops. They are mostly in the form of Excel files and images and collected using wireless sensors, drones, and cameras. This data will be stored for further use by the farmers to produce high yields.

Automated systems that analyze soil conditions are also used widely. Adding fertilizers to an already fertile piece of land can degrade the existing quality of soil in that region. This happens when the chemicals present in the fertilizers affect the elements that make the land fertile. In this scenario, it is very necessary to evaluate the soil conditions [23]. An idea of the amount of minerals and nutrients present in the soil can tell us how much more we must provide so that the plant will be free from nutrient deficiencies. Supplying large quantities of nutrients can degrade the quality of both soils and crops. The excess amount of minerals will act as toxins to the plants which can affect their growth. All these issues can be answered using automated systems. Using sensors and other technology-based devices, they collect the required data and will be used in the prediction systems. Another set of automated systems is used in disease and pest detection in crops. This is another significant area that should be given great care so as to provide a high-quality yield. The diseases affecting a plant could spread to other crops on the farm within a short span of time. This is the same for pest infestation also. Proper identification and remedies should be done at the earliest. Technology-driven devices and software packages are better options than the conventional ways of tackling these issues. A wide range of images and videos of the diseases and pests affected in the same farmland and other farmland with the same crops will be collected using drones, software systems, and other wireless sensor networks. Based on these datasets, an automated system will be developed, and the developers train them using the collected data [24]. There are various types of algorithms available for this purpose. Even now, studies and research are done in this area to produce more accurate results. After training, the system will be tested with a new set of data and if it shows good results will be used in the farms for the early detection and identification of diseases and pests [25]. These techniques can also be used to provide good care and protection to the crops in a less period. The given image (Fig. 1) shows the basic block diagram of a crop disease identification model.

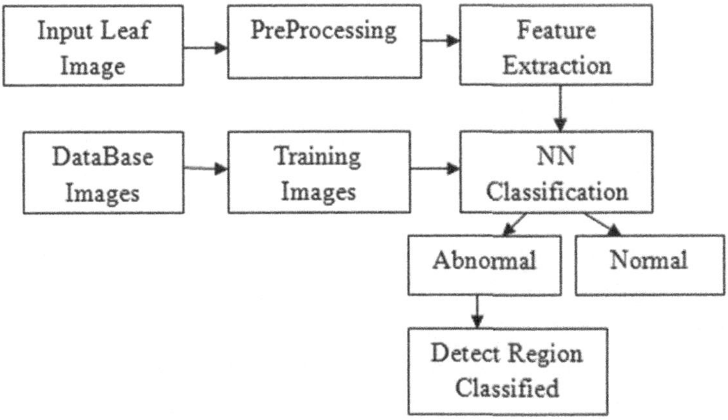

Fig. 1 Basic block diagram of a crop disease identification model. (Source: https://panteche-learning.com)

4.2 Production of High-Quality Yield

The production of good quality yield is a challenging factor in the case of small-scale and large-scale farming [26]. One major reason behind this is the dependence of yield production on various fields such as soil, climatic conditions, water resources, temperature, etc. Hence, a healthy combination of all these factors results in a good quality yield. Conventional farming techniques cannot be relayed in this scenario. With their limited facilities only, a certain feature can be predicted and is not of much use. Involving technology-based devices such as wireless sensors, drones, AI-based robots, and software can give more accurate results [27]. These advanced systems help us to track the crop conditions at various levels so that we can do the necessary activities accordingly. Wireless sensor networks are mainly used to monitor the environmental conditions that are necessary for agricultural crop cultivation. These sensor networks are designed in a manner that has a base system that controls all the activities and connects it to the Internet to share the acquired data. Drones, on the other hand, do multiple activities such as fertilizer distribution, data collection, water supply, etc., in large-scale farming. Using drones, we could see the very minute details of a crop even if it is a large farmland. This is helpful in the case of crop disease detection and classification [28]. Robots are other effective tools in smart farming. They can be used to reduce human labor and a lot of time [29]. They can be programmed and trained according to their needs. Software packages are developed based on the collected data to predict the yield and future results. Figure 2 is a center-pivot irrigation system which is very famous for its efficiency and the ability to irrigate uneven terrain uniformly.

Fig. 2 Center-pivot irrigation system. (Source: https://www.infosys.com)

4.3 Land Suitability, Yield Prediction, and Classification

Land and its surroundings are an important aspect of farming and crop cultivation. An efficient system with the ability to gather accurate information about agricultural land will be a useful tool for the farmers. Today, such prediction systems with advanced technologies are built using machine learning algorithms. In the olden days, to predict whether the respective land is suitable for cultivation, farmers didn't have many sources. Only after years of observation and experience they were able to make partially correct decisions. Artificial Intelligence and other technologies have brought a drastic change to the farming sector. Machine learning models built to predict the land suitability use real-time data such as soil composition, availability of water resources, temperature, humidity, other climatic factors such as rainfall, and the requirements of the vegetative crops that are planned to cultivate. There are various machine learning algorithms developed to predict such systems with high accuracy [30]. The advantages of these land suitability prediction models based on machine learning algorithms make them important in today's world. They foresee the quantity of yield produced and according to the amount a farmer can add more of it for a higher production without causing any damage to the environment. There is a high demand for such models, and prediction algorithms such as support vector machine, neural networks, K-means clustering, etc. are used to build them [31].

4.4 Crop Management and Harvesting

This is another area of the agricultural sector where machine learning algorithms are applied frequently. Crop management is an important aspect as it has a significant role in the economy of a nation. Therefore, it is mandatory to make sure that they are receiving proper monitoring at regular intervals of time. But it is always difficult to manage the crops in a large farmland. And manually doing this is even more tedious. As a solution to this, many computerized automated models are developed to manage the crops without any human intervention. They use machine learning algorithms and other technologies to build such systems [32]. Devices such as drones, unmanned aerial vehicles (UAVs), wireless sensor networks, robots, etc. are used to collect various types of data to monitor and manage the agricultural crops [33]. The data collected can be preserved for future uses and helps in analyzing the requirements of each crop during its growth. This reduces the overutilization of fertilizers and the exploitation of natural resources.

Many large-scale farming uses drones and helicopters for the distribution of fertilizers and water. Weeds that grow among the crops are also detected and eliminated using machine learning technologies. At each stage, the crops are provided with necessary requirements for their growth and development. Many systems are enhanced with special features which can detect and notify farmers of the presence of diseases and pests. There are also systems that indicate the harvesting or fruit ripening time so that the farmers can make the necessary preparations for it. Artificial Intelligence-based devices are used for harvesting which reduces the human labor and time. Machine-based harvesting is also a cost-effective process. This type of innovation using machine learning algorithms has made the farming sector more advanced and eco-friendlier at the same time. Even today, much research and studies are done in this field to make it more feasible and environmentally sound.

4.5 Seeds and Sapling Quality Prediction

Another challenging factor in crop production is the quality of the seeds used. Even if the farmers use land that is of high quality, there is no point if they sow a dormant or an unhealthy seed. So, it is very important to make sure that the seeds and sapling used are healthy and productive. It was an impossible task in the olden days. But today with the help of technologies such as machine learning, deep learning, etc., we can find the health of a seed within a short span of time. Predictive algorithms are used here, and the systems that give high accuracy could be used for real farming purposes. They improve the farming sector by providing a yield that is of high quality. Even the quantity of the crop production is also enhanced by using such models. Figure 3 is a general pictorial representation of the crop yield estimation in an automated model.

4.6 Reduce Ecological and Environmental Damages

Following a smart farming culture can benefit us in various ways. One main advantage is the elimination of environmental hazards that have occurred because of conventional farming. Agricultural pollution can be eliminated to a large extent through eco-friendly smart farming methods. Utilization of herbicides, an alternative to harmful pesticides, and less usage of fertilizers can reduce the emission of greenhouse gases. The traditional farming process consumes a lot of fresh waters, and this can affect the available water resources. Even if water comes under the category of renewable resources, we must realize the fact that improper usage of all these resources can make them disappear from this Earth itself; hence, it is important to follow sustainable farming culture. Providing water for irrigation based on crop needs and soil conditions can be effective to reduce water scarcity [34]. The use of wireless sensors and software helps us to monitor the plants to find the amount of water and other minerals required. This reduces the overutilization of water and fertilizers. And we will be able to supply an adequate amount of water and necessary fertilizers as per the analyses and records.

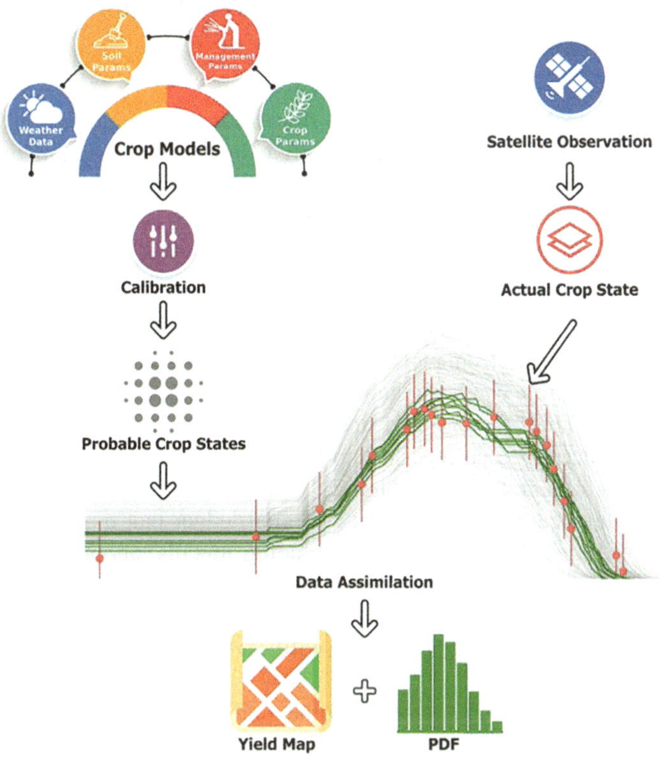

Fig. 3 Crop yield estimation model. (Source: https://www.geog.ucl.ac.uk)

4.7 For an Improved Decision-Making in Farming Sector

Smart farming and precision agriculture have already started paving our path to follow a systematic and planned farming. A farmer should improve his planning and decision-making skills as he becomes more market-oriented. One of the major challenges faced by the farming sector is the pressure on everyone to be updated and informed to make appropriate decisions. Using technologies such as GPS satellite, drones, sensors, etc., we can gather data from history to create a predictive model so that we can choose the suitable crops that can give higher yields. This form of decision-making is more accurate and provides better results. Gathering other details such as weather condition, soil nature, crop features, etc. can help us to provide them with their exact needs. As a result, computerized automated systems are developed widely for accurate prediction and improved decision-making in the agricultural sector [35].

5 Ensuring Transparency and Trust: Blockchain in Agriculture

In recent years, blockchain technology has drawn a lot of attention due to its potential to increase openness and trust in a variety of industries, including agriculture. The farming sector faces difficulties with regard to information asymmetry, traceability, and inefficient supply chains. Using blockchain in agriculture can solve these problems and have several advantages such as the following:

- Supply Chain Transparency: Blockchain can be used to provide a transparent and unchangeable record of every step in the agricultural supply chain. Every transaction and movement, from planting and harvesting through processing, packaging, and distribution, can be documented on the blockchain. This minimizes the possibility of fraud, forgery, and unauthorized alterations by ensuring that all parties have access to correct and up-to-date information about the products' whereabouts of manufacture and travel.
- Quality Assurance and Traceability: A digital ledger that maintains the whole lifespan of agricultural products, including details about the procedures employed, certificates gained, and any quality testing carried out, can be created using blockchain technology. Food safety may increase as a result of this transparency because any problems can be swiftly found and resolved.
- Accessibility and Consumer Trusts: Greater traceability and transparency can increase consumer trust in the items' authenticity and quality. This may open up new domestic and foreign markets where customers are willing to pay more for reliable items.
- Smart Contracts and Automation: Smart contracts are agreements that automatically carry out their obligations because they are written in code. Smart contracts can automate a number of operations in agriculture, including payment settlements, quality assurance, and compliance verification. For instance, if sensors or outside data show that a farmer's produce meets a given quality level, they can be automatically paid.
- Integration of technological expertise and a clear understanding of the specific challenges and opportunities within the agricultural sector can result in the successful implementation of blockchain technology which has the potential to revolutionize the agricultural industry by enhancing transparency, traceability, and trust throughout the supply chain.

6 Connecting Fields: 5G's Impact on Smart Farming

The advent of 5G technology promises unprecedented connectivity and low latency. This section will explore how 5G is poised to revolutionize smart farming by enabling real-time data collection, precision farming through remote-controlled machinery, and IoT integration.

The launch of 5G technology has the potential to significantly impact various industries, including agriculture. The possibilities provided by 5G can be very useful for smart farming, which uses cutting-edge technologies to optimize agricultural methods. Few impacts of 5G technology in smart farming are given below:

- Monitoring of real-time data: Gathering of real-time data and its transmission are made possible by 5G's high-speed and low-latency capabilities. The use of sensors, drones, and other IoT devices to monitor factors, like soil moisture, temperature, humidity, and crop health across their fields, provide precise information to farmers based on which they can take quick actions.
- Precision agriculture: By customizing activities to certain regions of a field, precision agriculture strives to maximize agricultural yields and resource efficiency. Precision remote control of machinery and equipment is made possible by 5G. For instance, real-time, minimally delayed operation of autonomous vehicles and drones enables precision planting, spraying, and harvesting.
- Remote operations and management: The utilization of cloud-based platforms and mobile applications enables farmers to remotely monitor and manage their operations. Farmers can access data, manage equipment, and make decisions from anywhere as a result of 5G's quick and dependable connectivity, which improves operational effectiveness.
- Livestock management: Wearable gadgets can be utilized in livestock husbandry to monitor animal behavior and health. Farmers can improve animal well-being and efficiency by real-time monitoring of vital signs and behavioral patterns to identify health issues early.
- Smart irrigation: Sensors powered by 5G can offer precise information on soil moisture levels, enabling irrigation with the right amount of water. This can minimize operational expenses, increase crop quality, and reduce water waste.
- Predictive analytics: 5G enables the real-time analysis of data from a variety of sources to produce insights and predictions. Machine learning algorithms can analyze sensor data, weather forecasts, and historical trends to produce recommendations for the best times to plant, how to prevent disease, and more.
- Supply chain management: Real-time connectivity provided by 5G can improve supply chain traceability and transparency. Every stage of the supply chain, from the farm to the consumer, may be watched, ensuring product quality, cutting waste, and fostering consumer trust.
- Agri-robotics: Robotic weeding, picking, and trimming are just a few of the advanced agri-robotics applications that 5G can support. Using 5G connectivity, these robots may be remotely controlled and steered, improving efficiency and lowering the need for manual labor.
- Virtual and augmented reality (VR/AR): They are immersive technologies that enhance our sensory experiences and interactions by blending the digital and physical worlds, in which 5G is well suited for due to its low latency. With augmented reality (AR) glasses, farmers can receive information about the field they are viewing in real time, such as crop health information or upkeep guidelines.

- Collaboration and knowledge-sharing: 5G makes it possible for farmers, professionals, and researchers to collaborate and share data easily. This helps people to share their finest ideas, perspectives, and solutions to agriculture-based problems.

Despite the significant advantages of 5G for smart farming, there are some challenges to take into account:

- Infrastructure: An extensive infrastructure and a proper network coverage are required for 5G installation which limits it to the urban areas due to the poor network in rural areas.
- Cost: For farmers, especially small-scale enterprises, installing 5G infrastructure and buying appropriate devices can be expensive.
- Data privacy and security: They are more important as more data is produced and sent, especially sensitive agriculture data.
- Spectrum and regulatory issues: Spectrum allocation and regulatory frameworks may have an impact on the availability and deployment of 5G networks.

Therefore, the low-latency connectivity and the high speed of 5G network have the potential to change smart farming enabling real-time data collection, remote monitoring, precision agriculture, and advanced analytics. The agricultural industry may experience a gain in productivity, resource efficiency, and sustainability as the technology develops and becomes more generally available.

7 Harvesting Insights: Big Data Analytics in Precision Agriculture

Big data analytics in precision agriculture is a significant advancement in modern farming practices which has the potential to transform raw agricultural data into actionable insights. By incorporating cutting-edge data analytics, precision agriculture, which adapts cultivation methods to specific circumstances, has undergone a revolution. This integration makes it easier to extract important insights from enormous and varied datasets, which results in improved agricultural processes. The acquisition and merging of data from multiple sources form the basis of precision agriculture. A huge amount of data is produced by sensors, drones, weather stations, machines, and satellite pictures. Big data analytics easily integrates these various data sources, giving farmers a thorough understanding of the state and dynamics of their farms. The basis for making informed decisions is this thorough knowledge.

The interpreting and processing abilities of big data analytics are essential in making sense of the gathered data. Huge amounts of data are searched through by sophisticated algorithms and models, which then reveal complex patterns, trends, and connections. It also reveals major relationships between crop yields and elements including soil type, irrigation intensity, and temperature variations [36]. These observations assist farmers aiming for increased output with beneficial guidance. The predictive ability of big data analytics in precision agriculture is among its most alluring features. The technology may predict future circumstances and results by examining historical and real-time data. The capacity to predict problems and opportunities gives farmers the flexibility to proactively modify their strategies and plans. Big data analytics also provides prescriptive advice in addition to predictions. These practical recommendations allow farmers to adjust planting timings, irrigation schedules, and fertilization methods. Precision agriculture is primarily driven by resource optimization, and big data analytics is essential to attaining this goal [37]. Waste is minimized, and environmental effect is decreased by adjusting the distribution of resources like water, fertilizers, and pesticides based on unique field conditions. The technology also assists in the early detection of pests, illnesses, and other hazards to crops. Monitoring many data sources enables quick responses that stop and lessen these problems. Another factor that sets apart precision agriculture driven by big data analytics is the customized decision-making. The technology permits the study of data that is unique to particular fields within a field. This level of specificity enables farmers to customize their choices to the distinctive qualities of each region, leading to improved efficiency and results [38]. The insights from big data analytics lead to cost savings in an economic perspective. The profit margins of farmers are boosted by greater resource use, higher yields, and reduced operational inefficiencies. Therefore, we can say that the integration of big data analytics in precision agriculture represents a fundamental change in the agricultural industry as it improves productivity, sustainability, and responsiveness which made precision agriculture a cornerstone of contemporary agricultural methods.

8 Orchestrating Farms: The IoT Revolution

The Internet of Things (IoT) has enabled smart devices and sensors to gather real-time data from agricultural fields. It is a network of interconnected machines, sensors, and systems that communicate and share information digitally. The IoT technology has transformed the conventional agricultural activities by facilitating real-time data gathering, analysis, and automation. As a result, many areas of farming are managed precisely and efficiently. The major areas enhanced by IoT technology are discussed below:

- Developments in irrigation using IoT: Irrigation has typically relied on predetermined schedules or manual monitoring, which can result in inefficient water use and unequal distribution. IoT uses sensors embedded in the ground to monitor soil moisture levels and send information to a central server. The precise amount of water required in various fields is then calculated using this data. Automated irrigation systems can be remotely managed and changed based on current information. This methodical technique reduces wasteful use of water, encourages ideal crop growth, and even guards against overwatering, which can result in soil erosion.

- Pest control using IoT: Agricultural sector faces an ongoing difficulty in controlling diseases and pests. IoT-enabled technologies provide a preventative strategy for pest management. Temperature, humidity, and other environmental elements that affect insect behavior can be monitored in the field by sensors. Potential insect outbreaks can be predicted using this data along with information from weather forecasts and historical trends [39]. IoT technologies also enable farmers to apply interventions like targeted pesticide spraying precisely where and when needed by providing them with notifications and recommendations. This decreases the need for extensive chemical use, lowers expenses, and has a minimal negative influence on the environment.
- Livestock management using IoT: Livestock and dairy farming are an important area of agricultural sector. Milk, dairy products, eggs, meat, etc. are essential in our day-to-day life. They also provide us with other beneficial things such as manure and wool and even help in farming. Therefore, it is very important to monitor the health and welfare of these animals. To improve dairy farming, we must receive products that are of high quality and quantity without affecting or disturbing their health and life. IoT technologies provide solutions such as real-time monitoring and better animal care for livestock husbandry [40]. These systems use wireless sensor networks, cameras, software, robots, etc. to monitor, analyze, and collect the data. Using this data that are collected, the farmers could identify the animals with poor health. They can also use robots to feed them and guide them to grazing in the fields. Artificial Intelligence-based devices are used in large farms to milk the cows and to reduce labor to a large extent. Farm animals can be monitored at any time from any location using drones and cameras. This type of tracking of farm animals helps farmers from losing them. Even the behavioral abnormalities of the cattle due to any diseases or pests also can be easily notified if the farmers follow a visual tracking system. This type of automated model helps to preserve the data and records easily and provides an insight into the financial and operational aspects. Farmers also can get familiarized with each animal through such automated models even if there are a large number of animals. Wearable gadgets with sensors can monitor an animal's health, activity level, and even behavioral habits. Farmers are able to remotely check on the health and welfare of their cattle by enabling the transmission of this data to cloud-based platforms [41]. Alerts are generated if anomalies are found, allowing for prompt intervention. IoT technology also helps feed distribution be optimized, ensuring animals get the proper quantity of nourishment. This method encourages healthier animals, more output, and lower mortality rates. There are various other fields that use IoT to flourish in today's farming practices. All these examples show how seamlessly integrating technology into farming practices can enhance the efficiency, sustainability, and productivity of the agricultural sector.

9 Digital Twins, Mobile Applications, and Cloud Computing for Smart Farming

A digital twin is a virtual representation of a physical object or system, such as a farm, a field, or even a single plant. In order to build a dynamic and interactive virtual counterpart to the actual agricultural environment, this method makes use of the power of cutting-edge technologies, such as sensors, Internet of Things (IoT) devices, and data analytics. Smart farming uses digital twins as crucial instruments for accurate decision-making and process optimization. Farmers can view data in real time gathered by sensors positioned throughout their fields and livestock. The digital twin is then fed with this information to produce a virtual depiction of the farm's conditions. The ability of digital twin approach for predictive modeling and analysis is one of its significant benefits. Farmers can learn more about how various factors, such as weather patterns, soil moisture levels, and planting methods may interact and affect crop growth by simulating these factors. Farmers may proactively modify their practices to enhance yields and resource efficiency, thanks to this foresight. The benefits of digital twin idea also extend to sustainability and risk management. Within the virtual environment, farmers can model the effects of pest outbreaks, droughts, or extreme weather conditions. As a result, they can create mitigation techniques and backup plans that they can quickly use when similar circumstances arise in real life.

The introduction of mobile applications revolutionized the agricultural sector drastically as it gives farmers immediate access to vital information, empowering them to make knowledgeable decisions and manage their businesses more successfully. These smartphone apps provide a variety of benefits across several facets of agricultural management in the context of smart farming. Through sensors and IoT devices, farmers can remotely monitor their fields, crops, and livestock to stay informed on important elements like soil moisture, temperature, and animal behavior. Rapid responses to developing situations and prospective problems are made possible by this rapid connectivity [42]. Additionally, these apps are essential for precision agriculture since they let farmers create precise field maps, specify planting zones, and optimize machinery routes. Such capabilities enable precision seeding, variable-rate fertilizer and pesticide delivery, and tailored irrigation, all of which boost agricultural yields and resource utilization. Mobile apps also speed up data gathering and analysis by enabling farmers to directly input data on planting plans, yields, and pest observations into their devices. Farmers are given the tools they need to make data-driven decisions so they may achieve greater results. In summary, mobility in agriculture is transforming how farmers run their businesses, promoting improved effectiveness, sustainability, and success in the context of contemporary farming.

The adoption of cloud computing technologies in agriculture has ushered in a new era of productivity and connection, revolutionizing conventional farming methods and paving the way for farming in the clouds. Delivery of computing services through the Internet, including storage, processing power, and data analytics, is referred to as cloud computing. This technology has several advantages for agriculture and has completely changed how farmers operate and make decisions. The capacity to store and access enormous volumes of data from anywhere at any time is one of the main benefits of cloud computing in agriculture. Cloud-based platforms allow farmers to upload and preserve information about soil characteristics, crop yields, weather patterns, and other topics. A comprehensive digital record of all agricultural activities is kept in this central location, enabling data-driven decision-making. Furthermore, having access to this information enables farmers to work with professionals, researchers, and other interested parties, exchanging knowledge and insights that help them make well-informed decisions. Precision agriculture also greatly benefits from the use of cloud computing. Farmers can produce precise field maps and remotely monitor conditions by combining data from numerous sources, including sensors, drones, and satellite photos. Cloud computing also improves agricultural risk management. By running models and simulations on cloud platforms, farmers can simulate various scenarios and make predictions about possible outcomes. For example, they can determine how different weather conditions affect agricultural growth or forecast the effects of shifting market pricing. Farmers may create risk mitigation plans and make well-informed decisions that can protect their revenue, thanks to this proactive approach.

The successful convergence of digital twins, mobile applications, and cloud computing has ushered in a new era of extraordinary creativity and efficiency in the quickly changing environment of modern agriculture. A wave of change in smart farming has been sparked by the dynamic interaction of digital twins, which provide virtual representations of the real world; mobile applications, which give farmers access to real-time insights and management tools; and cloud computing, which offers scalable data storage and processing capabilities. This effective trio enables farmers to adopt sustainable practices, allocate resources efficiently, increase productivity, and make educated decisions. A future where agriculture is not just smarter but also more resilient, productive, and tuned in to the changing requirements of an expanding globe is promised as these technologies continue to develop and converge.

10 Case Study in Leaf Disease Detection

The case study detailed here demonstrates the integration of technologies to provide the agricultural sector with more accurate results within a short span of time which can enhance the soil management and disease control strategies to another level.

Case study on disease detection: The applications of machine learning technologies in different fields are discussed in the previous sections. Advances related to vegetative crops and their high-quality production are a prominent area as it can benefit under any circumstances. One major problem that is influencing the production, quality, and quantity of the crops is the varieties of diseases and pests affecting the plants. This is damaging the food industry and agricultural sector as a whole.

The mutated variations of diseases are a real challenge to the farmers. Many automated systems are designed as a solution to tackle this issue. These models use machine learning technologies to detect and classify diseases accordingly. Each model is designed with many sophisticated algorithms at every step in order to generate an accurate result. Here is a system that is built using the latest machine learning algorithms to identify the diseases affecting the common vegetative crops. Figure 4 shows the general block diagram of an automated crop disease identification system.

These are the major steps in a disease detection model. The first and foremost activity is to gather the necessary data which must be fed to the system. In most cases, leaf images are used as data as they contain diseases, symptoms of diseases, and various other details. Pest and infections also affect other parts such as roots, stem, fruits, and flowers of the plants, and therefore, they can also be used as dataset. Once these images are collected, they must undergo preprocessing. It is done using various algorithms, resizing images without any data loss and with the help of many filtering techniques. This step is mainly taken to eliminate the presence of noise and other distortions that exist in the image. The unwanted portions of the image such as the background and other parts that don't contain the disease or symptoms also can be removed during this stage. This is an important step as further processes depend on the preprocessed image.

The preprocessed images are then segmented using various segmentation algorithms. These algorithms help us to segment the image into multiple partitions, and the formations of smaller sections are based on the similar characteristics of the pixel values. Watershed segmentation, canny edge detection, etc. are some of the conventional segmentation algorithms available. Recently, many hybrid algorithms are developed to obtain more accurate results. These segmentation algorithms help the system to extract the necessary features easily which results in proper feature extraction. They are extracted from the dataset fed into the system and are used to train the system to get familiarized with diseased and healthy leaf images. This

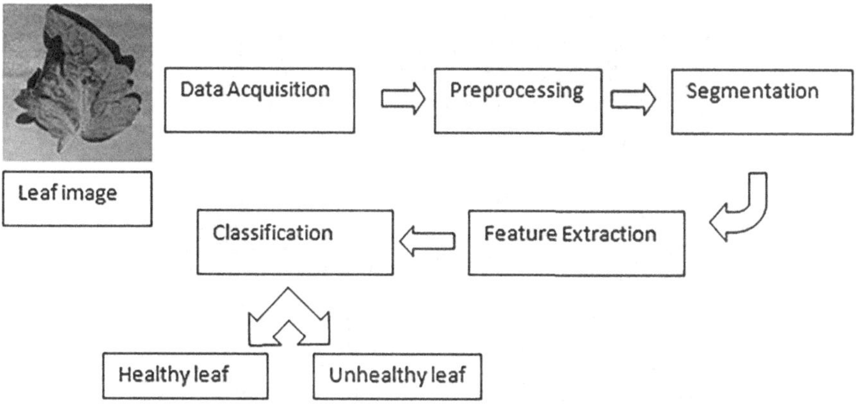

Fig. 4 The general block diagram of an automated crop disease identification system

helps the system to identify a diseased leaf image when tested using a similar image. There are algorithms to perfectly extract the features from a data, and this is a crucial step to develop an accurate model.

Here, we have taken two common rice leaf diseases such as brown spot disease and leaf smut which are great threats to rice production (Figs. 5 and 6, respectively). The images of diseased and healthy rice leaves were collected from here and preprocessed using the median filter to remove the noise. The unwanted background portions were also removed. After segmentation, the features that represent each disease were extracted to train the system. The features extracted for leaf smut are the black lesions present on the leaves and the dry, gray-colored leaf tips and the brown-colored large lesions for brown spot disease.

Fig. 5 Leaf smut disease

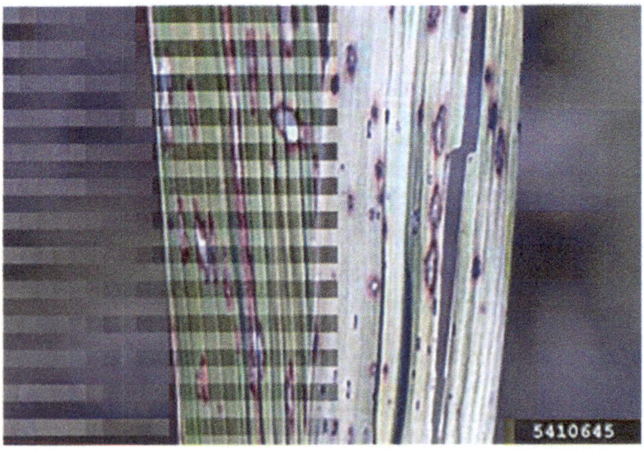

Fig. 6 Brown spot disease

10.1 Dataset

The dataset used here is collected from an online dataset repository known as UCI machine learning repository. A total of 470 images are taken and divided into a training and testing set of 422 images and 48 images, respectively.

10.2 Classification

The final step is the classification of images into the two categories, diseased or healthy. Again, there are various classification algorithms available to classify the input image. In our work, we have taken three supervised classification algorithms. Each one is trained and tested using our dataset, and the obtained experimental results along with the comparative study of each algorithm are shown here.

10.2.1 K-Nearest Neighbor Algorithm

It is one of the simplest algorithms that can be used in both classification and regression problems. This algorithm determines the K-nearest neighbors, and based on this neighbor weight, they determine the label of the samples. The aim of this algorithm is to classify the testing set by calculating the distance between the test samples and the training samples. The equation used here is

$$D\left(u, v_i\right) = \sum_{f \varepsilon F} \omega_f \delta\left(u_f, v_i\right) f \tag{1}$$

where D is the training set with v_i training samples and u is the testing sample. In this algorithm, the k values are based on the data values. After performing the classification, it is observed that the accuracy on the testing set is 91.6% when k value is taken as 1 and accuracy is 72.95% when the k value is 3.

10.2.2 Decision Tree

This is another classification algorithm that is used widely. Here, the algorithm partitions the dataset into two sections by taking a suitable value as the root. The splitting of the data continues till the values in each group become homogeneous. The decision tree algorithm is based on a greedy approach known as iterative dichotomies$_3$ (ID$_3$). In this method, the tree is constructed using concepts such as entropy and information gain which are borrowed from information theory. Here, entropy calculates the values that do not belong to the class. If all the values are of the same type, then it shows zero entropy. And when the values are different from that of the group, the entropy becomes positive. The given Eq. (2) shows the entropy value:

$$E = \sum_{c}^{i=1} - P_i \log_9 P_i \tag{2}$$

Here, *c* represents the number of classes.

The next node in the tree is selected using the information gain concept. The node with highest information gain is determined for this purpose. That is calculated using Eq. (3):

$$G(S,A) = E(S) - \Sigma \frac{|S_v|}{|S|} E(S_v) \tag{3}$$

In this equation, *A* denotes the known value, *G* is the gain, and *E* represents the Entropy. S_v is the subset of *A* as it has the value *v* in it. The test set gave an accuracy of 97.91% after performing classification using the decision tree algorithm.

10.2.3 Naïve Bayes Classifier

This is another commonly used classification algorithm that works according to the Bayes theorem. Naïve Bayes classifier is known as a probabilistic classifier as the prediction is based on the probability of an object. For this, first, the dataset is represented using a frequency table. A likelihood table is generated by finding the probabilities of the selected features, and the posterior probability is calculated using the Bayes theorem. The equation to select the best hypothesis is given by the following:

$$\hat{y} = amP(y) \prod_{n}^{i=1} \left(P(x_i \mid y)\right) \tag{4}$$

The accuracy obtained after performing Naïve Bayes classifier by the test set is 56% which is very less compared to that of K-nearest neighbor and decision tree classification algorithms.

10.3 Result

Table 1 shows the accuracy of the three classification algorithms when trained and tested using diseased and healthy rice leaf images.

The comparison of the accuracy obtained from the three classifiers is shown in the given graph (Fig. 7).

Table 1 Accuracy of the three classification algorithms

Classifiers	Accuracy (%)
KNN classifier	91.6
Decision tree	97.91
Naïve Bayes	56

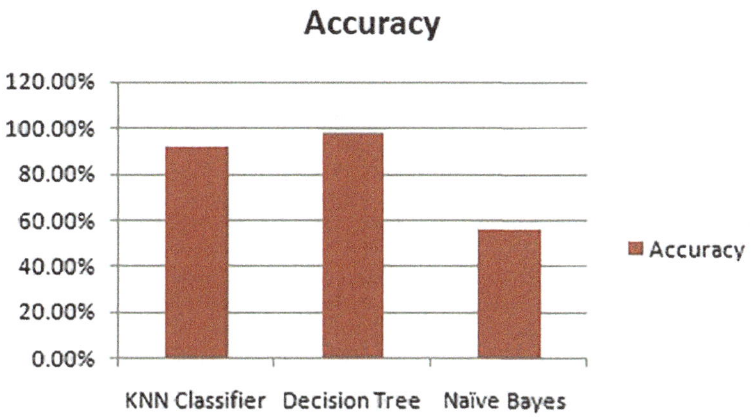

Fig. 7 Accuracy obtained from the three classifiers

11 Conclusion

Technological advances have transformed conventional farming practices to another level. Today, it is known as smart farming which uses intelligent robots and smart machines to propel the agricultural industry forward. A huge transformation has occurred since machine learning technologies invaded the agricultural sector. The rapidly increasing population is putting a tremendous pressure on the natural resources and grasslands for their food and other requirements. Due to this, many problems such as global warming, soil erosion, deforestation, climatic changes, etc. have been stimulated, making the situation even more serious. Fortunately, modern farming methods equipped with machine learning technologies are making it possible to meet the challenges faced by the industry. In this chapter, we have discussed the various applications and roles of machine learning technologies in smart farming. Many diagnostic applications powered by these highly advanced technologies are used by the farmers to retrieve complete information and find correct remedies to combat the diseases.

To innovate and improve the potential of advancing technology, agricultural experts and other technologists are expected to provide their maximum effort and enthusiasm. Besides, it is our duty to educate and spread the importance of precision agriculture. We must encourage everyone to embrace new technologies such as machine learning, Artificial Intelligence, etc., as they have the potential to increase yield and productivity.

References

1. S.K.S. Durai, M.D. Shamili, Smart farming using Machine Learning and Deep Learning techniques. Decis. Anal. J. **3**, 100041 (2022). https://doi.org/10.1016/j.dajour.2022.100041
2. N. Katyal, B. Jaganatha Pandian, A comparative study of conventional and smart farming, in *Emerging Technologies for Agriculture and Environment*, (Springer, Singapore, 2020), pp. 1–8. https://doi.org/10.1007/978-981-13-7968-0_1
3. M. Faridatul, M.S.G. Adnan, A. Dewan, Nexus of urbanization and changes in agricultural land in Bangladesh, in *Remote Sensing of Agriculture and Land Cover/Land Use Changes in South and Southeast Asian Countries*, (Springer, Cham, 2022), pp. 455–469. https://doi.org/10.1007/978-3-030-92365-5_26
4. P. Bhattacharyya, P. Santra, D. Mandal, B. Mondal, Concept and approaches of ecosystem services in agriculture, in *Pricing of Ecosystem Services in Agriculture: A Basis of Crop Insurance*, (Springer, Singapore, 2022), pp. 1–16. https://doi.org/10.1007/978-981-19-4416-1_1
5. S. De Alwis, Z. Hou, Y. Zhang, M.H. Na, B. Ofoghi, A. Sajjanhar, A survey on smart farming data, applications and techniques. Comput. Ind. **138**, 103624 (2022). https://doi.org/10.1016/j.compind.2022.103624
6. C. Lee, C. Lim, From technological development to social advance: A review of Industry 4.0 through machine learning. Technol. Forecast. Soc. Chang. **167**, 120653 (2021). https://doi.org/10.1016/j.techfore.2021.120653
7. C. Shang, F. You, Data analytics and machine learning for smart process manufacturing: recent advances and perspectives in the big data era. Engineering **5**(6), 1010–1016 (2019). https://doi.org/10.1016/j.eng.2019.01.019
8. S.S. Harakannanavar, J.M. Rudagi, V.I. Puranikmath, A. Siddiqua, R. Pramodhini, Plant leaf disease detection using computer vision and machine learning algorithms. Glob. Transit. Proc. (2022). https://doi.org/10.1016/j.gltp.2022.03.016
9. A. Singh, S. Prakash, A. Kumar, D. Kumar, A proficient approach for face detection and recognition using machine learning and high-performance computing. Concurr. Comput. Pract. Exp. **34**(3), e6582 (2022). https://doi.org/10.1002/cpe.6582
10. M. Rakhra, S. Sanober, N.N. Quadri, N. Verma, S. Ray, E. Asenso, Implementing machine learning for smart farming to forecast farmers' interest in hiring equipment. J. Food Qual. **2022** (2022). https://doi.org/10.1155/2022/4721547
11. B. Dudi, V. Rajesh, Optimized threshold-based convolutional neural network for plant leaf classification: A challenge towards untrained data. J. Comb. Optim. **43**(2), 312–349 (2022). https://doi.org/10.1007/s10878-021-00770-w
12. P.P. Kaur, S. Singh, Random forest classifier used for modelling and classification of herbal plants considering different features using machine learning, in *Mobile Radio Communications and 5G Networks*, (Springer, Singapore, 2022), pp. 83–94. https://doi.org/10.1007/978-981-16-7018-3_6
13. M.P. Vaishnnave, R. Manivannan, An empirical study of crop yield prediction using reinforcement learning. Artif. Intel. Tech. Wirel. Commun. Netw. **47-58** (2022). https://doi.org/10.1002/9781119821809.ch4
14. A.K. Tyagi, P. Chahal, Artificial intelligence and machine learning algorithms, in *Research Anthology on Machine Learning Techniques, Methods, and Applications*, (IGI Global, 2022), pp. 421–446. https://doi.org/10.4018/978-1-6684-6291-1.ch024
15. A.T. Schilling, P.P. Shah, J. Feghali, A.E. Jimenez, T.D. Azad, A brief history of machine learning in neurosurgery, in *Machine Learning in Clinical Neuroscience*, (Springer, Cham, 2022), pp. 245–250. https://doi.org/10.1007/978-3-030-85292-4_27
16. Y. Choi, K. Kim, S. Kim, D. Kim, Identification of odor emission sources in urban areas using machine learning-based classification models. Atmos. Environ. X **13**, 100156 (2022). https://doi.org/10.1016/j.aeaoa.2022.100156

17. M.L. Bangare, P.M. Bangare, E. Ramirez-Asis, R. Jamanca-Anaya, C. Phoemchalard, D.A.R. Bhat, Role of machine learning in improving tourism and education sector. Mater. Today Proc. **51**, 2457–2461 (2022). https://doi.org/10.1016/j.matpr.2021.11.615
18. M.U. Ahmed, I. Hussain, Prediction of wheat production using machine learning algorithms in northern areas of Pakistan. Telecommun. Policy **46**(6), 102370 (2022). https://doi.org/10.1016/j.telpol.2022.102370
19. A.M. Ali, M.A. Abouelghar, A.A. Belal, N. Saleh, M. Younes, A. Selim, S. Magignan, Crop yield prediction using multi sensors remote sensing. Egypt. J. Remote Sens. Space Sci. (2022). https://doi.org/10.1016/j.ejrs.2022.04.006
20. K. Alibabaei, P.D. Gaspar, T.M. Lima, R.M. Campos, I. Girão, J. Monteiro, C.M. Lopes, A review of the challenges of using deep learning algorithms to support decision-making in agricultural activities. Remote Sens. **14**(3), 638 (2022). https://doi.org/10.3390/rs14030638
21. Q. Chen, L. Li, C. Chong, X. Wang, AI-enhanced soil management and smart farming. Soil Use Manag. **38**(1), 7–13 (2022). https://doi.org/10.1111/sum.12771
22. N. Ohana-Levi, A. Ben-Gal, S. Munitz, Y. Netzer, Grapevine crop evapotranspiration and crop coefficient forecasting using linear and non-linear multiple regression models. Agric. Water Manag. **262**, 107317 (2022). https://doi.org/10.1016/j.agwat.2021.107317
23. A. Menaga, S. Vasantha, Smart Sustainable agriculture using machine learning and AI: A review. Ambient Commun. Comput. Syst. **447-458** (2022). https://doi.org/10.1007/978-981-16-7952-0_42
24. P.S. Thakur, P. Khanna, T. Sheorey, A. Ojha, Trends in vision-based machine learning techniques for plant disease identification: A systematic review. Expert Syst. Appl. **118117** (2022). https://doi.org/10.1016/j.eswa.2022.118117
25. V.B. Devi, R. Prabavathi, P. Subha, M. Meenaloshini, An efficient and robust random forest algorithm for crop disease detection, in *2022 International Conference on Communication, Computing and Internet of Things (IC3IoT)*, (IEEE, 2022), pp. 1–4. https://doi.org/10.1109/IC3IOT53935.2022.9767937
26. S. Kumar Apat, J. Mishra, K. Srujan Raju, N. Padhy, IoT-assisted crop monitoring using machine learning algorithms for smart farming, in *Next Generation of Internet of Things*, (Springer, Singapore, 2023), pp. 1–11. https://doi.org/10.1007/978-981-19-1412-6_1
27. D. Salifu, E.A. Ibrahim, H.E. Tonnang, Leveraging machine learning tools and algorithms for analysis of fruit fly morphometrics. Sci. Rep. **12**(1), 1–11 (2022). https://doi.org/10.1038/s41598-022-11258-w
28. M. Pei, M. Kong, M. Fu, X. Zhou, Z. Li, J. Xu, Application research of plant leaf pests and diseases base on unsupervised learning, in *2022 3rd International Conference on Computer Vision, Image and Deep Learning & International Conference on Computer Engineering and Applications (CVIDL & ICCEA)*, (IEEE, 2022), pp. 1–4. https://doi.org/10.1109/CVIDLICCEA56201.2022.9824321
29. A.D. Santos, I.C. de Lima Santos, J.G. Costa, Z. Oumar, M.C. Bueno, T.M.M. Mota Filho, J.C. Zanuncio, Canopy defoliation by leaf-cutting ants in eucalyptus plantations inferred by unsupervised machine learning applied to remote sensing. Precis. Agric. **1-17** (2022). https://doi.org/10.1007/s11119-022-09919-x
30. Z.F. Sokhangoee, A. Rezapour, A novel approach for spam detection based on association rule mining and genetic algorithm. Comput. Electr. Eng. **97**, 107655 (2022). https://doi.org/10.1016/j.compeleceng.2021.107655
31. S.M. Javidan, A. Banakar, K.A. Vakilian, Y. Ampatzidis, Diagnosis of grape leaf diseases using automatic K-means clustering and machine learning. Smart Agric. Technol. **3**, 100081 (2023). https://doi.org/10.1016/j.atech.2022.100081
32. Y. Resti, C. Irsan, M.T. Putri, I. Yani, A. Ansyori, B. Suprihatin, Identification of corn plant diseases and pests based on digital images using multinomial Naïve Bayes and K-nearest neighbor. Sci. Technol. Indonesia **7**(1), 29–35 (2022). https://doi.org/10.26554/sti.2022.7.1.29-35

33. A.E. Mezouari, A.E. Fazziki, M. Sadgal, Towards smart farming through machine learning-based automatic irrigation planning, in *Smart Sensor Networks*, (Springer, Cham, 2022), pp. 179–206. https://doi.org/10.1007/978-3-030-77214-7_8
34. S.K. Apat, J. Mishra, K.S. Raju, N. Padhy, The robust and efficient Machine learning model for smart farming decisions and allied intelligent agriculture decisions. J. Integr. Sci. Technol. **10**(2), 139–155 (2022) http://pubs.iscience.in/journal/index.php/jist/article/view/1463/814
35. S.R. Rose, A. Tamizharasi, J.S. Abishekkannao, A.S.S. Chowdary, B.V.K. Reddy, Reinforcement learning for rule extraction from labeled dataset, in *AIP Conference Proceedings*, vol. 2519, No. 1, (AIP Publishing LLC., 2022), p. 030085. https://doi.org/10.1063/5.0110939
36. A. Din, M.Y. Ismail, B. Shah, M. Babar, F. Ali, S.U. Baig, A deep reinforcement learning-based multi-agent area coverage control for smart agriculture. Comput. Electr. Eng. **101**, 108089 (2022). https://doi.org/10.1016/j.compeleceng.2022.108089
37. T. Pamuklu, A.C. Nguyen, A. Syed, W.S. Kennedy, M. Erol-Kantarci, IoT-aerial base station task offloading with risk-sensitive reinforcement learning for smart agriculture. IEEE Trans. Green Commun. Netw. (2022). https://doi.org/10.1109/TGCN.2022.3205330
38. M. Cordeiro, C. Markert, S.S. Araújo, N.G. Campos, R.S. Gondim, T.L.C. da Silva, A.R. da Rocha, Towards smart farming: Fog-enabled intelligent irrigation system using deep neural networks. Futur. Gener. Comput. Syst. **129**, 115–124 (2022). https://doi.org/10.1016/j.future.2021.11.013
39. M. Shehab, L. Abualigah, Q. Shambour, M.A. Abu-Hashem, M.K.Y. Shambour, A.I. Alsalibi, A.H. Gandomi, Machine learning in medical applications: A review of state-of-the-art methods. Comput. Biol. Med. **145**, 105458 (2022). https://doi.org/10.1016/j.compbiomed.2022.105458
40. D.M. Yousefi, A.M. Rafie, S.A.R. Al-Haddad, S. Azrad, A systematic literature review on the use of deep learning in precision livestock detection and localization using unmanned aerial vehicles. IEEE Access (2022). https://doi.org/10.1109/ACCESS.2022.3194507
41. S. Fuentes, C.G. Viejo, E. Tongson, F.R. Dunshea, H.H. Dac, N. Lipovetzky, Animal biometric assessment using non-invasive computer vision and machine learning are good predictors of dairy cows age and welfare: The future of automated veterinary support systems. J. Agric. Food Res. **10**, 100388 (2022). https://doi.org/10.1016/j.jafr.2022.100388
42. T. BlesslinSheeba, L.D. Anand, G. Manohar, S. Selvan, C.B. Wilfred, K. Muthukumar, B.T. Asfaw, Machine learning algorithm for soil analysis and classification of micro-nutrients in IoT-enabled automated farms. J. Nanomater. **2022** (2022). https://doi.org/10.1155/2022/5343965

Transforming Agriculture with Smart Farming: A Comprehensive Review of Agriculture Robots for Research Applications

T. R. Ashwini, M. P. Potdar, S. Sivarajan, and M. S. Odabas

1 Introduction

At present, the world population is around 7.4 billion, and by 2050, it would reach around 9.6 billion approximately [40]. Coming to Indian context, population is increasing day by day as it is around 1.44 billion now. The FAO of the UN has projected that 60% extra increment of food per annum may be needed by the middle of the current century [101]. Keeping this in mind, there is an urgent need to boost the farm produce to fulfill the target demand [41]. The solution lies in either enhancing food production from the available land area or reducing food demand [51]. However, industrialization, as well as urbanization, demands for production of more food grains because the value-added and processed food need more raw materials. Under the circumstances, there is no magic stick to achieve the goal, but technological interventions during recent time offered mankind to achieve the same. In this regard, it may be stated that by properly adopting the latest technologies applicable to the farming sector, the so-called "impossible" things can be achievable, which is the strength of the recent technological development.

T. R. Ashwini (✉)
Department of Agronomy, VIT School of Agricultural Innovations and Advanced Learning (VAIAL), Vellore Institute of Technology, Vellore, Tamil Nadu, India
e-mail: ashwini.tr@vit.ac.in

M. P. Potdar
Department of Agronomy, University of Agricultural Sciences, Dharwad, Karnataka, India

S. Sivarajan
Department of Agri Engineering, VIT School of Agricultural Innovations and Advanced Learning (VAIAL), Vellore Institute of Technology, Vellore, Tamil Nadu, India

M. S. Odabas
School of Agriculture, Ondokuz Mayis University, Samsun, Turkey

S. Balasubaramanian et al. (eds.), *Intelligent Robots and Drones for Precision Agriculture*, Signals and Communication Technology,
https://doi.org/10.1007/978-3-031-51195-0_3

In agriculture, the introduction of robotics and smart agriculture technologies has aided data gathering and advanced analytics, allowing farmers to make better farming decisions and save money on inputs and labor. As farmers accepted more technology in the pursuit of higher yields throughout the years, the concept that "bigger is better" has come to dominate farming, making small-scale enterprises unprofitable. However, advancements in robotics and sensing technology are threatening to destabilize the current agrarian model. "Intelligent robots can change the costly robots and become feasible for small and marginal farmers" says Robotics Engineer George Kantor at Carnegie Mellon University in Pittsburgh, Pennsylvania. Blackmore (2017) believes that agriculture may be more efficient and sustainable by switching to a robotic agriculture system.

Although many innovations are in use, most of them are in the prototype stage. "The big companies are not investing in agricultural robots because of lack of business," says Blackmore (2017). Agriculture 4.0 is a new wave of precision agriculture that began in the early 2010s, thanks to the advancement of many technologies such as low-cost sensors and actuators with high-performance cellular communication with high bandwidth, microprocessors with a cheap cost, cloud-based ICT systems, and big data analytics. Finally, researchers and agriculture pioneers are thinking to bring one more revolution in agriculture mainly based around robotics and artificial intelligence in Agriculture 5.0. By integrating digital tools, sensors, and control technologies, agricultural robotic design and development are accelerating in modern agriculture [68, 69]. These advancements include anything from digitizing plants and fields by gathering precise and detailed data (temporal and spatial) in real time to completing complex nonlinear control tasks for the navigation of robots. The use of robots and smart farming technology in digital farming is generating a growing interest in automation, transforming field jobs into high-tech industrial tasks that are attracting investors, professional engineers, and businesses. Many of these robots are in prototype stage. They can currently do a variety of farming tasks, such as crop reconnaissance and phenotyping [7, 11], seeds and sowing [22, 76], automated robotic weeding and targeted spraying [6, 92], real-time crop monitoring and fertilizing [82], robotic irrigation [16], selective harvesting [9, 27, 46, 89], pruning [102], and pollination [90].

2 Robots

The term "robot" comes from the Czech word *robota*, which means "forced labor." It refers to a mechanical entity created by humans and typically controlled through an electromechanical system. These devices utilize software engineering to simplify complex tasks. Agricultural robotics involves the use of automation in the field of life sciences, encompassing areas such as agriculture, forestry, and fisheries. In farming processes, intelligent robots are deployed to perform appropriate actions, at precise locations and times, with optimal efficiency. Implementing automation in agriculture has led to significant advancements and cost and time savings for farmers. Although the movie *Runaway* in 1984 portrayed agrorobots as scouts removing insects from corn leaves, the field of robotics in agriculture is still relatively young. In fact, the first agricultural field robot was not introduced until a later period [44]. In this review paper, an extensive literature survey has been performed to identify robotic applications in various agricultural operations right from planting to harvesting throughout the crop growing season. A total of 102 journal articles have been surveyed to analyze the importance of robotics in agriculture and understand the challenges involved. The limitations in this technology and suggestions for future work have also been discussed.

2.1 Working of Robots

Robots possess remarkable capabilities in terms of mobility and perception. To navigate and operate effectively in unfamiliar environments, robots rely on multiple sensors. These sensors transmit valuable information to the controller in the form of electronic signals, providing crucial data about the robot's surroundings. By employing specialized sensors, robots can acquire information that surpasses the limitations of human senses. The controller, often referred to as the robot's "brain," plays a pivotal role in governing its actions. It acts as a computerized command center, enabling the robot to execute programmed tasks and facilitating connections with other systems. This connectivity allows robots to collaborate with other machines, processes, or even fellow robots, enhancing their functionality and versatility. Actuators or drives serve as the "engine" of the robot, generating the necessary motion for various tasks. These actuators can be hydraulic, pneumatic, DC, stepper, or servo motors, depending on the specific requirements of the robot's design. The robot arm, comprising the shoulder, elbow, wrist, and fingers, closely resembles the structure of a human arm. It facilitates the precise positioning of end effectors and sensors, enabling them to fulfill their predetermined functions effectively. Lastly, the end effectors represent the robot's final connection point. These components serve as interfaces with the surrounding environment, allowing the robot to interact with objects and perform tasks. End effectors encompass a wide range of tools and devices tailored to specific applications, expanding the robot's capabilities in various contexts.

3 Current Status of Agricultural Robots [44]

The agricultural industry has witnessed significant advancements in robotics and automation, including the MF-Scamp robot developed by Blackmore. This robot performs scouting, weeding, and harvesting tasks, saving labor time and costs. However, its implementation may pose challenges for small-scale farmers due to increased costs. The Danish Institute of Agricultural Sciences has developed the

Autonomous Plant Inspection (API) platform, which uses RTK GPS for autonomous field reconnaissance, resulting in a 75% reduction in herbicide usage. Hohenheim University's Sub Canopy Robot ISAAC 2 assesses crop health using sensors and GPS technology, though complete weed eradication is still being pursued. Deepfield Robotics' BoniRob performs various agricultural tasks and has demonstrated over 90% effectiveness in weed eradication. However, it remains limited to laboratory usage due to high costs. The lettuce bot in California utilizes robotics, computer vision, and machine learning to recognize lettuce plants and eradicate weeds. Although it has implications for organic farming, its inability to utilize organic fertilizers restricts its application. These agricultural robots offer efficiency and cost benefits but require further development and cost reduction. Small-scale farmers should consider the impact and viability for their operations.

CROPS, a project by the European Union, involves clever robots designed to detect and evaluate the maturity of fruits in crops. These robots can navigate through the fields, locate ripe fruits hidden behind leaves, and use grasping mechanisms to detach them. Additionally, they can perform targeted spraying of foliage by analyzing specific sites. However, challenges remain in recognizing hidden fruits due to their diverse forms and sizes. HortiBot, developed in Denmark, is a robust tool carrier primarily used for high-tech plant nursing, with a focus on controlling weeds. It reduces labor costs by performing repetitive tasks such as mechanical weeding and optimizes herbicide usage, thus lowering herbicide costs. Nonetheless, HortiBot has drawbacks such as high short-term expenses, the need for skilled operators, and limited suitability for small farms. Despite these limitations, it offers innovative technology to enhance farmers' productivity. AgBot II, an Australian prototype, aids farmers in decision-making regarding herbicides, insecticides, fertilizers, and irrigation. It utilizes sensor networks, historical data, satellites, and drones to assist farmers in farm management decisions.

India's AgriBot, developed by students at BITS Hyderabad, performs essential farming functions, including harvesting, spraying, sowing, and weed removal. Its design is based on image processing, and motor control is achieved through a relay. AgriBot aims to increase productivity, speed, and precision while reducing labor costs in India's agrarian economy. Vitirover, a solar-powered robot from New Zealand, operates at a speed of 500 meters per hour. It effectively removes grass and weeds within a short distance from grape vines using sensors and a GPS system. The use of technology-friendly gadgets in grape vines is vital for farmers. These agricultural robots demonstrate advancements in automation and technology, offering potential solutions to labor-intensive tasks, improving productivity, and reducing costs in various farming practices worldwide.

4 Applications of Robotic Technology in Agriculture

4.1 Seeds and Sowing

Sunitha et al. [76] developed an agricultural robot as a superior substitute for traditional manual seed sowing and expensive tractor technology. The robot can autonomously plow the land, localize its path, and traverse without human intervention. It utilizes cost-effective dc locomotion motors, a Raspberry Pi CPU for efficient picture processing and motor control. Due to wetland limitations, the limited area can be plowed through robots, and the seeder sows the seeds in the same pattern. The proposed system combines plowing and seed sowing, offering an affordable and efficient solution for farmers. Divya et al. [22] found that the equipment performed well for dry clay soil, with a seeding accuracy of 94.8% compared to seeding on a flat surface. In sandy soil, the accuracy was 82.8%, and in very coarse soil, it was 72.4%. Ryu and Han [64] developed a vision-assisted robotic transplanter with a success rate of 98% for seedling transplantation. The end effector positioning accuracy was within 1.0 mm, sufficient for the task. Griepentrog et al. [32] proposed a kinematic model to calculate seed positions, while Swapnil et al. [77] created a robot for autonomous plowing and seed dispensing, reducing labor and seed waste.

4.2 Crop Scouting and Phenotyping

Obtaining timely and reliable data for plant phenotyping is a crucial task that requires leaf-level physiological and chemical trait measurements [67]. The interaction between complex plant features and the environment is at the core of plant phenotyping [26]. Quantitative assessments of plant phenotypes during the growing season are essential [93]. However, measurement manually is time-consuming, challenging, and error-prone. To address this, robotics has emerged as a promising solution, enabling minimal involvement of humans and monitoring automatically. Over the past decade, numerous automated high-throughput field-based phenotyping platforms have been developed. These platforms utilize sensors like RGB, depth or hyperspectral cameras, light curtains, and infrared radiometers to traverse short-row crops [2]. Furthermore, an autonomous field survey mobile robot platform with a customized manipulator and gripper has been created for transporting imaging sensors and GPS devices, enabling autonomous navigation and data collection in greenhouse and open-field agriculture environments. Researchers have also integrated various multispectral imaging devices and LiDAR sensors into adapted mobile robot platforms for automatic monitoring and the creation of reconstructed 3D point clouds, which can generate computer images of trees and plants [68, 69].

Measurement of physiological and chemical traits at the leaf level is crucial in plant phenotyping for monitoring plant health. Manual measurement is time-consuming and error-prone. Atefi et al. [7] developed a robotic system with a

MATLAB GUI to automate leaf phenotyping in maize and sorghum. The robot accurately predicted chlorophyll, water content, and potassium, but not nitrogen and phosphorus. Maize had a 78% grasping success rate, while sorghum had 48%. The robot's data complements image-based phenotyping, which requires manual measurements to establish correlations with picture data [29, 54, 56]. Sorghum is a valuable biofuel feedstock, but manually assessing its growth attributes limits research scalability. A high-throughput robotic phenotyping device was developed to capture side-view stereo images and measure stem diameter [11]. Image-derived features correlated well with manual measurements. To quantify plant surface area, the point cloud data was transformed into a triangle mesh using the greedy projection triangulation method. Moving least squares smoothed noisy and overlapping surfaces [49, 63]. Sorghum's architectural features such as plant height [65], leaf area index [55], and leaf angle [62, 80] have been shown to strongly influence biomass yield.

4.3 Robotic Weeding

Novel weeding technologies are being developed to reduce manual effort and herbicide usage in farming. These robots can work in harsh environments with limited space and uneven surfaces. The Danish Farm Research Authority has created a four-wheel-drive weed-seeking robot that replaces manual hoeing. To reduce the need for herbicides, the navigation system is used by the intelligent hoe to identify the crop rows. Vision-guided intra-row cultivators eliminate the need for chemical weed control by mechanically targeting weeds within sugar beet rows. Additionally, autonomous mobile robots with two independently driven wheels are being utilized for various agricultural operations.

By using image interpretation features, weed detection cameras mounted on weeders aid in the identification and distinction of crop and weed. The robot's weeding arm/weeding equipment removes weeds or sprays herbicide on them without harming the crop plants (Anon 2017). Several experiments using various discriminating and classification strategies have been conducted to detect weeds automatically. To segment individual weed leaves, Manh et al. [48] employed parametric deformable templates, Sokefeld et al. [74] used Fourier descriptors and shape parameters to identify more than 20 weed species, and Søgaard [103] used feature-based models to assess 19 weed plants.

Researchers have utilized artificial neural networks [17, 31] and image processing techniques [30, 87] for weed classification in agriculture. Manual weed scouting takes about 0.7 man-hours per year per hectare [58]. Autonomous field scouting using GPS and GIS technologies can cover 4.32 hectares per hour. Herbicide savings of 30% to 75% are possible with decision assistance systems for patch spraying [33, 103]. Selective weed detection and post-processing enhance chemical application. Mechanical and chemical methods are used for weed removal, with row-based removal requiring high-speed sensing and mechanisms for effective treatment.

Several weed control robots have been developed, focusing on strategies such as mechanical weeding [57], flaming [21], selective chemical spraying [45], and electric discharging [72]. Mechanical weeding is gaining more attention due to its pesticide-free approach [57]. Researchers have used real-time crop row detection and computer vision systems to guide hoes and rotary hoes for inter-row and selective in-row weed eradication, achieving removal rates of up to 53% [78, 83]. Autonomous robots with multiple tools, including hoes, brushes, and springs, are also being developed for intra-row and in-row weeding [53]. Integrated weed management systems aim to maximize weed treatment efficiency by combining different strategies [18, 98]. Automated weed and crop classification has achieved over 90% accuracy in distinguishing and removing weeds, regardless of delays, topography, or crop growth stages [42, 92].

In direct-seeded scenarios with high crop emergence rates and low to moderate weed densities, the combination of special context, plant form, and color proved to enhance the resistance of the robotic weed management system to variations in plant appearance and weed species [6]. However, during a field test on an organic sugar beet field at the first true leaf stage, the robot only removed 1% of the sugar beets and 41–53% of the weeds. Of the weeds not removed, 31% were growing too close to crop plants, and 18% were in areas where sugar beet seeds did not germinate. A visual navigation system implemented in a simulated paddy field allowed a weeding robot to move at a speed of 156.07 mm/s while effectively following the rice seedling line [99, 100]. Robotic weeding has demonstrated automatic, precise, and efficient control of weeds near or within crop rows [28]. CNNs, a deep learning technique, particularly models like Inception v3, GoogLeNet, and DenseNet, have shown success in crop/weed detection and classification, even in uncontrolled illumination conditions [23, 36, 37, 50, 104]. The Drop on Demand (DoD) robotic system has effectively controlled all weeds in field trials,pote reducing herbicide use by tenfold and serving as an alternative to conventional spraying [81]. By utilizing the Department of Defense technology, herbicide consumption can be minimized by over 90%, reducing environmental and health concerns while potentially eliminating the need for manual in-row weeding. Robotic systems are also exploring targeted spraying to confine herbicide application to weeds and mechanical methods to eradicate weeds without herbicides [52, 92].

4.4 Nutrient Management

Fertilizer technology plays a crucial role in agricultural development [79] by addressing issues such as escalating costs of inorganic fertilizers and decreased soil fertility [4]. Integrated nutrient management and need-based application of fertilizers have gained importance among farmers. Nitrogen, an essential macronutrient for plant growth, is often limited in crop yield [5]. Effective nitrogen fertilization management is vital, but detecting nitrogen insufficiency early is challenging [14]. In recent decades, machine vision and digital image processing techniques have been developed in greenhouse engineering, offering low-cost image acquisition equipment for noncontact phyto-monitoring. These technologies enable the determination of plant health status [3, 82], detection of nutrient deficiencies [75, 82], and identification of viral and fungal diseases [20, 60, 66]. Various imaging approaches, including visible range cameras, thermography [88], chlorophyll fluorescence [38], and spectral sensors [12, 97, 105], provide valuable data. Combining artificial vision and pattern recognition systems has been suggested for assessing nutrient concentrations, while machine vision-enabled robotic systems allow real-time plant health and growth monitoring [70, 82]. Such advancements not only increase plant production efficiency and quality but also reduce nitrogen fertilizer usage, minimizing environmental risks and cutting production costs.

4.5 Fertilizers Will Be Spread by Flying Robots

A flying robot keeps an eye on the crops' progress. The robot can fly autonomously and apply fertilizer on its own, thanks to camera equipment and an automatic fertilizing mechanism in the front.

4.6 Robotic Irrigation

The mechatronic sprinkler is a robotic irrigator designed to provide customized water and chemigation rates to specific areas, including field corners. This technology addresses the issue of water scarcity in arid environments and dry seasons by efficiently applying water where it is needed. The sprinkler operates autonomously, utilizing solar power as an energy source [96] or employing other innovative methods such as RFID, iris systems, or human excitation [34, 73]. Continuous improvements are being made to reduce human involvement in irrigation processes. The system described by Khriji et al. [39] utilizes wireless sensor networks for farmers, while Bodunde et al. [16] present an adaptive sprinkler irrigation robot based on Zigbee communication. With a tank capacity of approximately 5 L, the sprinkler can empty in less than 100 s, completing a full sprinkling cycle in around 2 min and 30 s. This Zigbee is cost-effective and has a high network capacity, long life span, and minimal damage to farmland, plants, and pipes during installation.

4.7 Pollination

Researchers have been working on the development of autonomous pollinators that can effectively spray crops, similar to targeted weed-spraying robots [1, 13, 43, 47]. These advancements aim to enhance fruit cultivation efficiency by providing a viable alternative to natural kiwifruit pollination, which has become increasingly challenging. The high costs associated with labor and inefficient use of expensive pollen have made it difficult for many farmers to adopt alternative methods. To address these issues, precision agriculture is focusing on the development of intelligent robotic devices. In a recent study by Williams et al. [91], a groundbreaking kiwifruit pollination robot was created and tested. The results showed that the robotic system was able to target and pollinate 79.5% of flowers at a speed of 3.5 km/h while consuming a pollen amount comparable to that of a commercial Cambrian operator. Additionally, the study found that kiwifruits pollinated at a speed of 1 km/h by the robot exhibited the same high quality as commercially farmed ones.

4.8 Pruning

Researchers have proposed a robotic system for automatic grape vine pruning [106], using trinocular stereo cameras, computer vision, artificial intelligence, and a robotic arm. The system captures images, creates a 3D model, selects pruning targets, and performs cuts. It achieves pruning times comparable to humans, but reliability needs improvement for commercial viability [102].

4.9 Selective Harvesting

Traditional fruit and vegetable picking is labor-intensive and costly, prompting a need for automated harvesting. However, current technology has only achieved a 33% success rate in picking sweet pepper fruit, taking an average of 94 s per fruit [35]. Selective harvesting, which involves harvesting specific sections of the plant-based on quality criteria, requires the ability to recognize quality factors before harvest and to harvest without damaging the remaining crop [15]. Agriculture's unstructured environment necessitates complex machine learning approaches [8, 61], but advancements in machine vision, deep learning, sensing, and end effector manipulation have facilitated the use of robots in this field. Although various harvesting robots have been developed for crops like melons [24, 25], oranges [59], cucumbers [84–86], tomatoes [19, 95], sweet peppers [10], and strawberries [94], commercialization is limited. Recent studies have shown success rates ranging from 26% to 86% in different fruit-harvesting robots [71]. Accurate and real-time fruit detection in the canopy is crucial for successful commercial harvesting, and while recognizing fruit in 2D photos has become reliable, accurately pinpointing fruit and canopy structure in three dimensions remains challenging [8, 72].

5 Other General Agricultural Robots

5.1 *Demeter (Used for Harvesting)*

Demeter is an autonomous crop-cutting robot named after the Roman goddess of agriculture. It can chop wheat and alfalfa without human intervention, working tirelessly and increasing productivity. Equipped with cameras, it can differentiate between cut and uncut crops. Demeter can be controlled remotely or taught a specific path to follow with its precise onboard sensors and computer control. The system offers three levels of automation, starting with "cruise control" for steering and controlling the harvester. The "drone" feature enables one operator to control multiple harvesters, and finally, a fully autonomous machine can harvest an entire field independently.

5.2 *Robot for Weed Control*

Novel weeding technologies are being developed to reduce manual effort and herbicide usage in farming. These robots can work in harsh environments with limited space and uneven surfaces. The Danish Farm Research Authority has created a four-wheel-drive weed-seeking robot that replaces manual hoeing. Vision-guided intra-row cultivators eliminate the need for chemical weed control by mechanically targeting weeds within sugar beet rows. Additionally, autonomous mobile robots with two independently driven wheels are being utilized for various agricultural operations.

5.3 *Forester Robot*

This innovative robot is designed specifically for tasks such as wood cutting, tree care, tree pruning, and extracting pulp and hardwood in forests. It utilizes specialized jaws and axes to efficiently chop branches. With its six-legged movement system, the forester robot autonomously coordinates its legs, while a human operator controls its navigation through the forest.

5.4 *Fruit-Picking Robot*

Fruit-harvesting robots have been an established concept since the early 1980s, revolutionizing crop harvesting. To fully develop this technology, collaboration is needed from high-tech industries, agricultural groups, and farm equipment manufacturers. These robots must be capable of picking ripe fruit without damaging the tree's branches or leaves while also being mobile enough to access all areas of the tree. They require advanced intelligence and humanlike interaction with their surroundings, utilizing touch, sight, and image processing. By using video image capture and color detection, the robots can distinguish between fruit and leaves. If fruit is obstructed, an air jet can clear the way for better visibility. The robot arm, wrapped in rubber, features five degrees of freedom for precise movement. The gripper mechanism, powered by motors, hydraulics, or pneumatics, applies enough pressure to detach the fruit without crushing it. Significant progress has been made in France with end effectors that can harvest apples and citrus, resulting in robots collecting over 75% of the crops

5.5 Micro-flying Robot

Scientists worldwide are reverse engineering insect mechanics to develop miniature robots for various applications. The tiniest microrobot, entered in the genius book of world records, has propellers enabling high-flying capabilities. Its impressive skills include precise landing on potato chips and swift takeoff by flapping wings. These experiments aim to explore operational fields, such as scanning battlefields, locating victims in rubble, and capturing images in agricultural fields. The mini robot's potential use in agriculture includes combating weeds and insects, promising advancements in crop protection.

Some of the robots discussed here are field robots, while others are mobile robots. They alter their appearance to meet the needs of the situation. Mobile robots are those that can move around in relation to a medium. The entire system is in sync with its surroundings.

6 Conclusion

Agricultural robots have been developed to address various challenges in the farming sector. These robots utilize DC motors, which offer efficient performance and cost-effectiveness, particularly when it comes to plowing and seed planting tasks. One significant advantage of autonomous robotic weeding devices is their potential to decrease reliance on agrochemicals such as herbicides and pesticides. By enabling real-time treatment, these robots contribute to reducing pollution and promoting sustainability in agriculture. In the context of greenhouse crops, machine vision-equipped robotic systems can be employed for online plant health and growth monitoring. These robots assess leaf physiological and chemical properties, offering a means to complement image-based high-throughput plant phenotyping. This integration of technologies holds promise for enhancing plant phenotyping capabilities. The use of Zigbee-based wireless plant irrigation robots overcomes the limitations of fixed sprinkler systems and saves valuable space in agricultural operations. By efficiently and autonomously managing irrigation, these robots contribute to optimizing water usage and improving crop yield.

Farmers stand to benefit significantly from the adoption of farming robots. For instance, they can save approximately 20% on cereal scouting, 12% on sugar beet weeding, and 24% on inter-row weeding tasks. Additionally, agricultural robots can facilitate harvesting through real-time data sensing, processing, and utilization of robotic vehicles. Precision monitoring and harvesting with the assistance of agri-robots can help alleviate labor shortages and rising labor costs in the industry while also ensuring high food quality during the harvesting process. As the global population continues to grow, the farming sector faces increasing pressure to meet food demands. Agricultural robots have emerged as a potential solution to address manpower shortages while enhancing productivity. These robots integrate various technologies such as machine vision, image processing, and mechatronics into a single platform, enabling autonomous agricultural operations and offering a promising option for the future of farming.

References

1. G.J. Amador, D.L. Hu, Sticky solution provides grip for the first robotic pollinator. Chem **2**(2), 162–164 (2017)
2. P. Andrade-Sanchez, M.A. Gore, J.T. Heun, et al., Development and evaluation of a field-based high-throughput phenotyping platform. Funct. Plant Biol. **41**(1), 68–79 (2014)
3. A. Arefi, A.M. Motlagh, K. Mollazade, et al., Recognition and localization of ripen tomato based on machine vision. Aust. J. Crop. Sci. **5**(10), 11–44 (2011)
4. T.R. Ashwini, G. Math, H. Babalad, et al., Effect of different residue-based composts and biofertilizers on uptake of nutrients and protein yield of pigeonpea (*Cajanus Cajan* (L.) Millsp.). J. Pharmacog. Phytochem. SPI, 284–287 (2018a)
5. T.R. Ashwini, G. Math, V. Venkatachalapathi, Integrated nutrient management with different residue-based composts on growth, yield and economics of pigeonpea (*Cajanus cajan* (L.) Millsp.). Multilogic Sci. **8**, 19–21 (2018b)
6. B. Åstrand, A.J. Baerveldt, An agricultural mobile robot with vision-based perception for mechanical weed control. Auton. Robot. **13**(1), 21–35 (2002)
7. A. Atefi, Y. Ge, S. Piha, et al., In vitro human like robotic phenotyping of leaf traits in maize and sorghum in green house. Comp. Electron. Agril. **163**, 1–10 (2019)
8. C.W. Bac, E.J. Henten, J. Hemming, et al., Harvesting robots for high-value crops: State-of-the-art review and challenges ahead. J. Field Robot. **31**(6), 888–911 (2014)
9. C.W. Bac, T. Roorda, R. Reshef, et al., Analysis of a motion planning problem for sweet-pepper harvesting in a dense obstacle environment. Biosyst. Eng. **146**, 85–97 (2016)
10. W.C. Bac, J. Hemming, B.A.J. Tujil, et al., Performance evaluation of harvesting robot for sweet pepper. J. Field Robot. **34**(6), 1123–1139 (2017)
11. Y. Bao, L. Tang, M.W. Breitzman, et al., Field based robotic phenotyping of sorghum plant architecture using stereo vision. J. Field Robot. **36**, 397–415 (2018)
12. G. Bareth, H. Aasen, J. Bendig, et al., Low weight and UAV-based hyperspectral full-frame cameras for monitoring crops: Spectral comparison with portable spectroradiometer measurements. Photogram Fernerkundung-Geoinform **1**, 69–79 (2015)

13. S. Berman, V. Kumar, R. Nagpal, Design of control policies for spatially inhomogeneous robot swarms with application to commercial pollination, in *2011 IEEE International Conference on Robotics and Automation (ICRA)*, (IEEE, 2011), pp. 378–385
14. T.M. Blackmer, J.S. Schepers, G.E. Varvel, et al., Nitrogen deficiency detection using reflected shortwave radiation from irrigated corn canopies. Agron. J. **88**(1), 1–5 (1996)
15. S. Blackmore, B. Stout, M. Wang, et al., Robotic agriculture – The future of agriculture mechanization? in *5th European Conference on Precision Agriculture*, (2005), pp. 1–9
16. O.P. Bodunde, U.C. Adie, O.M. Ikumapayi, et al., Architectural design and performance evaluation of a ZigBee technology-based adaptive sprinkler irrigation robot. Comput. Electron. Agril. **160**, 168–178 (2019)
17. T.F. Burks, S.A. Shearer, J.R. Heath, et al., Evaluation of neural-network classifiers for weed species discrimination. Biosyst. Eng. **91**(3), 293–304 (2005)
18. R. Chikowo, V. Faloya, S. Petit, et al., Integrated weed management systems allow reduced reliance on herbicides and long-term weed control. Agric. Ecosyst. Environ. **132**(3–4), 237–242 (2009) (8) (PDF) Robotic weed control using automated weed and *crop classification*. Available from: https://www.researchgate.net/publication/339086261_Robotic_weed_control_using_automated_weed_and_crop_classification. Accessed June 27 2023
19. Y.C. Chiu, P.Y. Yang, S. Chen, Development of the end-effector of a picking robot for greenhouse-grown tomatoes. Appl. Eng. Agric. **29**, 1001–1009 (2013)
20. G. Corkidi, K.A. Balderas-Ruíz, B. Taboada, et al., Assessing mango anthracnose using a new three-dimensional image-analysis technique to quantify lesions on fruit. Plant Pathol. **55**(2), 250–257 (2006)
21. A. Datta, Knezevic, Flaming as an alternative weed control method for conventional and organic agronomic crop production systems: A review. Adv. Agron. **118**, 399–428 (2013)
22. C.H. Divya, H. Ramakrishna, G. Praveena, Review article: Seeding and fertilization using an automated robot. Int. J. Curr. Res. **5**(3), 461–466 (2013)
23. M. Dyrmann, R.N. Jørgensen, H.S. Midtiby, et al., RoboWeedSupport – Detection of weed locations in leaf occluded cereal crops using a fully convolutional neural network. Adv. Anim. Biosci. **8**(2), 842–847 (2017). https://doi.org/10.1017/S2040470017000206
24. Y. Edan, Design of an autonomous agricultural robot. Appl. Intell. **5**(1), 41–50 (1995)
25. Y. Edan, D. Rogozin, T. Flash, et al., Robotic melon harvesting. IEEE Trans. Robot. Autom. **16**(6), 831–835 (2000)
26. S. Foix, G. Alenyà, C. Torras, 3D Sensor planning framework for leaf probing, in *2015 IEEE/RSJ International Conference on Intelligent Robots and Systems (IROS)*, (2015), pp. 6501–6506
27. K.G. Fue, W.M. Porter, E.M. Barnes, et al., An extensive review of mobile agricultural robotics for field operations: Focus on cotton harvesting. Agric. Eng. **2**(150), 174 (2020)
28. J. Gai, L. Tang, B.L. Steward, Automated crop plant detection based on the fusion of color and depth images for robotic weed control. J. Field. Robot. **37**(1), 35–52 (2020)
29. Y. Ge, G. Bai, V. Stoerger, et al., Temporal dynamics of maize plant growth, water use, and leaf water content using automated high throughput RGB and hyperspectral imaging. Comput. Electron. Agric. **127**, 625–632 (2016)
30. R.J. Godwin, P.C.H. Miller, A review of the technologies for mapping within-field variability. Biosyst. Eng. **84**(4), 393–407 (2003)
31. P.M. Granitto, P.F. Verdes, H.A. Ceccatto, Large-scale investigation of weed seed identification by machine vision. Comput. Electron. Agric. **47**(1), 15–24 (2005)
32. H.W. Griepentrog, M. Nørremark, H. Nielsen, et al., Individual plant care in cropping systems, in *Proceedings of the 4th European Conference on Precision Agriculture, Berlin. A Method for High Accuracy Geo-referencing of Data From Field Operations*, (2003) Available from: https://www.researchgate.net/publication/257270212_A_method_for_high_accuracy_geo-referencing_of_data_from_field_operations. Accessed June 27 2023
33. T. Heisel, S. Christensen, A.M. Walter, Whole-field experiments with site-specific weed management, in *ECPA 02: Proceedings of the Second European Conference on Precision*

Agriculture, Odense, Denmark, 11–16 July, 1999, Part 2, ed. by J.V. Stafford, (1999), pp. 759–768
34. N. Hema, R. Aswani, M. Malik, Plant watering autonomous mobile robot. Int. J. Robot. Autom. **1**(3), 152–162 (2012)
35. J. Hemming, J. Ruizendaal, J. Hofstee, et al., Fruit detectability analysis for different camera positions in sweet-pepper. Sensors **14**(4), 6032–6044 (2014)
36. Z. Huang, S. Wane, S. Parsons, Towards automated strawberry harvesting: Identifying the picking point. Lect. Notes Comput. Sci. (including Subser. Lect. Notes Artif. Intell. Lect. Notes Bioinformatics) **10454 LNAI**, 222–236 (2017a)
37. J. Huang, V. Rathod, C. Sun, et al., Speed/accuracy trade-offs for modern convolutional object detectors, in *IEEE CVPR*, (2017b), p. 4
38. S.J. Jeong, D. Schimel, C. Frankenberg, et al., Application of satellite solar-induced chlorophyll fluorescence to understanding large-scale variations in vegetation phenology and function over northern high latitude forests. Remote Sens. Environ. **190**, 178–187 (2017)
39. S. Khriji, D.E. Houssaini, W. Jmal, et al., Precision irrigation based on wireless sensor network. IET Sci. Meas. Technol. **8**(3), 98–106 (2014)
40. A. King, Technology: The future of agriculture. Nature, 544–S21 (2017)
41. N.Y. Kondo, P.P. Nishitsuji, K.C. Ling, et al., *Visual Feedback Guided Robotic Cherry Tomato Harvesting* (The American Society of Agricultural and Biological Engineers, St. Joseph, 1996)
42. C. Kunz, J.F. Weber, G.G. Peteinatos, et al., Camera steered mechanical weed control in sugar beet, maize and soybean. Precis. Agric. **19**(4), 708–720 (2018)
43. H. Kurosaki, H. Ohmori, M. Takaichi, Development of an automatic fruit-set-reagent spraying robot for tomato plants to promote uniform fruit ripening, in *International Symposium on Advanced Technologies and Management Towards Sustainable Greenhouse Ecosystems: Greensys*, vol. 9, (2011), pp. 931–936
44. H.L. Kushwaha, J.P. Sinha, T.K. Khura, et al., Status and scope of robotics in agriculture, in *International Conference on Emerging Technologies in Agricultural and Food Engineering, 27–30, December, 2016*, (Agricultural and Food Engineering Department, IIT Kharagpur, 2016), pp. 1–10
45. W.S. Lee, D. Slaughter, D. Giles, Robotic weed control system for tomatoes. Precis. Agric. **11**, 95–113 (1999)
46. Z. Li, F. Miao, Z. Yang, et al., Factors affecting human hand grasp type in tomato fruit-picking: A statistical investigation for ergonomic development of harvesting robot. Comput. Electron. Agric. **157**, 90–97 (2019)
47. H. Maghsoudi, S. Minaei, B. Ghobadian, et al., Ultrasonic sensing of pistachio canopy for low-volume precision spraying. Comp. Electron. Agril. **112**, 149–160 (2015)
48. A.G. Manh, G. Rabatel, L. Assemat, et al., Weed leaf image segmentation by deformable templates. J. Agril. Eng. Res. **80**(2), 139–146 (2001)
49. Z.C. Marton, R.B. Rusu, M. Beetz, On fast surface reconstruction methods for large and noisy point clouds, in *2009 IEEE International Conference on Robotics and Automation (ICRA)*, (IEEE, Kobe/Piscataway, 2009), pp. 3218–3223
50. C. McCool, J. Beattie, J. Firn, C. Lehnert, et al., Efficacy of mechanical weeding tools: a study into alternative weed management strategies enabled by robotics. IEEE Robot. Auto Lett. **3**(2), 1184–1190 (2018). https://doi.org/10.1109/lra.2018.2794619
51. A.T. Meshram, A.M. Vanalkar, Technology for agriculture to increase food production and quality; A review. Int. J. Res. Appl. Sci. Eng. Tech. **45**(98), 2321–2653 (2018)
52. A. Michaels, S. Haug, A. Albert, Vision-based high-speed manipulation for robotic ultra-precise weed control, in *2015 IEEE/RSJ International Conference on Intelligent Robots and Systems (IROS)*, (IEEE, 2015), pp. 5498–5505
53. Naio, (2018), https://www.naio-technologies.com/en/agricultural-equipment/weeding-robot-oz/

54. E.H. Neilson, A.M. Edwards, C.K. Blomstedt, et al., Utilization of a high-throughput shoot imaging system to examine the dynamic phenotypic responses of a C_4 cereal crop plant to nitrogen and water deficiency over time. J. Exp. Bot. **66**, 1817–1832 (2015)
55. S.N. Olson, K. Ritter, W. Rooney, et al., High biomass yield energy sorghum: Developing a genetic model for C4 grass bioenergy crops. Biofuels Bioprod. Biorefin. **6**(6), 640–655 (2012)
56. P. Pandey, Y. Ge, V. Stoerger, et al., High throughput in vivo analysis of plant leaf chemical properties using hyperspectral imaging. Front. Plant Sci. **8**, 1348 (2017)
57. E. Pannacci, B. Lattanzi, F. Tei, Non-chemical weed management strategies in minor crops: A review. Crop Prot. **96**, 44–58 (2017)
58. S.M. Pedersen, S. Fountas, B.S. Blackmore, Agricultural-Applications and economic perspectives. Precis. Agric. **7**(4), 295–308 (2008)
59. A. Plebe, G. Grasso, Localization of spherical fruits for robotic harvesting. Mach. Vis. Appl. **13**(2), 70–79 (2001)
60. J. Qin, T.F. Burks, M.A. Ritenour, W.G. Bonn, Detection of citrus canker using hyperspectral reflectance imaging with spectral information divergence. J. Food Eng. **93**(2), 183–191 (2009)
61. C.W. Roldan, E.J. Henten, J. Hemming, et al., Harvesting robots for high-value crops: State-of-the-art review and challenges ahead. J. Field Robot. **31**(6), 888–911 (2014)
62. W.L. Rooney, J. Blumenthal, B. Bean, et al., Designing sorghum as a dedicated bioenergy feedstock. Biofuels Bioprod. Biorefin. **1**(2), 147–157 (2007)
63. R.B. Rusu, S. Cousins, 3d is here: Point cloud library (pcl), in *2011 IEEE International Conference on Robotics and Automation (ICRA)*, (IEEE, Shanghai/Piscataway, 2011), pp. 1–4
64. K.H. Ryu, J.S. Han, Development of robotic transplanter for bedding plants. J. Agric. Eng. Res. **78**(2), 141–146 (2000)
65. M.G. Salas Fernandez, P.W. Becraft, Y. Yin, et al., From dwarves to giants? Plant height manipulation for biomass yield. Trends Plant Sci. **14**(8), 454–461 (2009)
66. N. Schor, A. Bechar, T. Ignat, et al., Robotic disease detection in greenhouses: combined detection of powdery mildew and tomato spotted wilt virus. IEEE Robot. Automat. Lett. **1**(1), 354–360 (2016)
67. D. Shah, L. Tang, J. Gai, et al., Development of a mobile robotic phenotyping system for growth chamber-based studies of genotype x environment interactions. IFAC-PapersOnLine **49**, 248–253 (2016). https://doi.org/10.1016/j.ifacol
68. R.R. Shamshiri, I.A. Hameed, M. Karkee, et al., Robotic harvesting of fruiting vegetables: A simulation approach in V-REP, ROS and MATLAB, in *Proceedings in Automation in Agriculture-Securing Food Supplies for Future Generations, 2018*, (InTech, 2018a)
69. R.R. Shamshiri, C. Weltzein, I.A. Hameed, et al., Research and development in agricultural robotics: A perspective of digital farming. Int. J. Agric. Biol. Eng. **11**(4), 1–14 (2018b)
70. F.D.F.D. Silva, P.H.C. Luz, L.M. Romualdo, M.A. Marin, A.M.G. Zúñiga, V.R. Herling, O.M. Bruno, A diagnostic tool for magnesium nutrition in maize based on image analysis of different leaf sections. Crop Sci. **54**(2), 738–745 (2014)
71. A. Silwal, J.R. Davidson, M. Karkee, C. Mo, Q. Zhang, K. Lewis, Design, integration, and field evaluation of a robotic apple harvester. J. Field Robot. **34**(6), 1140–1159 (2017)
72. D. Slaughter, D. Giles, D. Downey, Autonomous robotic weed control systems: A review. Comput. Electron. Agric. **61**(1), 63–78 (2008)
73. L.R. Sobenko, A.P. deCamargo, T.A. Botrel, et al., An iris mechanism for variable rate sprinkler irrigation. Biosyst. Eng. **175**, 115–123 (2018)
74. M. Sokefeld, R. Gerhards, W. Kuhbauch, Automatic identification of weed species using digital image processing. J. Plant Dis. Prot. **14**, 143–152 (1994) (8) (PDF) Weed and crop discrimination using image analysis and artificial intelligence methods. Availablefrom: https://www.researchgate.net/publication/221990357_Weed_and_crop_discrimination_using_image_analysis_and_artificial_intelligence_methods. Accessed June 27 2023
75. D. Story, M. Kacira, C. Kubota, A. Akoglu, et al., Lettuce calcium deficiency detection with machine vision computed plant features in controlled environments. Comput. Electron. Agric. **72**(2), 238–243 (2010)

76. K.A. Sunitha, G.S. Suraj, P.N. Sowrya, et al., Agricultural robots designed for seeding mechanism. Front. Automob. Mech. Eng. **197**, 1–6 (2017)
77. A. Swapnil, B.A. Rajashri, M.A. Kajal, Agricultural robot for automatic seeding. Int. J. Innov. Res. Sci. Eng. Technol. **7**(3), 2531–2535 (2018)
78. N.D. Tillet, T. Hague, Computer-vision-based hoe guidance for cereals – An initial trial. J. Agric. Eng. Res. **69**, 225–236 (1999)
79. D. Tilman, Global environmental impacts of agricultural expansion: The need for sustainable and efficient practices. Proc. Natl. Acad. Sci. U. S. A. **96**, 5995–6000 (1999). https://doi.org/10.1073/pnas.96.11.5995
80. S.K. Truong, R.F. McCormick, W.L. Rooney, et al., Harnessing genetic variation in leaf angle to increase the productivity of Sorghum bicolor. Genetics **201**(3), 1229–1238 (2015)
81. T. Utstumo, F. Urdal, A. Brevik, et al., Robotic in-row weed control in vegetables. Comp. Electron. Agril. **154**, 36–45 (2018)
82. K.A. Vakilian, Massah, A farmer assistant robot for nitrogen fertilizing management of greenhouse crops. Comput. Electron. Agril. **139**, 153–163 (2017)
83. D. Van der Schans, P. Bleeker, L. Molendijk, et al., *Practical Weed Control in Arable Farming and Outdoor Vegetable Cultivation Without Chemicals*, PPO Publication 532 (Applied Plant Research Wageningen University, Lelystad, 2006), p. 77
84. E.J. Van Henten, J. Hemming, B.A.J. Van Tuijl, et al., An autonomous robot for harvesting cucumbers in greenhouses. Auton. Robot. **13**(3), 241–258 (2002)
85. E.J. Van Henten, B.A.J. van Tuijl, J. Hemming, et al., Field test of an autonomous cucumber picking robot. Biosyst. Eng. **86**(3), 305–313 (2003)
86. E.J. Van Henten, B.A.J. Van Tuijl, G.J. Hoogakker, et al., An autonomous robot for de-leafing cucumber plants grown in a high-wire cultivation system. Biosyst. Eng. **94**(3), 317–323 (2006)
87. E. Vrindts, J. De Baerdemaeker, H. Ramon, Weed Detection Using Canopy Reflection. Precis. Agric. **3**, 63–80 (2002)
88. X. Wang, W. Yang, A. Wheaton, et al., Automated canopy temperature estimation via infrared thermography: A first step towards automated plant water stress monitoring. Comput. Electron. Agric. **73**(1), 74–83 (2010)
89. H.A. Williams, M.H. Jones, M. Nejati, et al., Robotic kiwifruit harvesting using machine vision, convolutional neural networks, and robotic arms. Biosyst. Eng. **181**, 140–156 (2019a)
90. H. Williams, M.H. Jones, M. Nejati, et al., Autonomous pollination of individual kiwifruit flowers: Toward a robotic kiwifruit pollinator. J. Field Robot. (2019b). https://doi.org/10.1002/rob.21861
91. H. Williams, M. Nejati, S. Hussein, et al., Autonomous pollination of individual kiwifruit flowers: Toward a robotic kiwifruit pollinator. J. Field Robot. **37**, 246–262 (2020)
92. X. Wu, S. Aravechia, L. Phillipp, et al., Robotic weed control using automated weeds and crop classification. J. Field Robot. **37**, 322–340 (2020)
93. D. Xiao, L. Gong, C. Liu, et al., Phenotype-based robotic screening platform for leafy plant breeding. IFAC-Papers OnLine **49**, 237–241 (2016). https://doi.org/10.1016/j.ifacol.2016.10.044
94. Y. Xiong, C. Peng, L. Grimstand, et al., Development and field evaluation of a strawberry harvesting robot with a cable-driven gripper. Comput. Electron. Agril. **157**, 392–402 (2019)
95. H. Yaguchi, K. Nagahama, T. Hasegawa, et al., Development of an autonomous tomato harvesting robot with rotational plucking gripper. IEEE Int. Conf. Intell. Robot Syst., 652–657 (2016)
96. L. Yang, Y. Tianguang, Design of the solar energy watering robot. Int. J. Mater. Mech. Manuf. **3**(3), 201–204 (2015)
97. Y.H. Yeh, W.C. Chung, J.Y. Liao, et al., Strawberry foliar anthracnose assessment by hyperspectral imaging. Comput. Electron. Agric. **122**, 1–9 (2016)
98. S.L. Young, Beyond precision weed control: A model for true integration. Weed Technol. **32**(1), 7–10 (2018)

99. C. Zhang, H. Gao, J. Zhou, et al., 3D robotic system development for high-throughput crop phenotyping. IFAC-PapersOnLine **49**(16), 242–247 (2017a)
100. J. Zhang, L. He, M. Karkee, et al., Branch detection with apple trees trained in fruiting wall architecture using stereo vision and regions-convolutional neural network (R-CNN), in *2017 ASABE Annual International Meeting*, (American Society of Agricultural and Biological Engineers. CROPS (2014). CROPS: Clever Robots for Crops, 2017b), p. 1. http://www.crops-robots.eu/. Accessed 1-9-2014
101. FAO, Google, https://www.fao.org/state-of-food-agriculture/2019/en/ (2019)
102. T. Botterill, S. Paulin, R. Green, et al., A robot system for pruning grape vines (2016). https://onlinelibrary.wiley.com/doi/10.1002/rob.21680
103. H. T. Søgaard., Weed classification by active shape models. Biosyst. Eng. **91**(3), 271–281 (2005). https://doi.org/10.1016/j.biosystemseng.2005.04.011
104. A. Milioto, P. Lottes, C. Stachniss., Real-time Blob-wise sugar beets vs weeds classification for monitoring fields using convolutional neural networks. ISPRS Annals of the Photogrammetry Remote Sensing and Spatial Information Sciences IV-2/W341-48 (2017). https://doi.org/10.5194/isprs-annals-IV-2-W3-41-2017
105. C. Römer, M. Wahabzada, A. Ballvora, F, et al., Early drought stress detection in cereals: simplex volume maximisation for hyperspectral image analysis. Funct. Plant Biol. **39**(11), 878–890 (2012). https://doi.org/10.1071/FP12060
106. G. Dryden, Viticulture monitoring report. Technical Report ISBN 978-0-478-437060-7, Ministry of Primary Industries and New Zealand Winegrowers. (2014)

Empirical Analysis of Crop Yield Prediction Using Hybrid Model

E. Chandra Blessie, Sundaravadivazhagan Balasubaramanian, and V. Kumutha

1 Introduction

Driven by technological advancements and data-driven methodologies, smart farming has emerged as a vital role in modern agriculture. An integral aspect of smart farming is crop yield prediction, a vital element that empowers farmers and stakeholders in agriculture to make well-informed decisions, optimize resource allocation, and safeguard food security in an ever-evolving global landscape. In the age of precision agriculture, where data analytics, sensor technologies, and machine learning reign supreme, crop yield prediction has evolved into an indispensable instrument for elevating productivity, sustainability, and overall efficiency within the agricultural sector. This introductory exploration dives deep into the importance, obstacles, and hopeful prospects of crop yield prediction [1] in the context of smart farming, illuminating its capacity to bring about a transformative shift in how we cultivate and harvest crops worldwide.

In the era of smart farming, machine learning techniques have evolved as a powerful tool for various agriculture fields. Among them is the prediction of crop yield

E. C. Blessie (✉)
Department of Computing (Artificial Intelligence and Machine Learning), Coimbatore Institute of Technology, Coimbatore, Tamil Nadu, India
e-mail: blessie@cit.edu.in

S. Balasubaramanian
Department of Information Technology, University of Technology and Applied Science-AL Mussanah, Al Mussanah, Oman

V. Kumutha
Department of Computer Science, Avinashilingam Institute for Home Science and Higher Education for Women, Coimbatore, Tamil Nadu, India

S. Balasubaramanian et al. (eds.), *Intelligent Robots and Drones for Precision Agriculture*, Signals and Communication Technology,
https://doi.org/10.1007/978-3-031-51195-0_4

[2]. Several ML algorithms can be utilized for predicting crop yields in the context of smart farming, and they can be classified into various categories. The support vector regressor (SVR) technique is a supervised machine learning technique utilized for solving classification and regression tasks. SVRs [3] offer a robust and adaptable framework for forecasting crop yields, especially when working with datasets that display intricate, nonlinear connections among variables. Train the SVR by utilizing historical data, where the chosen features serve as inputs and the target variable is the crop yield. The SVR's objective is to establish an optimal decision boundary that effectively distinguishes various yield levels. Following training, employ the trained SVR to forecast crop yields for forthcoming seasons or timeframes, incorporating pertinent data like upcoming weather forecasts, environment factors, and soil conditions as input. The K-nearest neighbor (KNN) algorithm represents a straightforward supervised machine learning technique capable of addressing both classification and regression tasks. Predicting crop yield, an unknown feature value, can be accomplished by utilizing the values of the nearest known neighbors. This is achieved by calculating the Euclidean distance between data points to envisage crop yield depending on given constraints. A supervised learning technique decision tree (DT) may be implemented to solve both regression and classification issues. The structure of this classifier resembles a tree, with every leaf node representing the outcome, branches demonstrating decision rules, and core nodes reflecting dataset properties. In the training phase, the DT algorithm identifies the optimal feature for segmentation of data using a metric like entropy or Gini impurity. These metrics measure the level of disorder or diversity within the data subsets. The aim is to pinpoint the features that give maximum information gain or minimum impurity post-split. Within the domain of decision trees, when predicting a target for a dataset, the task initiates at the root of the tree. The root feature values are compared with the attributes of the record. Based on this assessment, the respective branch associated with that data is pursued, guiding the traversal to the subsequent node.

Ensemble algorithms [4] also show a crucial function in forecasting crop yield by combining the predictions of multiple machine learning models to enhance accuracy and robustness. Some of the notable ensemble algorithms used in this context are gradient boosting (GB), AdaBoost, and extreme gradient boosting (XGB). Gradient boosting stands as a potent boosting technique, amalgamating multiple weak learners to form a formidable learner. In this method, each subsequent method is trained to diminish the loss function, like mean squared error or cross entropy, of the former method applying gradient descent. The method computes the gradient of the loss procedure in relation to the calculations made by the current ensemble for each process [5]. Subsequently, it prepares a new weak model to diminish this gradient. The calculations generated by this novel technique are incorporated into the ensemble, and this iterative procedure continues till a predetermined terminating condition is achieved. AdaBoost, short for Adaptive Boosting, belongs to the ensemble of boosting classifiers. Its primary objective is to enhance classifier accuracy by integrating several weak classifiers. Adaptive Boosting is an iterative ensemble approach, where a strong classifier is constructed by amalgamating multiple initially weak performing classifiers. This incorporation aims to create a high-accuracy robust classifier. The fundamental principle of AdaBoost revolves around adjusting the weights of classifiers and training the data samples iteratively. This iterative process ensures precise predictions, especially for atypical observations.

XGBoost [6] is an open-source software library that implements highly optimized distributed machine learning algorithms utilizing the gradient boosting framework. XGBoost is a scalable and remarkably precise implementation of gradient boosting, designed to maximize computational efficiency and enhance the performance of machine learning models. Specifically engineered to harness the potential of computing power for boosted tree algorithms, XGBoost revolutionizes the building of trees by leveraging parallel processing instead of sequential methods like GBDT. Employing a level-wise strategy, it efficiently evaluates the quality of potential splits across the training set by scanning through gradient values and utilizing partial sums.

This study presents an empirical analysis focusing on utilizing ML algorithms and ensemble-based approaches for predicting crop yields. By employing ensemble techniques, which combine the predictions of various ML models, we aim to harness the collective wisdom of these models to achieve more accurate and robust predictions.

In this analysis, we explore diverse ML algorithms including support vector regressor (SVR), K-nearest neighbor (K-NN), and decision tree regressor (DTR) to forecast the crop yield in the first phase. During the second phase, we investigate ensemble methods like gradient boosting (GB), AdaBoost, and extreme gradient boosting (XGB), which integrate the predictions from multiple base models, leveraging the strength of each individual model to enhance overall prediction accuracy.

The dataset used in this analysis is carefully investigated and preprocessed to ensure relevance and accuracy. It encompasses a wide array of features crucial for crop yield prediction. Through comprehensive experimentation and evaluation, we seek to demonstrate the comparative performance of individual ML models and ensemble methods. The prediction is done based on the evaluation metrics. By improving prediction accuracy, farmers and stakeholders can make better-informed decisions, optimize resource management, and ultimately work toward a more sustainable and productive agricultural sector.

Results indicate that ensemble methods significantly improve prediction accuracy compared to individual models. The ensemble approach mitigates biases present in single models, leading to more reliable and precise crop yield predictions. Moreover, this study demonstrates the significance of feature selection and hyperparameter tuning in achieving optimal model performance.

This article is well organized as follows: Sect. 2 introduces a review of predicting crop yield by both machine learning and ensemble learning algorithms. Section 3 defines the existing ML and Sect. 4 about Ensemble methodologies in detail followed by the hybrid approach in Sect. 5. The empirical examination of the crop production forecast is examined in Sect. 6. The conclusion is given in Sect. 7.

2 Literature Review

Purushotam Naidu et al. [7] proposed a hybrid machine learning ensemble model for crop prediction. The ensemble model used is voting technique. The model presented in this study is structured utilizing a set of five diverse learning algorithms: support vector machine (SVM), Naïve Bayes, random forest, K-nearest neighbor (KNN), and logistic regression. The results demonstrate that the proposed hybrid ensemble model gives improved prediction accuracy than other algorithms. Van Klompenburg et al. [8] have done a thorough systematic literature review (SLR) to extract and amalgamate the approaches and attributes applied in the prediction of crop yield. Based on their study on deep learning algorithms, they found out that these algorithms are better than machine learning algorithms.

Mishra et al. [9] studied on different week regressors such as SVR, ridge, lasso, and linear and proposed a strong predictor to enhance the predictive performance of the learning problem. They demonstrated that the forecasting model shows individual regressors may lack strength, but the boosted model exhibits robust regression capabilities. Priya et al. [10] emphasized their work on predicting crop yield using the random forest algorithm based on available data, specifically focusing on Tamil Nadu's real data. The study involved constructing models and evaluating their performance with sample data. The results suggest that the random forest algorithm proves to be effective for precise crop yield prediction. Banu et al. [11] presented crop yield prediction by leveraging historical data encompassing weather patterns, soil characteristics, rainfall metrics, and past crop yields. Employing a machine learning algorithm, particularly the random forest, enables an analysis of crop growth concerning prevailing climatic conditions and biophysical alterations, leading to enhanced solutions for the agricultural system. The study also involves the creation of a web application to forecast overall crop yield as well as yield for specific crops. Furthermore, the system offers recommendations to farmers regarding the appropriate fertilizer for their chosen crops.

Agarwal and Tarar [12] focused on an approach based on deep learning techniques which gives improved accuracy by taking into consideration the climatic and soil condition of the land. While LSTM and RNN are used as deep learning algorithms in aforementioned work, SVM is used as a machine learning algorithm. Nosratabadi et al. [13] propounded innovative crop yield prediction techniques leveraging hybrid machine learning approaches. The study assesses the effectiveness of two hybrid models: artificial neural networks-imperialist competitive algorithm (ANN-ICA) and artificial neural networks-gray wolf optimizer (ANN-GWO) for accurately predicting crop yields. Sunitha Devi et al. [14] presented an inventive deep learning strategy proficient in effectively capturing and fusing spatial and temporal attributes, representing a significant advancement over conventional methodologies that frequently struggle with these facets. This method accurately predicts crop produces by a notably low error rate, capitalizing on the resilience with distinctive hybrid architecture blending WaveNet and LSTM. This introduces a novel standpoint to agricultural yield predictions. The proposed approach outperformed the existing approaches.

3 Machine Learning Methodologies

3.1 Decision Tree Regressor

The decision tree is a supervised learning technique utilized to both regression and classification tasks, although it is often supported for addressing classification problems. Operating as a tree-structured classifier, its internal nodes symbolize dataset features, branches signify decision rules, and every leaf node signifies an outcome.

The splitting process initiates at the root node and progresses throughout a branched tree structure, culminating at a leaf node (terminal node) holding the algorithm's prediction or final result. The construction of decision trees typically follows a top-down approach, selecting the most suitable variable at each stage to split the item set. Each sub-tree within the decision tree model can be depicted as a binary tree, in which a decision node is subdivided into two nodes depending on specific criteria. In this paper, we will consider regression trees, a type of decision tree where the target variable or terminal nodes can encompass continuous values, often real numbers. Conversely, when the target variable is capable of assuming a discrete set of values, these trees are termed classification trees.

In decision tree regression [15], the algorithm examines an object's features and constructs a tree-like model to predict future data, generating meaningful continuous output. Continuous output implies that the result is not limited to specific, distinct numbers or values but rather encompasses a broader, uninterrupted range. The framework of DTR given in Fig. 1 depicts the step-by-step procedure.

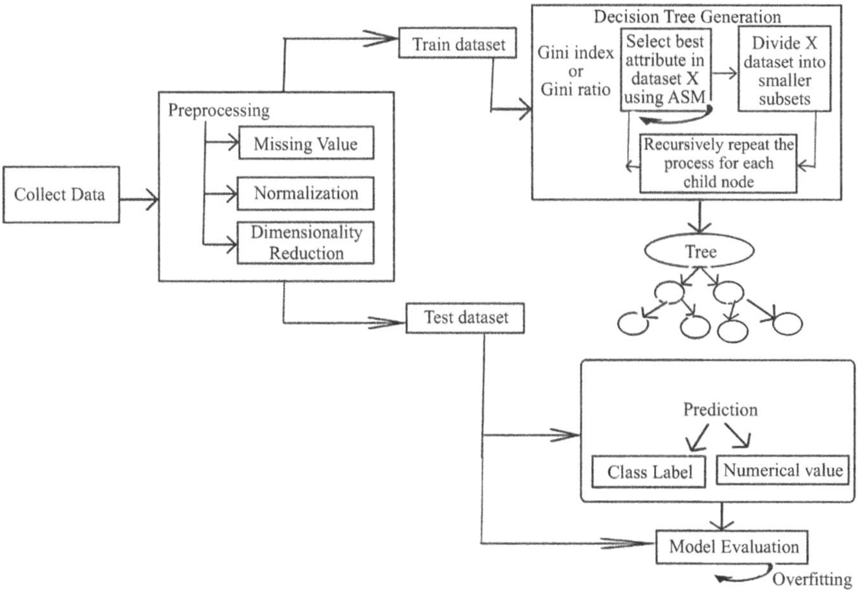

Fig. 1 Framework of decision tree regressor

Step-by-Step Procedure for Decision Tree Algorithm

1. Collect the data, and apply preprocessing techniques such as filling in missing values, dimensionality reduction, and normalization.
2. Select an attribute selection metric to identify the optimal feature for dividing the data at each node.
3. Choose the most suitable feature as the root node, guided by the selected attribute selection measure.
4. Partition the dataset into subgroups based on the values of the chosen feature (root node), where each subgroup corresponds to a branch stemming from the root node.
5. a. Within each subset (branch), iterate through attribute selection to identify the most appropriate feature for the subsequent tree level.
   b. Continue this recursive process till terminating condition is obtained (e.g., attaining a highest depth, no notable improvement in impurity reduction, or attaining a lowest count of representatives per leaf).
6. Define stopping criteria to cease tree expansion, mitigating the risk of overfitting. These criteria may encompass an extreme allowable depth, a limited quantity of samples per leaf, or a minimal enhancement in impurity reduction.
7. Allocate a class label or value to every leaf node by considering the predominant class or the mean (in the case of regression) of the samples present within that node.
8. Perform pruning to mitigate overfitting by eliminating redundant branches or nodes that do not notably enhance predictive accuracy.
9. Visualize the decision tree to grasp its structure and comprehend how features are utilized for classification or regression.
10. Use the trained decision tree to make predictions of class labels (for classification) or numerical values (for regression) for unseen data.
11. Assess the model's performance by employing suitable metrics like precision, accuracy, recall, mean squared error, F1-score (for classification), or R-squared (for regression).

Features of Decision Tree Regressor

- Decision tree regressors can handle both categorical and numerical features, determining the data split based on the specific feature type.
- The structure of decision trees enables high interpretability and ease of understanding. It provides a clear view of how the model formulates predictions based on the input features, making it intuitive for users to comprehend.
- Decision trees have the capability to capture intricate, nonlinear associations between features and the target variable, enhancing their versatility for addressing a diverse set of regression problems.
- Methods such as pruning, imposing a cap on the tree's maximum depth, and establishing a minimum amount of samples per leaf are implemented to mitigate overfitting and craft a more generalized model.

- Decision tree regressors exhibit robustness to outliers within the dataset, showing little impact on their performance, thus proving their resilience in the presence of noisy data.

Advantage of Decision Tree

- Intuitive and user-friendly.
- Capable of handling both categorical and numerical data.
- Robust against outliers, minimizing the need for extensive data preprocessing.
- Predicting using a decision tree is quick, requiring a logarithmic time complexity in the count of data points for every prediction.
- A decision tree is the most promising approach to establish relationships between variables and identify the most impactful variable.

Disadvantage of Decision Tree

- Susceptible to overfitting.
- Demand assessment of performance metrics.
- Careful parameter tuning is essential.
- Prone to biased tree creation if certain classes dominate the dataset.
- Decision trees are not mainly suited for effectively handling continuous numerical variables.

3.2 K-Nearest Neighbor (K-NN) Algorithm

K-nearest neighbors (K-NN) are considered as the most straightforward machine learning algorithms, built upon the values of supervised learning. The K-nearest neighbor (K-NN) algorithm functions depending on the hypothesis of likeness between a new data instance and the existing cases, assigning the new instance to the category that closely matches the available categories. By retaining and analyzing all available data, the K-NN algorithm facilitates straightforward classification of new data points, efficiently placing them into the most fitting category. The K-nearest neighbor (K-NN) algorithm is versatile, suitable for both regression and classification tasks, although it finds primary application in classification. Being nonparametric, K-NN avoids assumptions about the underlying data distribution, enhancing its adaptability and effectiveness.

In the training stage, K-NN algorithms preserve the complete training dataset for reference. During prediction, K-nearest neighbors (K-NN) compute the distance among the input data point and each training instances, employing a specified distance measure, namely, Euclidean distance. Following this, the algorithm recognizes the K closest neighbors to the input data point depending on these distances. In classification circumstances, the algorithm designates the prevailing class label among the K neighbors as the forecasted label for the initial given values. In regression, it computes either the mean or the weighted mean of the target values from the K neighbors to anticipate the value for the initial given values. Figure 2 depicts the step-by-step procedure for KNN.

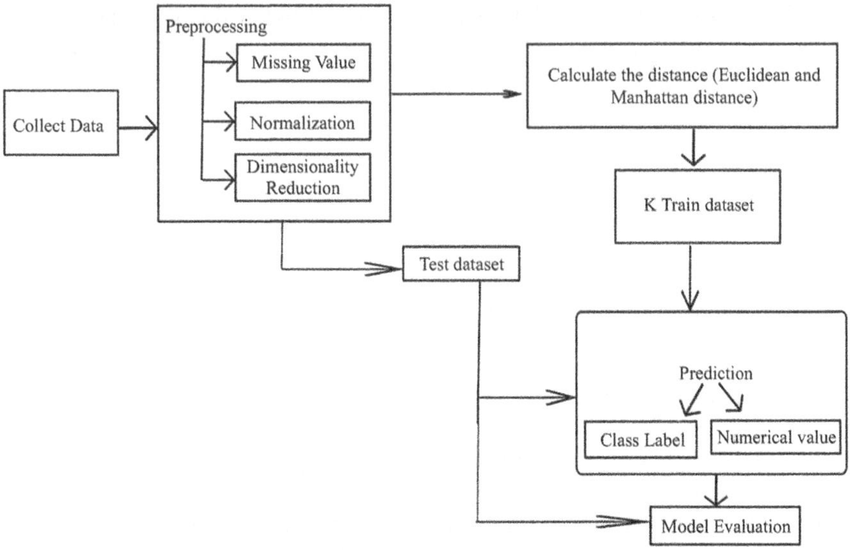

Fig. 2 Framework of K-nearest neighbor

Step-by-Step Procedure for KNN

1. Load and preprocess the data.
2. Select K, representing the quantity of closest neighbors, which is a critical hyper-parameter affecting the model's performance, one and the other classification and regression tasks.
3. Calculate the distance (like Euclidean or Manhattan distance) between the target data point (the one intended for classification or prediction) and every data point within the training set.
4. Choose K training instances that exhibit the shortest distances to the new data point, identifying them as the K-nearest neighbors.
5. In classification problem, identify the most frequent class label among the K-nearest neighbors, and allocate it to the newly generated values. For regression, compute the mean of the target values belonging to the K-nearest neighbors, and allocate it to the recent data point.
6. In classification, the forecasted class label for the new data point is determined by superiority voting. In regression, the predicted value for the new values is the average derived from the K-nearest neighbors.
7. Evaluate the K-NN model's performance by employing suitable metrics like accuracy, mean squared error, or other relevant measures based on the specific task, whether it's classification or regression.
8. When there's a requirement to classify or predict new data points, iterate through steps 4–7 for each of the new data points.

Features of KNN

- K-nearest neighbors (KNN) operate as a supervised learning technique, utilizing labeled input datasets to forecast the outputs of data points.
- This algorithm is among the simplest in machine learning, allowing for easy implementation across diverse problem domains.
- Primarily relies on feature similarity, where KNN evaluates a data point's similarity to its neighboring points and assigns it to the class it closely resembles.
- K-nearest neighbors (KNN) do not involve learning a specific model.
- It predicts outcomes based on the similarity between an input sample and each training instance.

Advantages of KNN

- Implementation is straightforward.
- Demonstrates robustness to noisy training data.
- Enhances effectiveness with a larger training dataset.
- It can adeptly manage scenarios with multiple classes.
- Its versatility lies in its suitability for both regression and classification applications.

Disadvantages of KNN Algorithm

- Requires determination of the value of K, which can be complex at times.
- Involves high computation costs due to calculating distances between data points for all training samples.
- Predictions become sluggish with high values of N.
- The algorithm is responsive to irrelevant features.
- Need substantial memory storage.

The input features (x1,x2,......xn) are fed into the model, and SVR often employs a kernel function (e.g., linear, polynomial, radial basis function) to map the input features into a higher-dimensional space as shown in Fig. 3. SVR aims to minimize the cost function by optimizing the parameters. This optimization involves finding the optimal weight vector u and bias b.

Step-by Step Procedure of SVR

1. Gather the dataset, perform preprocessing techniques such as scaling and filling the missing values, and perform feature engineering and feature selection.
2. Select Kernel like polynomial, linear, or radial basis function (RBF), and fine-tune the hyperparameters.
3. Instantiate the SVR model with the selected kernel and hyperparameters.
4. Train and assess the SVR model using the training data and test data, respectively.
5. Fine-tune the model by optimization technique.
6. Do prediction on unseen data.

Features of SVR

- Support vector machines (SVMs) excel in high-dimensional feature spaces, rendering them suitable for intricate real-world problems.

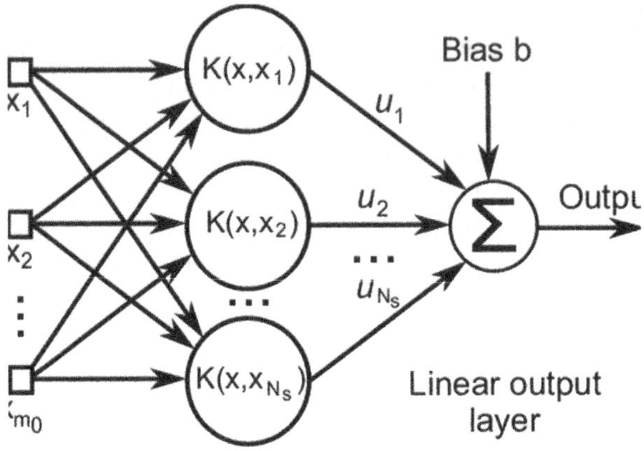

Fig. 3 Framework of support vector regressor

- Support vector machines (SVMs) are proficient in handling both binary and multiclass classification tasks with effectiveness.
- The kernel trick enables SVMs to manage nonlinear relationships by implicitly projecting the input data into a higher-dimensional feature space without the need for explicit transformation computations.
- SVMs demonstrate lower susceptibility to overfitting, primarily due to their objective of maximizing the margin, especially when suitable regularization parameters are chosen.
- SVMs usually involve a small number of hyperparameters for tuning, namely, the regularization criterion, and also the choice of kernel, simplifying their use and implementation.

Advantages of SVR

- Efficient in high-dimensional scenarios.
- Memory-efficient by using a subset of training points (support vectors) for the decision function.
- Flexibility in choosing varied kernel functions for decision-making, involving the option to describe custom kernels.
- Resistance to overfitting.
- Support vector machines (SVMs) demonstrate strong performance even with small training datasets.

Disadvantages of SVR

- Support vector machines (SVMs) can incur high computational costs, particularly with large datasets. Training time and memory demands escalate considerably with an increase in the number of training samples.
- Support vector machines (SVMs) involve crucial parameters like the regularization parameter and the selection of the kernel function. The performance of SVMs can be highly influenced by these parameter settings. Inadequate tuning may result in suboptimal outcomes or extended training durations.

- Support vector machines (SVMs) yield binary classification outputs and do not directly estimate class probabilities.
- Support vector machines (SVMs) have the ability to generate intricate decision boundaries, especially when nonlinear kernels are employed. However, this complexity can pose challenges in interpreting the model and grasping the fundamental patterns within the data.

4 Ensemble Learning Algorithms

4.1 Boosting Methods

Boosting has emerged as a popular method for addressing binary classification tasks. By transforming weak learners into strong learners, these algorithms significantly enhance predictive capabilities. The fundamental principle underlying boosting involves constructing an initial model on the training dataset and subsequently building additional models to correct errors made by the previous ones. This iterative process continues until errors are minimized, resulting in accurate predictions for the dataset. The three main boosting algorithms are AdaBoost, gradient boosting, and extreme gradient boosting.

4.1.1 AdaBoost

The AdaBoost algorithm, also known as Adaptive Boosting, is a boosting technique employed as an ensemble technique in the realm of machine learning. Its designation as Adaptive Boosting stems from the reassignment of weights to each instance, with notably higher weights allocated to instances that were classified incorrectly. AdaBoost, an ensemble technique, combines several weak classifiers to enhance classifier accuracy. AdaBoost operates as an iterative ensemble method, sequentially creating a robust classifier by amalgamating multiple weak classifiers. This amalgamation is strategically designed to transform initially weak performers into a high-accuracy strong classifier. The fundamental concept driving AdaBoost revolves around setting classifier weights and training data samples in each iteration to prioritize accurate predictions, particularly for atypical observations.

Step-by-Step Procedure for AdaBoost

1. AdaBoost begins by giving equal weights to all training samples. Each sample has an initial weight of 1/N, where N is the total number of samples.
2. A model is constructed using a subset of the data.
3. Predictions are created for the entire dataset using this model.
4. Bias is computed by evaluating the predictions with the original values.
5. In the subsequent model creation, elevated weights are assigned to data points with incorrect predictions.

6. The determination of weights can be based on the error value; for instance, higher error results in a greater weight assigned to the observation.
7. This iterative process continues until the error function stabilizes or the maximum limit of estimators is reached.

The depicted diagram in Fig. 4 illustrates the creation of the initial model, with subsequent identification of errors by the algorithm. Misclassified records from the first model are utilized as input for generating the next model.

This iterative procedure continues until the specified condition is satisfied. As illustrated, "n" models are constructed by incorporating errors from the preceding model. This exemplifies the fundamental operation of boosting. The models denoted as 1, 2, 3,..., N represent distinct models akin to decision trees. All variants of boosting models operate on this common principle.

Features of AdaBoost

- The weak learners employed in AdaBoost are decision stumps, which are decision trees with a single split.
- AdaBoost operates by assigning higher weights to instances that are challenging to classify and lower weights to those already effectively handled.
- AdaBoost algorithms are versatile and can be operated for classification issues as well as regression drawbacks.
- AdaBoost constructs a sequence of weak learners in a stepwise manner, with each subsequent weak learner placing increased emphasis on rectifying the errors of its predecessors. It allocates higher weights to misclassified samples, enabling subsequent models to concentrate on challenging-to-classify instances.
- AdaBoost effectively addresses the bias-variance trade-off through the amalgamation of weak models. By integrating a range of weak learners, it mitigates bias and consequently lowers the overall error.

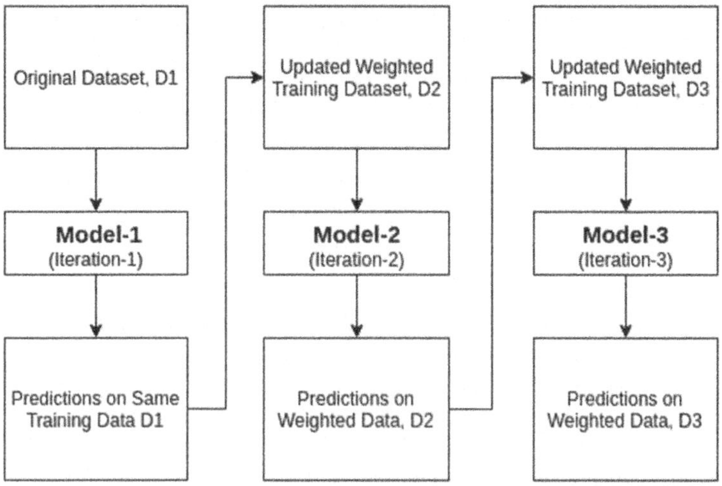

Fig. 4 Framework of AdaBoost

Advantages of AdaBoost

- AdaBoost enhances the accuracy of weak machine learning models.
- AdaBoost is resistant to overfitting as it processes every model sequentially and assigns weights to them, promoting better generalization.
- It also provides the flexibility to be united with any machine learning algorithm, requiring no additional parameter tuning.
- It demonstrates versatility as it could be utilized with both text and numeric data.
- Ability to manage intricate data and feature interactions.

Disadvantages of AdaBoost

- AdaBoost relies on high-quality training data as it is highly sensitive to noisy data and outliers.
- AdaBoost may encounter challenges with imbalanced datasets where one class has a considerably larger number of samples than others.
- AdaBoost can incur high computational costs, particularly with large datasets and complex models. Consequently, the training process might be prolonged and necessitate additional resources.
- AdaBoost can be easy to interpret.
- AdaBoost might assign higher weight to features highly correlated with the target variable. This potential bias in the model can result in inaccurate outcomes.

4.1.2 Gradient Boosting

A powerful boosting technique called gradient boosting joins several weak learners to produce robust and strong learners. Using gradient descent, each successive model in this process is prepared to minimize the cost function of the previous technique, which might be anything from mean squared error to cross entropy. In every iteration, the method determines the gradient of the loss function with respect to the predictions made by the current ensemble and then trains a new weak method to minimize this gradient. This new model's forecasts are supplemented to the ensemble, and the process is repetitive until a predetermined stopping threshold is met. The residual errors from the previous predictor are used as labels for each predictor during training. Gradient-boosted tree is a technique that uses CART (classification and regression trees) as its foundation learner. Figure 5's schematic illustrates gradient-boosted trees being competent for regression problems.

The residual errors (r1) in the training set are then determined using the predictions labeled y1(hat) from Tree1. Subsequently, the feature matrix X and the enduring error r1 from Tree1 are used as labels for training Tree2. The new residual, r2, is ascertained using the anticipated outcomes, r1(hat). Till all N trees in the ensemble are trained, this iterative procedure is continued. A key component of this method is shrinkage, which is the process of increasing each tree's prediction thereby the learning rate (eta), that spans from 0 to 1. This helps strike a balance among the eta and the amount of perceptors; a reduced learning rate requires

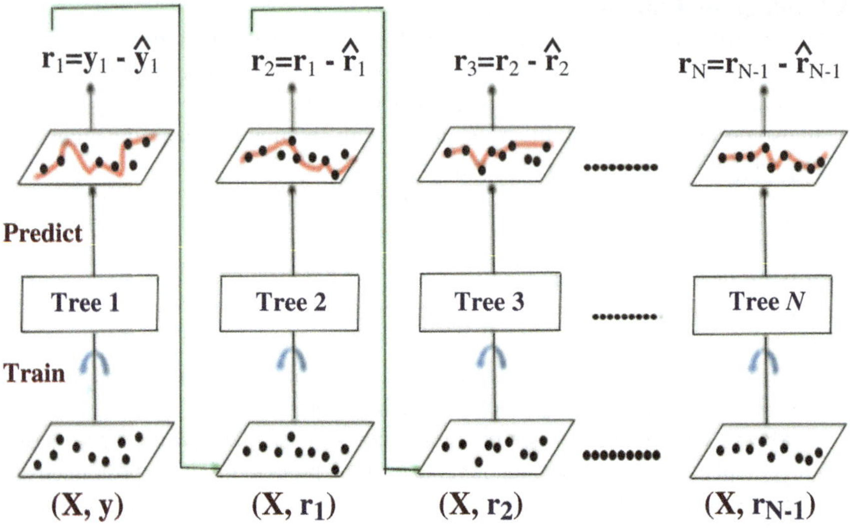

Fig. 5 Framework of gradient boosting

compensatory increase in estimators to achieve desired method performance. With all trees trained, predictions can be performed.

Features of Gradient Boosting

- Decision trees are utilized as the weak learner in gradient boosting.
- The gradient descent process is utilized to minimize the loss during the addition of trees.
- Gradient boosting, being a greedy algorithm, can rapidly overfit a training dataset.
- The predictions of each tree are sequentially aggregated.
- Gradient boosting algorithm operates based on three primary components: weak learners, the loss function, and an additive model.

Advantages of Gradient Boosting

- Fast in training process especially on large dataset.
- Efficient in handling categorical features.
- This algorithm is proficient not only in handling numerical datasets but also in efficiently managing categorical data.
- Due to its ensemble learning approach, interpreting and managing data with the gradient boosting model are more straightforward.
- It provides the flexibility to utilize a range of loss functions (e.g., regression loss, classification loss), enabling customization based on the specific problem at hand.

Disadvantages of Gradient Boosting

- The inclination to address every error from predecessor nodes can lead to overfitting of the model.
- Training complete gradient boosting models on CPUs can be computationally expensive and time-consuming.
- Very difficult to interpret the final model.
- Sensitive to outliers.
- Training the model might demand additional time and resources compared to simpler algorithms.

4.1.3 XGBoosting

XGBoost is a powerful ensemble learning algorithm in the field of data science. XGBoost, a gradient boosting algorithm extensively utilized in the field of data science, represents an implementation of gradient boosting engineered for exceptional efficiency, adaptability, and portability. XGBoost operates by combining numerous weak learners into potent learner. A weak learner refers to a machine learning model that exhibits slight improvement over random guessing. However, through their combination, these weak learners synergize to create a strong learner with significantly enhanced accuracy. XGBoost functions by training huge decision trees. Every tree is trained on a particular subset of the data, and the predictions from these individual trees are aggregated to produce the ultimate prediction.

Figure 6 shows the framework of XGBoost. It constructs decision trees using a depth-first approach. It greedily splits the data at each node, optimizing the objective function to find the best split.

Features of XGBoost Algorithm

- XGBoost (extreme gradient boosting) employs a range of regularization techniques to mitigate underfitting or overfitting, ultimately enhancing the model's performance.
- For each node, XGBoost uses parallel processing.
- XGBoost (extreme gradient boosting) alleviates the need for imputing missing values, as the model inherently handles them. By default, XGBoost autonomously determines whether these values should belong to the right or left node during the learning process.
- XGBoost employs cache optimization techniques to efficiently manage and utilize computational resources.
- XGBoost possesses the capability to identify and extract insights from nonlinear patterns within data.

Advantages of XGBoost

- It exhibits higher efficiency compared to alternative machine learning algorithms.

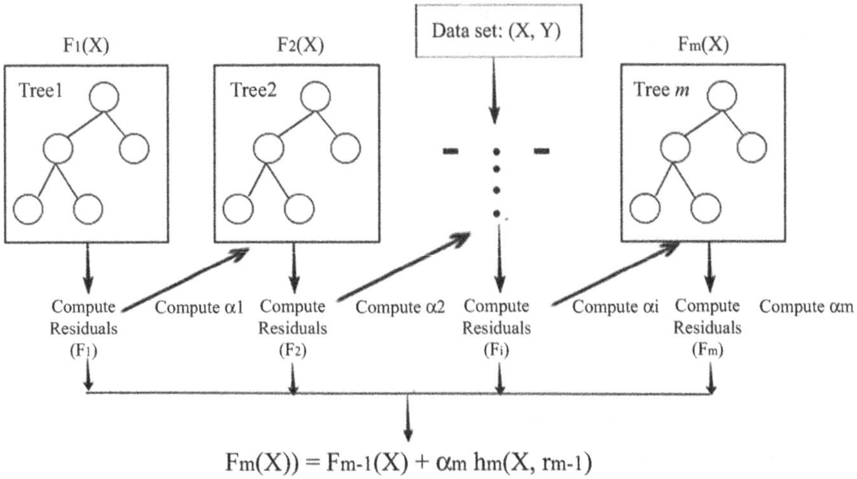

Fig. 6 Framework of XGBoost

- The XGBoost algorithm operates at a high execution speed, providing rapid and efficient output thanks to its parallel computation capabilities.
- XGBoost is equipped with a built-in functionality to manage missing values. When encountering a missing value at a node, XGBoost attempts both left- and right-hand splits, investigating the path that results in greater loss for every node.
- XGBoost is proficient in managing extensive datasets, parallelizing computations, and mitigating overfitting.
- It shows efficient execution speed.

Disadvantages of XGBoost

- XGBoost is sensitive to outlier.
- Difficult in visualization and interpretation.
- It's more demanding to fine-tune due to the abundance of hyperparameters.
- Without proper tuning of XGBoost parameters, overfitting is prone to occur.
- It overfits the model if the model is not stopped too early.

5 Hybrid Model

A hybrid model that integrates the predictive capabilities of decision tree regressor, K-nearest neighbors (KNN), and support vector regressor (SVR) through ensemble algorithms like AdaBoost, gradient boosting (GBoost), and XGBoost offers a powerful approach for robust prediction. The decision tree regressor excels in capturing nonlinear relationships and intricate feature interactions. Conversely, K-nearest neighbors (KNN) utilize proximity-based learning, considering nearby data points. Support vector regressor (SVR) is particularly adept at managing complex patterns and high-dimensional spaces. When their strengths are combined using ensemble methods like AdaBoost, GBoost, and XGBoost, the model achieves superior predictive accuracy by leveraging the diversity and complementary capabilities of these base learners.

AdaBoost, as an example, iteratively fine-tunes instance weights, prioritizing challenging cases and enhancing the overall model performance. On the other hand, GBoost and XGBoost enhance predictions by progressively incorporating weak learners, with each one rectifying prior errors, culminating in a precise and resilient predictive model. This hybrid strategy effectively amalgamates the unique strengths of each algorithm, producing a potent and adaptable predictive model suitable for diverse applications, such as crop yield prediction. The general framework of our crop yield prediction system is given in Fig. 7.

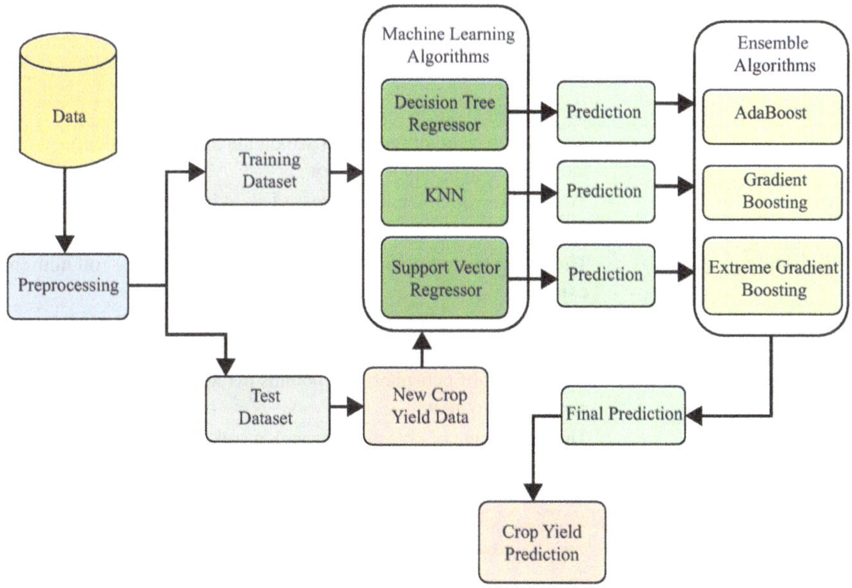

Fig. 7 The general framework of our crop yield prediction system

6 Experimental Analysis

6.1 Dataset Description

The crop yield prediction dataset consists of 756 rows and 12 columns. It also includes several factors that can potentially influence or are related to crop yield. These factors are represented by the columns in the dataset (Table 1).

These elements incorporate soil nutrient levels; temporal, geographic, cultivated, and harvested areas; and crop yield. Thorough analysis and understanding of these components can assist in predicting crop yield and enhancing agricultural practices to achieve improved productivity. The scatterplot in Fig. 8 shows crop yield divided by years categorized by the states. Each point on the plot would represent a data entry, with one axis representing the year and the other axis representing the crop yield.

The box plot showing crop yield vs. state given in Fig. 9 offers a visual representation of the distribution and statistical summary of crop yields for different states in the dataset. Each box represents the crop yield distribution within a specific state. The central line within each box represents the median yield for the respective state, providing a measure of the typical yield. The top and bottom edges of the box specify the third quartile (Q3) and the first quartile (Q1), correspondingly. This provides insight into the range where the bulk of crop yields fall within a specific state.

Table 1 Dataset description for crop yield prediction

Factors	Description
State	The specific state or region for which the data is recorded, indicating a geographical factor
Year	The year in which the data is recorded, indicating a temporal factor
Nitrogen (%)	The percentage of nitrogen in the soil, a soil nutrient factor
Nitrogen (pounds/acre)	The amount of nitrogen in pounds per acre, indicating soil nutrient content
Phosphorus (%)	The percentage of phosphorus in the soil, another soil nutrient factor
Phosphorus (pounds/acre)	The amount of phosphorus in pounds per acre, representing soil phosphorus content
Potash (%)	The percentage of potash in the soil, a soil nutrient factor
Potash (pounds/acre)	The amount of potash in pounds per acre, indicating soil potash content
Area planted (acres)	The area in acres that was planted for cultivation, representing the cultivated land area
Harvested area (acres)	The area in acres that was harvested, indicating the actual harvested land area
Lint yield (pounds/harvested acre)	The lint yield in pounds per harvested acre represents the crop yield

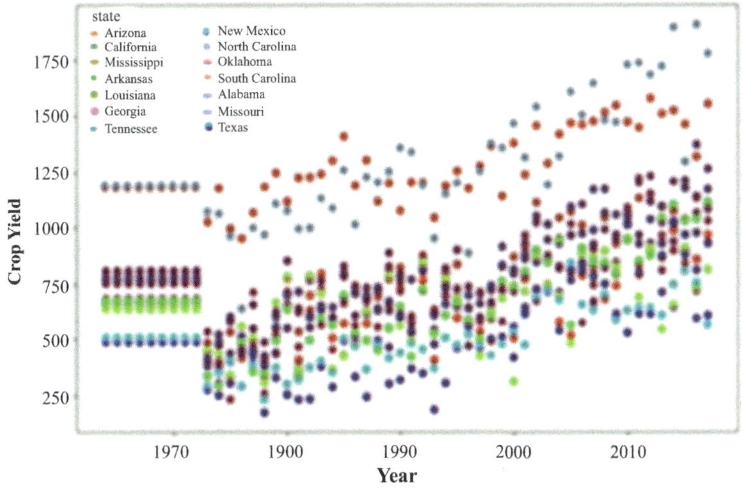

Fig. 8 Crop yield divided by years categorized by the states

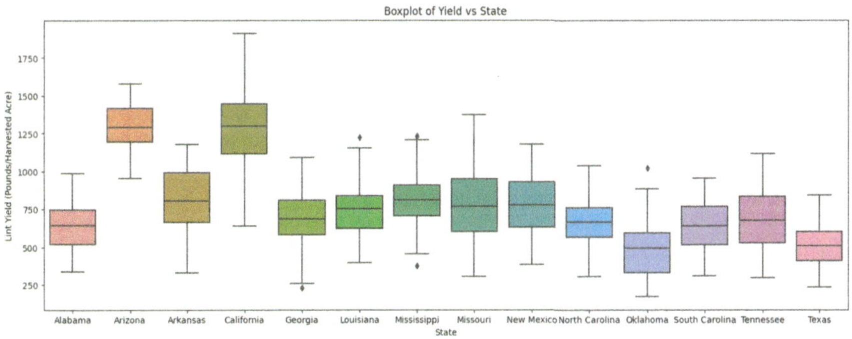

Fig. 9 The box plot showing crop yield vs. state

This visual representation facilitates a comparative examination of crop yields across distinct states, presenting disparities in yield distribution and pinpointing states with potential high or low yields. It offers a brief outline of the yield distribution, assisting in the evaluation of state-wise crop productivity and guiding agricultural decision-making.

The bar chart given in Fig. 10 presents a visual comparison of crop yields across the initial ten states in the dataset. Each bar characterizes a state and the height of the bar associated with the crop yield in pounds per acre for the particular state. This visualization allows for a quick assessment of the relative crop productivity among these states.

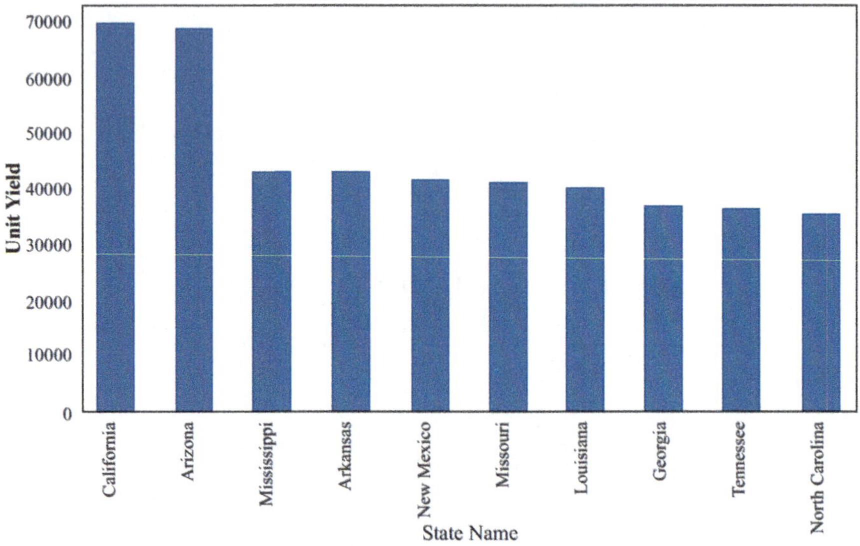

Fig. 10 Comparison of crop yields

Table 2 Parameter list for regression techniques

Machine learning and ensemble techniques used	Parameters
K-nearest neighbor (KNN)	Value of K, metrics = *"minkowski"*
Support vector regressor (SVR)	Kernel type = "rbf", Degree (Degree of the polynomial kernel function)
Decision tree regressor (DTR)	Criterion, splitter, max_depth
AdaBoost	n_estimators = 100, base_estimator = model3, learning_rate = 1
Gradient boosting (GB)	n_estimators = 10,000, learning_rate = 1
Extreme gradient boosting (XGB)	n_estimators = 5000, max_depth = 7, eta = 0.1, subsample = 0.7, colsample_bytree = 0.8, booster = 'gbtree', base_score

6.2 Parameter Discussion

In this experiment, the parameters with their values taken for result analysis are given in Table 2.

6.3 Result and Discussion on Machine Learning Model

We conducted experiment on crop yield dataset using machine learning algorithms such as K-nearest neighbors (KNN), support vector regressor (SVR), and decision tree regressor (DTR). The result shows that decision tree regressor outperforms the other two ML algorithms. The prediction done by DTR used AdaBoost, XGBoost, and gradient boosting to predict a target variable. We measured the performance of each model using mean squared error (MSE), a common metric used to evaluate regression models. Table 1 shows the MSE values of ML algorithms.

The decision tree regressor (DTR) yields lower MSE of 20922.607374918778 when compared to other two algorithms. This demonstrated that DTR is the best model with the lowest mean squared error (MSE) among the three models.

6.4 *Result and Discussion on Hybrid Model*

From Table 3, we see that the ML algorithm decision tree regressor (DTR) performs well for crop yield prediction. DTR is used as base learner in ensemble methods like AdaBoost, gradient boosting, and XGBoost. DTR is chosen as the weak learner in AdaBoost because of its capabilities to capture nonlinear relationship in the data. Gradient boosting also uses DTR as weak learner as it fits the weak learners sequentially to the residual errors of the previous models. XGBoost enhances the traditional gradient boosting by adding regularization. The number of estimators used in the ensemble learning technique is 1000, and the base estimator used is DTR. Table 2 shows that the gradient boost regressor performs well with a mean squared error of 26600.09. It captures intricate patterns in the relationship between soil types and crop yield. AdaBoost achieves a good balance between bias and variance. In this experiment for crop prediction using soil types, XGBoost regressor exhibited the best performance with the lowest mean squared root. The result suggests that XGBoost is the most promising choice for accurately predicting crop yield based on soil types (Figs. 11, 12 and 13).

Table 3 MSE values of all approaches

Algorithms	MSE
K-nearest neighbors	27923.50
Support vector regressor	97451.28
Decision tree regressor	20922.60
AdaBoost regressor	14463.70
Gradient boost regressor	26600.08
XGBoost regressor	13193.10

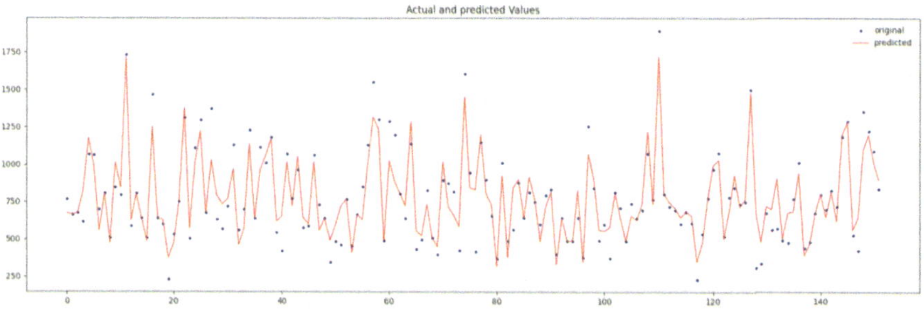

Fig. 11 Real and projected values using AdaBoost

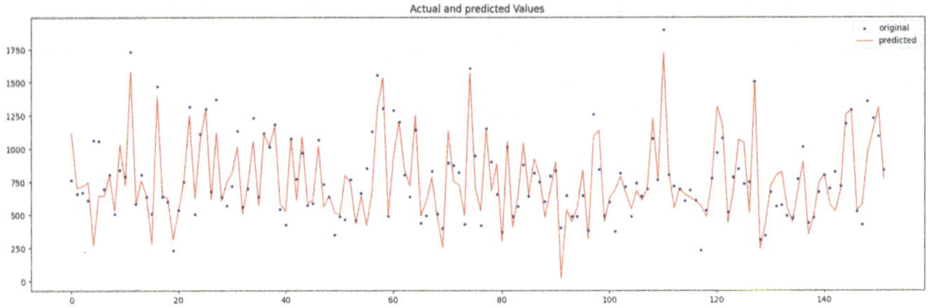

Fig. 12 Real and projected values using gradient boosting

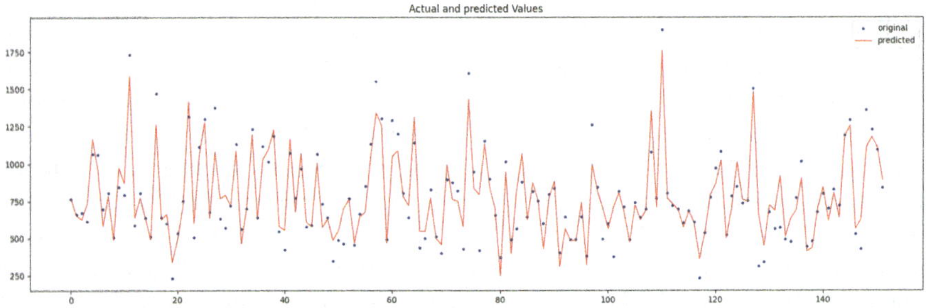

Fig. 13 Real and projected values using XGBoosting

7 Conclusion

In conclusion, the main objective of forecasting the optimal crops for cultivation by farmers was achieved. A hybrid model was built by using ensemble techniques to minimize the mean squared error which ultimately increases the accuracy. This study infers that the current research significantly improves prediction compared to existing methodologies. The empirical analysis on crop yield prediction using the AdaBoost, gradient boosting, and XGBoost (AdaBoost-gradient-XGB) hybrid model has proven to be highly effective and promising. This hybrid approach combines the strengths of AdaBoost, gradient boosting, and XGBoost algorithms, resulting in an ensemble model with superior predictive performance. The AdaBoost algorithm, known for its ability to sequentially correct errors and improve predictive accuracy, significantly contributes to the hybrid model. By focusing on challenging instances and adjusting instance weights iteratively, AdaBoost enhances the model's robustness. Gradient boosting, on the other hand, sequentially adds weak learners to correct errors from previous iterations, further refining the predictive capabilities of the hybrid model. XGBoost, with its efficiency and capability to handle complex patterns, enriches the ensemble by improving predictive accuracy.

This hybrid model excels in handling a diverse range of features, including numerical and categorical variables, making it versatile and applicable to various agricultural datasets. It effectively addresses the issues associated with crop yield prediction by leveraging the complementary strengths of these individual algorithms. The empirical analysis demonstrates that the AdaBoost-gradient-XGB hybrid model outperforms standalone algorithms, offering more accurate and reliable crop yield predictions. Also, XGB outperforms AdaBoost and gradient boosting algorithms. The integration of ensemble learning and hybridization in this manner presents a valuable approach to enhancing predictive modeling in the domain of agriculture.

References

1. R. Maclin, D. Opitz, Popular ensemble methods: An empirical study. J. Artif. Intell. Res. **11**, 169–198 (2011)
2. A. Chlingaryan, S. Sukkarieh, B. Whelan, Machine learning approaches for crop yield prediction and nitrogen status estimation in precision agriculture: A review. Comput. Electron. Agric. **151**, 61–69 (2018). https://doi.org/10.1016/j.compag.2018.05.012
3. Y. Zhang, Y. Zhu, S. Lin, X. Liu, Application of least squares support vector machine in fault diagnosis, in *ICICA 2011, Part II*, CCIS, ed. by C. Liu, J. Chang, A. Yang, vol. 244, (Springer, Heidelberg, 2011), pp. 192–200
4. D. Kocev, C. Vens, J. Struyf, S. Džeroski, Ensembles of multi-objective decision trees, in *European Conference on Machine Learning*, (Springer, 2007), pp. 624–631
5. A. Taherkhani, G. Cosma, T.M. McGinnity, AdaBoost-CNN: An adaptive boosting algorithm for convolutional neural networks to classify multi-class imbalanced datasets using transfer learning. Neurocomputing **404**, 351–366 (2020)
6. E. Bauer, R. Kohavi, An empirical comparison of voting classification algorithms: Bagging, boosting, and variants. Mach. Learn. **36**, 105–139 (1999)
7. K. Purushotam Naidu, P. Krishna Subba Rao, M.H.M. Krishna Prasad, Machine learning based hybrid ensemble model using majority voting technique for crop prediction. Webology **18**(5), 2282–2293 (2021). ISSN: 1735-188X
8. T. van Klompenburg, A. Kassahun, C. Catal, Crop yield prediction using machine learning: A systematic literature review. Comput. Electron. Agric. Elsevier **177**, 105709 (2020)
9. S. Mishra, D. Mishra, G.H. Santra, Adaptive boosting of weak regressors for forecasting of crop production considering climatic variability: An empirical assessment. J. King Saud Univ. Comput. Inf. Sci. **32**(8), 949–964 (2020)
10. P. Priya, U. Muthaiah, M. Balamurugan, Predicting yield of the crop using machine learning algorithm. Int. J. Eng. Sci. Res. Technol. **9**, 2 (2018)
11. R. Banu, N. Harshavardhan, S. Bharath, K.P. Dileepa, V.V. Rao, Random forest machine learning algorithm for crop yield prediction. Int. J. Creat. Res. Thoughts (IJCRT) **11**(5), 966–968 (2023) ISSN: 2320-2882
12. S. Agarwal, S. Tarar, A hybrid approach for crop yield prediction using machine learning and deep learning algorithms. J. Phys. Conf. Ser. **1714**(1) (2021). https://doi.org/10.1088/1742-6596/1714/1/012012
13. S. Nosratabadi, F. Imre, K. Szell, S. Ardabili, B. Beszedes, A. Mosavi, Hybrid machine learning models for crop yield prediction. Neural Evol. Comput. (2020). https://doi.org/10.48550/arXiv.2005.04155
14. B. Sunitha Devi, N. Sandhya, K. Shahu Chatrapati, Hybrid deep WaveNet-LSTM architecture for crop yield prediction. Multimed. Tools Appl. (2023). https://doi.org/10.1007/s11042-023-16235-7
15. S. García, A. Fernández, F. Herrera, Enhancing the effectiveness and interpretability of decision tree and rule induction classifiers with evolutionary training set selection over imbalanced problems. Appl. Soft Comput. **9**(4), 1304–1314 (2009)

Digital Twins and Predictive Analytics in Smart Agriculture

S. Clement Virgeniya

1 Introduction

Michael Grieves gave the concept of digital twins in 2002 [1]. Almost all fields like manufacturing, aerospace, healthcare, defense, transportation, construction, education, and obviously agriculture are making its footprint in digital twin. It is actually a replica involving all geometrical components and material properties and connected to real world through IoT and sensors [2].

In the aspect of healthcare industry, medical procedures were virtually created. This reduces the stress of medical practitioners in doing complex surgical operation. Analytics twinned with this virtual representation helps to further analyze and predict in advance. Moreover, separate distinct record is maintained for every single individual containing all the medical records starting from ECG, X-rays, MRI if any, and blood sample records. This helps in forecasting future outcomes in health of individuals. Many instruments for medical industry developed, test the product, and check its reliability. It creates models for disorder. This helps better learning about disorder strength and provides proper ailment before involving in-person activity. Nowadays, many medical students use this during their training period which gives them practical experience. It is also used in managing resources in hospitals and visualizing medicine utilizations, bed and room availability, etc. Although there are several booms, security of patient data, privacy, and other ethical issues need to be considered while implementing in healthcare.

In the aspect of education, digital twins and predictive analytics together have great achievement in today's education. They provide customized learning approach for students depending on their strength and weakness by collecting data, and this

S. C. Virgeniya (✉)
PG Department of Computer Science, Dr. Umayal Ramanathan College for Women, Karaikudi, Tamil Nadu, India

S. Balasubaramanian et al. (eds.), *Intelligent Robots and Drones for Precision Agriculture*, Signals and Communication Technology,
https://doi.org/10.1007/978-3-031-51195-0_5

data is analyzed for future educational needs. It helps to design course curriculum, and it is also used in future enhancement in course syllabi. Also, it checks for resources such as availability of classrooms, Internet, etc. It helps to identify weak students and provide strategies to improve their learning. Although there are many pros including decision-making by institutions and student support, there is every possible risk for security of student data.

Digital twin in smart cities helps to design and plan urban areas. It creates 3D model for developing cities for constructing roads, buildings, bridges, tunnels, etc. It reduces traffic by having collaboration with sensors and IoT platform. It can also decrease accident and control traffic congestion. They are widely used in monitoring the environmental condition of our country. It includes managing noise levels, weather forecasting, and monitoring air and water quality index. An alert about seismic waves is also given. Hence, digital twins in conjunction with predictive analytics give a comprehensive view of urban environment.

Digital twins are used widely in aerospace and automotives. It helps in designing vehicles, aircrafts, etc. It creates prototypes of aircrafts before actual implementation. It is also helpful in monitoring the vehicles and aircraft in real time.

In agriculture, there is always an increasing demand for food. This is because of increasing population and lack of proper utilization of food supplies and due to the attack of pests and weedicides to plants. Furthermore, environmental factors and climatic changes also affect agriculture growth. Keeping this situation in mind, digital twin arose in agriculture known as smart farming or smart cultivation. Digital twin along with predictive analytics is actually revolutionizing the entire agriculture sector. Digital twin is divided into physical and virtual space. Physical virtual space involves many steps. First and foremost, it involves the following:

1. Preparing soil where soil is plowed, leveled, and manured
2. Sowing involving appropriate selection of seeds
3. Manuring
4. Irrigation
5. Weeding at appropriate time
6. Harvesting when the crop is fully matured

This physical stream is very complex, whereas virtual space has a different way of approaching this issue as shown in Fig. 1.

First step is monitoring soil. There are a number of sensors which are used to monitor the pH and moisture level of soil and also pollutant in soil. These sensors also check the nutrient present in soil. Yield and health of plants depend upon soil quality. All the sensors are connected through IoT, and real-time data is collected from soil sensors at regular interval of times. Apart from data storing in cloud, real-time data analytics is also done with deep leaning and machine learning algorithms to give future insights of data. This is termed as predictive analytics.

Involving machineries like tractors gives knowledge to and fro and helps in controlling machinery. Digital twin uses robots in agriculture. It is used to model design and develop prototypes. Sensors are also involved in managing water supply to

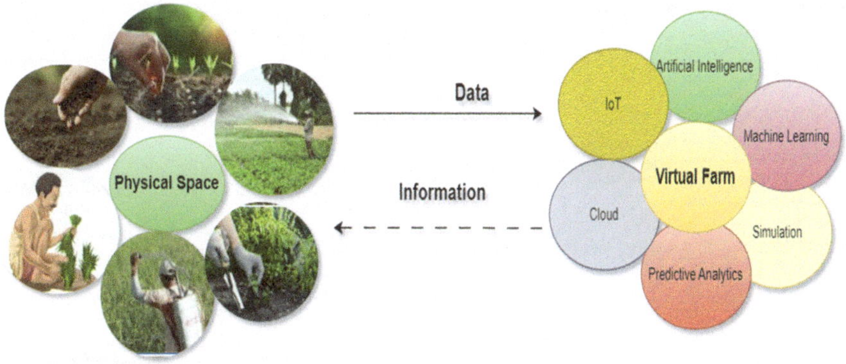

Fig. 1 Digital twin in agriculture

crops. Weather forecasting sensors check the climatic condition and water the plants appropriately.

The present study focuses on smart farming involving cloud, IoT, and predictive analytics. Many farmers were not aware of digitalization of agriculture. They rely on old methodologies and techniques for finding crop diseases like manually seeing affected crops after the occurrence of disease, not aware of soil nutrient and pH level. This results in decreasing the yield and profit of farmers. Few years back, it also created a situation where many farmers created suicide due to decreased yields in their farmland [3].

However, with the advent of digitalization, the previous scenario is completely revoking. Moreover, it could be prevented in advance if prior knowledge is given to farmers. This is the main aim of digital twin technology. Checking plant disease and inspecting soil and water nutrient level, temperature, moisture level, pH level, etc. could be completely made automatic and digitalized with less man power with the help of IoT, cloud, and predictive analytics [4].

Moreover, drones [5] are also used in keeping the plants healthy. They are used in monitoring the health of plants, monitoring the water supply of plants, identifying weeds, monitoring the presence of cattle, etc. [6]. Drones are equipped with GPS, cameras, and sensors as depicted in Fig. 2. In the coming years, 80% of farmers will use drones in their fields.

2 Literature Survey

Sun et al. [7] discussed digital twins in healthcare industry in particular in the musculoskeletal system. The authors focused on its usage in medical industry before using in real time. Inspite of its limitations in the availability of data, fusion of data, simulation, which restricted the usage of Digital twin. The authors proposed new

Fig. 2 Drones in agriculture [5]

trends in big data and IoT technologies which created real-time monitoring for humans especially elders.

Chakshu et al. [8] proposed inverse analysis approach which helps in monitoring and preventing medical conditions. However, there are some issues with clinical environment where only one type of heart disease is addressed and deep learning is applied on virtual database which dwindles accuracy of model. The authors also gave a solution to tackle the same using transfer learning.

Vikhman et al. [9] gave the prospective of digital twins in education. The author laid emphasis on grasping the process of involving digital twins in education. The author discussed social effects of implementing digital twin. He also discussed the complexity and unexpected effects and other sociotechnological consequences.

Nikolaev et al. [10] developed a simulation-driven product. The authors introduced a module for PG students for designing, prototyping, and testing complex systems like unmanned aerial vehicle.

Sepasgozar et al. [11] developed a digital pedagogy. The instructors could see the students' performance and give feedback instantly. The students were given online mixed reality modules to help them solve real-life problems like how a tunnel boring machine works, knowledge on running an excavator, etc.

Deng et al. [12] gave a systematic review of implementing digital twins in smart cities. The authors elaborated the techniques, applications, and theories concerned with digital twins in smart cities. A self-perceiving, self-determining, self-organizing, self-executing, and adaptive platform is built via mapping technology, IoT discernment, computing, simulation, and deep learning [12].

White et al. [13] said that digital twins improve the manufacturing process. With the advent of IoT and sensors, 3D model of a city can be made online before bringing them into practice. Thus, this open public model gathers feedback where the public can give their suggestions and can report problems in their area. It is also used in flood expulsion planning.

Verdouw et al. [14] analyzed how digital twins could advance smart cultivation. The authors stated that smart farming would no longer need close proximity for monitoring the crops. It not only represents actual states but also analyzes historical states and makes future decisions.

Pylianidis et al. [15] identified two distinctive characteristics of digital twins in agriculture: First, many digital twins involve directly or indirectly living systems and perishable products. However, their integration with the physical twin can be difficult. The second one lies in the spatiotemporal dimension of their operation.

Mohamed et al. [16] gave the importance of smart agriculture. The authors used IoT in connecting agriculture fields and all other applications. They also integrated IoT with UAV and robots and discussed the limitation of implementing in developing countries.

Farooq et al. [17] discussed the concept of IoT in agriculture. The authors threw light on smart farming including cloud, big data, and analytics. They also studied the importance of security in storing agriculture data.

3 Digital Twin Steps

The first and foremost step is setting sensors in the fields. The sensor appropriate to the field is chosen and installed in the fields. Placement of sensors should be well planned considering the depth, location, and gap between each sensor. The overall work is illustrated into different stages in Fig. 3.

The various data-capturing sensors and their purpose are listed in Table 1.

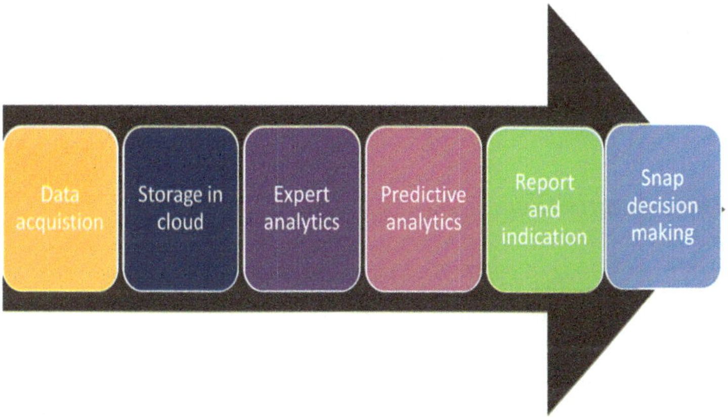

Fig. 3 Stages in digital twin technique

Table 1 Sensor and its purpose

Name of the sensor	Purpose
Soil moisture sensor	Retrieves moisture content of soil
Soil pH	Gets pH values of the soil
Light intensity	Checks whether sunlight is present or not
CO2	CO2 level is observed
Air, temperature, and humidity sensor	Air, temperature, and humidity are observed
Fire sensor	Fire sensor alarms if there exists fire
GPS	Tracks location
Water level	Checks water level
Gas sensor	Checks for hazardous gas
Soil nutrient sensor	Checks the nutrient level of soil

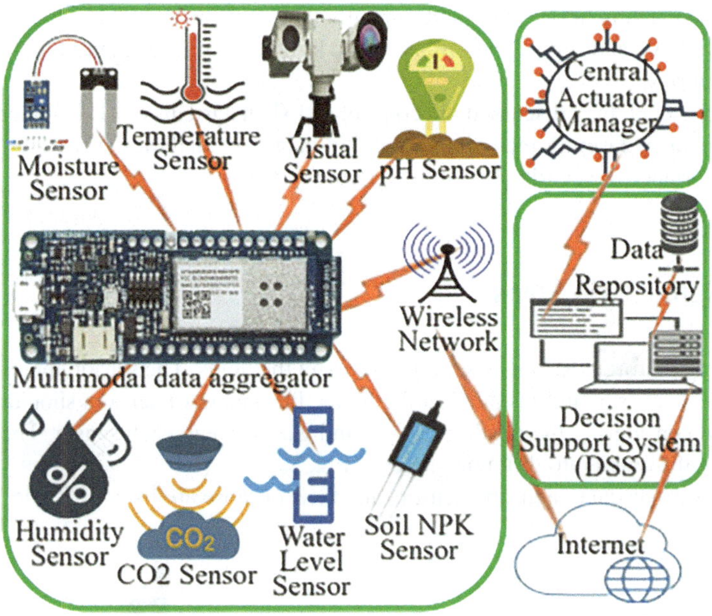

Fig. 4 Data acquisition from different sensors [19]

3.1 *Data Acquisition*

IoT platform plays a major role in acquiring data [18]. A model of data acquisition from various sensors is depicted in Fig. 4 [19].

The sensors are activated via actuators. It is maintained by central actuator manager. The data from these sensors is further taken into cloud. They are either connected using WI-FI connection or through low-power wide area network which transfers data. Sensors have reliable power supply either through batteries, solar panels. etc. Sensors are connected through microcontroller like Arduino or low-cost Raspberry Pi. These microcontrollers are involved in receiving data from sensors and also responsible for storage.

3.2 Storage in Cloud

There are different types of storage systems ranging from relational databases, NoSQL, cloud, warehouses, Hadoop distributed file system, and time series database. Cloud-based system like Amazon, Azure, and Google Cloud is economical and easy to use. A well-defined schema is first designed so that structured data makes it easier to store and retrieve. Appropriate APIs are used to store data in real-time streams.

3.3 Expert Analytics

It includes data preprocessing, feature extraction, and visualization. Data preprocessing and feature extraction are important steps in bringing data visualization better and accurately. Depending on the structure of data, it is normalized. Invalid or duplicate streams give wrong decision. So, it is eliminated.

It is always good to evaluate the performance of data after preprocessing. Secure encryption technique is followed to maintain the security of data, since farmers are unaware about data hacking.

3.4 Predictive Analytics

Predictive analytics is a class of supervised learning algorithms [4]. There are two main types, namely, supervised and unsupervised learning algorithms [20] with classification and regression problems. Since it is a categorical regression, algorithms like linear regression and logistic regression are used [21] in predicting the output.

Algorithms under predictive analytics are listed in Fig. 5. Before undergoing analytics, it is actually compared with historical data that is already present. Analyzing future trends from past data already present is known as predictive analytics. In this paper, five different algorithms depicted are used. It is observed that depending upon the data present, the accuracy and performance of the algorithm vary.

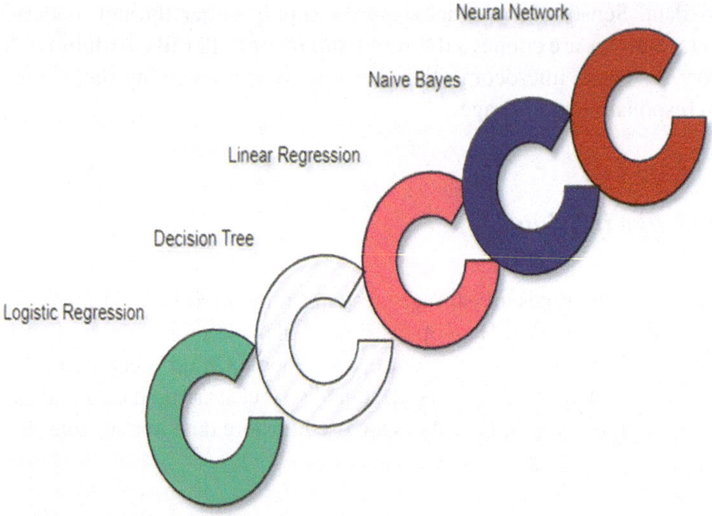

Fig. 5 Predictive analytics algorithms

3.5 Report and Indication

Various data visualization tools are available to indicate the status of the crops now and then. Apps help farmers to know the status of the crops immediately. Sensors placed at intervals in the fields (Fig. 6) sense data at hourly or daily basis and send to the experts to analyze. Experts provide insights of data using predictive analytics algorithms and report the same visually so that farmers could understand.

Initially, the data collected is *raw*. This means it won't make any sense if a farmer sees that data. It should be preprocessed; outliers have to be removed and structured in such a way that it can be easily predicted with the predictive analytics algorithms illustrated in Fig. 5.

3.6 Snap Decision-Making

It helps farmers make appropriate decision at the right time. If water level increases, the pH value is imbalanced, and the pests attack or if there is fire nearby, it alerts farmers about its presence and take decisions. Figure 6 shows the overall smart work in agriculture to decision-making via apps. The data collected from sensor reading through IoT is stored in cloud in real streams.

In parallel, historical data is also present in database. Experts analyze the data, preprocess, extract valuable features, and apply machine learning algorithms to find crop status. Finally, status of crops is visualized to farmers via apps, and hence they could make appropriate decision about plants' health condition.

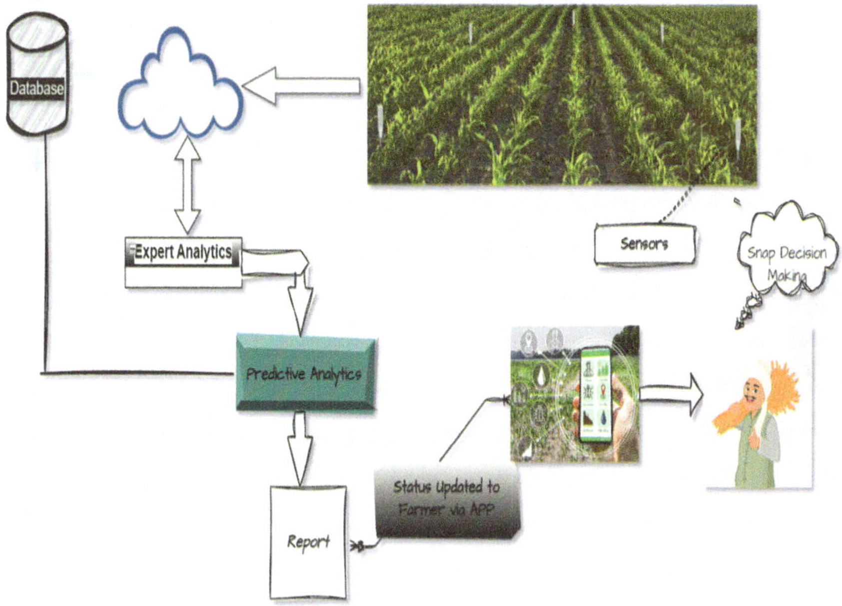

Fig. 6 Smart work in agriculture

4 Experimental Study

Experimental results show a detailed view of applying predicting analytics algorithms in predicting crop disease. Five different algorithms as depicted in Fig. 5 are used, and neural network algorithm gives better performance.

Neural network algorithm used here is convolutional neural network (CNN or ConvNet). Since images of crops are sensed in a grid-like structure, CNN is used. CNN's architecture has three layers, namely, convolutional layer, pooling, and fully connected layer. CNN is a product of two matrices. One matrix is known as kernel and other is receptive. The kernel's height and width are small and depth is large [22] (Fig. 7).

Two-dimensional image called activation map is produced during forward pass, and sliding size of the kernel is called a stride. Convolution layer [22] is calculated using Eq. (1):

$$W_{out} = \frac{W - F + 2P}{S} + 1 \tag{1}$$

where

Spatial size is represented as F.

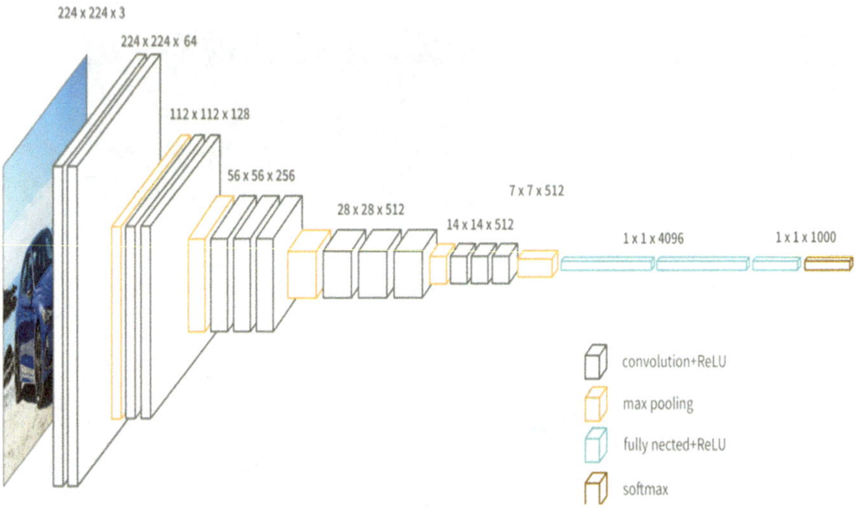

Fig. 7 Architecture of a CNN [23]

Stride is represented as S.
Amount of padding is represented as P.
W_{out} is represented as the size of output volume.

The operation of CNN is given in Fig. 8.

CNN is based on three key points: sparse interaction, parameter sharing, and equivariant representation [22]. Sparse interaction, where a portion of input is involved for interaction. This is made possible by making the kernel smaller. Only few important and meaningful pixels are chosen among thousands of pixel (depicted in Fig. 8). Through this, only lesser amount of memory is required, and it also increases the accuracy of the model.

Pooling layer replaces the output by nearby inputs. In this layer, spatial size is small with less computations. Pooling operation is performed on every slice. Slicing operation is portrayed in Fig. 9. Max pool is very commonly used. If an activation map is of size $W \times W \times D$, then the pooling layer is calculated using Eq. (2):

$$W_{out} = \frac{W - F}{S} + 1 \tag{2}$$

In a fully connected layer, all the neurons are connected to one another preceding and succeeding layer.

In the present study, cotton leaves collected from Kaggle are being considered for applying predictive learning algorithms. It consists of diseased and fresh cotton leaves. This repository is used as historical dataset for comparing with sensor data arriving at real streams. Data is preprocessed, and predictive learning algorithms are applied to find the diseased plant.

The metrics used to evaluate the algorithms are given from Eqs. (3) to (6):

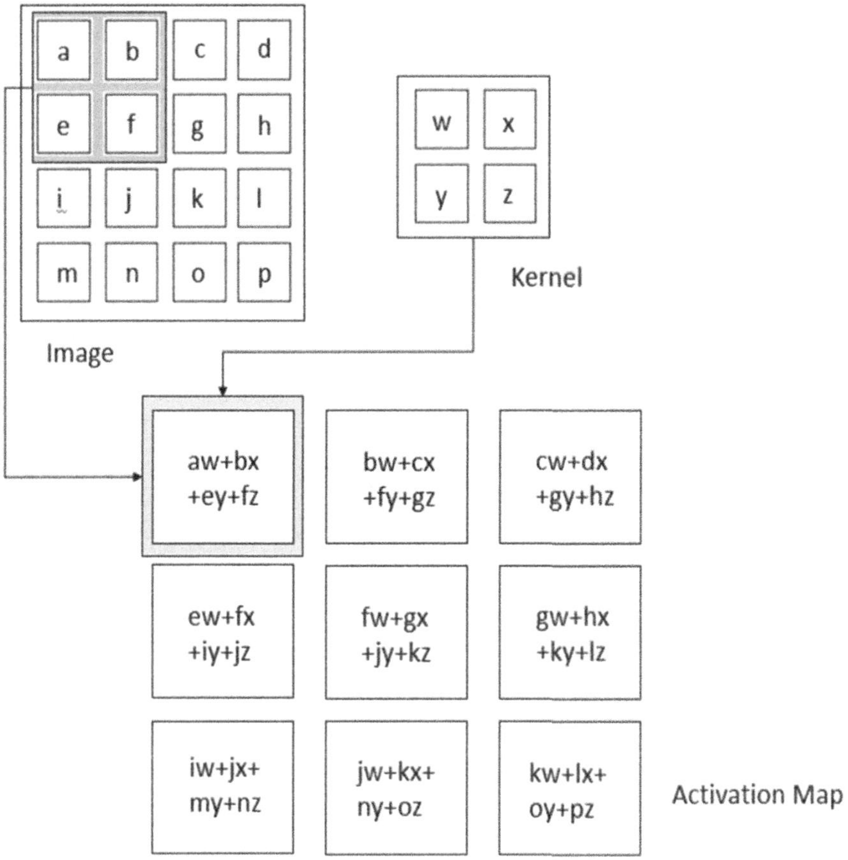

Fig. 8 Convolution operation [24]

$$\text{Precision} = \frac{\text{TP}}{\text{TP} + \text{FP}} \tag{3}$$

$$\text{Recall} = \frac{\text{TP}}{\text{TP} + \text{FN}} \tag{4}$$

$$\text{F1score} = 2 \times \left(\frac{\text{Precision} \times \text{recall}}{\text{Precision} + \text{recall}} \right) \tag{5}$$

$$\text{Accuracy} = \frac{\text{TP} + \text{TN}}{\text{TN} + \text{TP} + \text{FN} + \text{FP}} \tag{6}$$

Table 2 describes the training and validation accuracy of plants, and Fig. 10 represents the same visually for data stored in database.

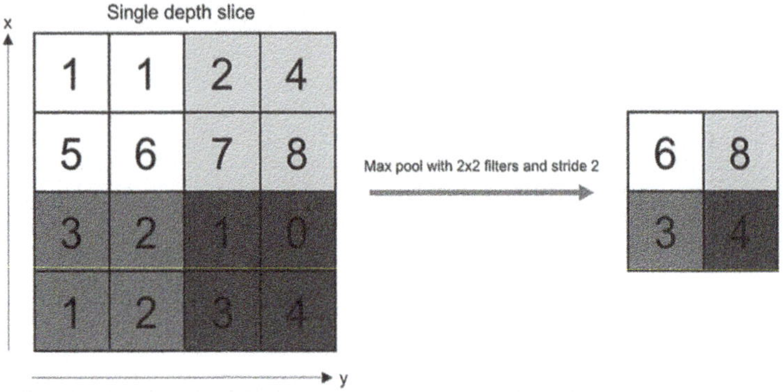

Fig. 9 Pooling operation. (Source: O'Reilly Media)

Table 2 Accuracy of predictive learning algorithms

Algorithm	Training accuracy (in %)	Validation accuracy (in %)
Decision tree	99.04	91.08
Logistic regression	89	84
Linear regression	88.24	84.23
Neural network	99.67	94.87
Naïve Bayes	66	67

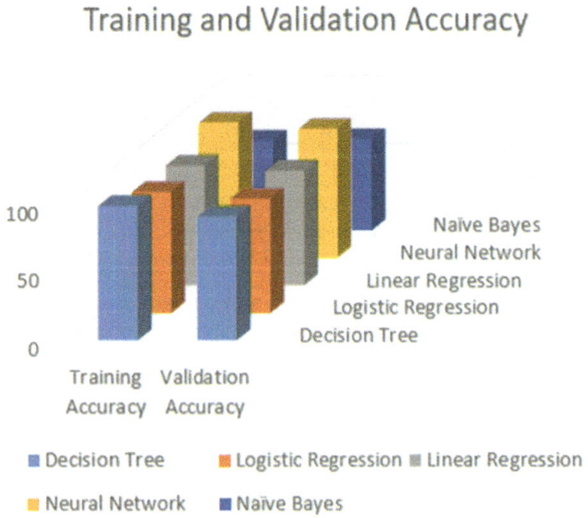

Fig. 10 Training and validation accuracy

It is observed that decision tree algorithms give an accuracy of 99.04% for training and 91.08% for validation. Logistic regression gives an accuracy of 89% and 84% for validation. Linear regression gives an accuracy of 88.24% for training and 84.23% for validation. Neural network gives an accuracy of 99.67% for training and 94.87% for validation. Naive Bayes gives an accuracy of 66% for training and 67% for validation.

5 Conclusion

The present study depicted the digital environment in setting smart agriculture-based environment. It also threw light on data acquisition from sensors, storage, processing, expert analytics, report generation, and finally informing farmers through apps.

In this paper, cotton leaves are tested for diseased and healthy leaves. Five different algorithms are being implemented, and neural network-based convolutional neural network achieves higher accuracy. Likewise, data from other sensors are also taken and analyzed. Due to insufficiency of data availability, the proposed work targets cotton leaf disease alone. Yet there are a number of other issues to be considered like monitoring soil moisture level, pH level, nutrient level, etc. But due to the lack of data, they are under process.

Smart agriculture is a developing area in developed and developing countries. This system would be more useful for farmers if it comes in the market. This system could be affordable for croft lands since affording sensors to the entire land is cost-effective. Many areas practice such technology placing sensors at distance.

As the popularity of smart agriculture is growing, many industries are involving them in their work. If this comes into reality, many farmers would be benefitted, reducing their manual work, saving their time and money, and increasing the yield. It also helps farmers to make more informed decision about the plant growth and availability.

For this, farmers need to be given general awareness about how to implement and activate sensors and apps and how to incorporate those in their field. Prior training before implementation will help farmers to do better. If this is possible, then suicide due to less yield would never happen in our country.

References

1. https://timesofindia.indiatimes.com/readersblog/education-insights/digital-twins-in-education-an-insight-38045/
2. T. Defraeye et al., Digital twins are coming: Will we need them in supply chains of fresh horticultural produce? Trends Food Sci. Technol. **109**, 245–258 (2021)
3. https://en.wikipedia.org/wiki/Farmers%27_suicides_in_India

4. S.C. Virgeniya, E. Ramaraj, Predictive analytics using rule-based classification and hybrid logistic regression (HLR) algorithm for decision making. Int. J. Sci. Technol. Res. **8**(10), 1509–1513 (2019)
5. https://www.freepik.com/photos/agriculture-drone
6. F. Veroustraete, The rise of the drones in agriculture. EC Agric. **2**(2), 325–327 (2015)
7. T. Sun, X. He, X. Song, L. Shu, Z. Li, The digital twin in medicine: A key to the future of healthcare? Front. Med. **9**, 907066 (2022). https://doi.org/10.3389/fmed.2022.907066
8. N.K. Chakshu, I. Sazonov, P. Nithiarasu, Towards enabling a cardiovascular digital twin for human systemic circulation using inverse analysis. Biomech. Model. Mechanobiol. **20**(2), 449–465 (2021)
9. V.V. Vikhman, M.V. Romm, "Digital twins" in education: Prospects and reality. Vysshee Obrazovanie v Rossii = High. Educ. Russ. **30**(2), 22–32 (2021)
10. S. Nikolaev, M. Gusev, D. Padalitsa, E. Mozhenkov, S. Mishin, I. Uzhinsky, Implementation of "digital twin" concept for modern project-based engineering education, in Proceedings of the IFIP International Conference on Product Lifecycle Management, Turin, Italy, 2–4 July 2018, pp. 193–203
11. S.M.E. Sepasgozar, Digital twin and web-based virtual gaming technologies for online education: A case of construction management and engineering. Appl. Sci. **10**(13), 4678 (2020)
12. T. Deng, K. Zhang, Z.-J.M. Shen, A systematic review of a digital twin city: A new pattern of urban governance toward smart cities. J. Manag. Sci. Eng. **6**(2), 125–134 (2021)
13. G. White et al., A digital twin smart city for citizen feedback. Cities **110**, 103064 (2021)
14. C. Verdouw et al., Digital twins in smart farming. Agric. Syst. **189**, 103046 (2021)
15. C. Pylianidis, S. Osinga, I.N. Athanasiadis, Introducing digital twins to agriculture. Comput. Electron. Agric. **184**, 105942 (2021)
16. E.S. Mohamed et al., Smart farming for improving agricultural management. Egypt. J. Remote Sens. Space Sci. **24**(3), 971–981 (2021)
17. M.S. Farooq et al., A survey on the role of IoT in agriculture for the implementation of smart farming. IEEE Access **7**, 156237–156271 (2019)
18. S.C. Virgeniya, E. Ramaraj, IoT and big data for ECG signal classification-a quick decision system, in *2021 2nd International Conference on Communication, Computing and Industry 4.0 (C2I4)*, (IEEE, 2021)
19. P. Tripathy, A. Tripathy, A. Agarwal, S. Mohanty, MyGreen: An IoT-enabled smart greenhouse for sustainable agriculture. IEEE Consum. Electron. Mag. **10**, 57–62 (2021). https://doi.org/10.1109/MCE.2021.3055930
20. https://machinelearningmastery.com/supervised-and-unsupervised-machine-learning-algorithms/
21. https://www.upgrad.com/blog/types-of-regression-models-in-machine-learning/
22. https://towardsdatascience.com/convolutional-neural-networks-explained-9cc5188c4939
23. https://towardsdatascience.com/an-introduction-to-convolutional-neural-networks-eb0b60b58fd7
24. I. Goodfellow, Y. Bengio, A. Courville, *Deep Learning* (MIT Press, 2016)

Soil Classification and Crop Prediction Using Machine Learning Techniques

Tilottama Goswami, Divyajyothi Mukkatira Ganapathi, and Prakriti Goswami

1 Introduction

Soil is the most precious natural resource on Earth which plays a major role in ecosystem sustainability. Both food and fabric needed by the world for survival depend on successful agriculture. For successful agriculture, farmers need to be acquainted with knowledge of soil characteristics that best suit a particular crop to increase productivity. Adoption of modern agricultural techniques for soil preparation, crop selection, and seed selection can improve crop yield production, leading to a more sustainable agricultural development. Soil obtained from the Earth's crust comes in different particle sizes, shapes, and compositions.

Classification of soil makes it much more convenient to study them and arrive at a certain generalized conclusion pertaining to crop cultivation. Any classification system must provide the expected engineering properties of a soil after detailed investigation and experiments. This serves as a language of communication between engineers across the globe. Classification is done to have a commonly accepted pedagogy of soils universally.

T. Goswami
Department of Information Technology, Vasavi College of Engineering, Hyderabad, Telangana, India

D. M. Ganapathi (✉)
Department of Information Technology, University of Technology and Applied Sciences – Al Musanna, Al Musanna, Sultanate of Oman
e-mail: divya@act.edu.om

P. Goswami
Department of Environmental Science, School of Earth Sciences, Central University of Rajasthan, Ajmer, India

S. Balasubaramanian et al. (eds.), *Intelligent Robots and Drones for Precision Agriculture*, Signals and Communication Technology,
https://doi.org/10.1007/978-3-031-51195-0_6

The lowest category of soil classification is called *soil series*. A soil series serves as the foundation for the placement of that soil series within the soil family and provides a record of the soil qualities required to create soil interpretations. The features of all soils belonging to a specific soil series would be uniform across the terrain [1]. Machine learning algorithms can be used to categorize the soil series data [2]. In order to identify appropriate crops based on geographic factors that are suitable for the soil series of a specific place and its climatic circumstances, the findings of such classification can also be integrated with crop databases. Consequently, the soil dataset and the crop dataset can be used for classification. The datasets consist of chemical and geographical attributes of soil and crops. This will contribute to precision agriculture which will help farmers get an informed farming strategy based on the soil health parameters like rainfall, climate, fertilizer, nitrogen content in soil, humidity, temperature, etc. Numerous variables, including meteorological, geographic, organic, political, and economic considerations, have an impact on plant crop output. Farmers may find it difficult to cultivate numerous crops, especially if they are not familiar with market prices. The soil series data may then be classified using machine learning classification and regression algorithms. Next, suitable crops can be predicted, and finally the crop yield can be regressed. ML includes a lot of practical ways for figuring out the input and output link in yield and crop prediction. For a range of tasks in agriculture, including yield prediction, smart irrigation, crop disease prediction, crop selection, and weather forecasting, machine learning techniques are applied. The following is how the chapter is structured. Section 2 discusses various taxonomies for classifying dirt. The crop forecast is briefly discussed in Sect. 3. The importance of remote sensing and machine learning methods in this field is discussed in Sect. 4. Section 7 brings the chapter to a close.

2 Soil Classification

Various soil classification systems are available which are going to be discussed in the following paragraphs.

2.1 Soil Classification Based on UNESCO Soil Map and USDA System

The authors in [3] aggregated various soil classification systems used in India since ancient times and the factors they based their classification systems on. Agriculture has always been the traditional occupation of Indians, and a vast amount of knowledge regarding soil nature and fertility has been passed on among agriculturists for generations. This study talks about the native classification that had been used, which was based on three main factors – soil fertility, climate, and revenue system of the land.

Classification based on soil fertility mainly included two types of soil, "urvara" and "anurvara," which mean fertile and sterile, respectively. They were further divided into different classes based on crop suitability and the source of their irrigation. On the other hand, climate-based classification focused on the nature of the soil and its characteristics in respective climatic conditions. Classification of soil is mainly based on dry, wet, and moderate (neither dry nor wet) land conditions. Citations from ancient texts like "Vishvavallabha" and "Jala-Bhumi: Krishi-Sukti" also helped enumerate the significance of soil characteristics like color and taste. Lastly, the revenue system focused on the productivity of land to generate revenue. This led to considering factors such as slope, texture of the soil, availability of water, crop yield, etc.

Subsequent efforts to classify the soil diversity of India in modern times led to the formation of the first soil map. This was further revised and simplified to a simple three-level classification where numerous soil families could be grouped together, making it easier for research purposes like the international correlation with foreign soil systems and soil mapping.

The paper [4] studied extensively on the huge diversity of soils present in the Indian subcontinent. India, with its diverse physical features and physiographic regions, has the largest variety of soils compared to other similar-sized countries in the world. Initiatives for soil classification in India date back to the nineteenth century, when soils were divided into four major groups – red soil, black soil, Indo-Gangetic alluvium, and laterite soil. The World Soil Map Project then generated the FAO/UNESCO soil map of the world, where the revised soil map of India consisted of 23 major classes.

The need of remapping soil was required for the purpose of sustainable resource management and efficient use and conservation of soil. India was divided into seven major zones, namely, northern, southern, western, eastern, northeastern, central, and islands. The soil resources in these regions were analyzed on the basis of water-holding capacity, nutrient capacity, base saturation, etc. Soil families were observed, and the cause for changes was analyzed. An increase in the number of soil families was an indicator of soil variation. The northeastern region (NER) was found to have a greater number of soil families per million hectares compared to other zones due to hilly conditions having different soil-forming processes and different physiographic features. Other factors like microbial population were also recorded to be found more in rain-fed ecosystems followed by mountain ecosystems. The current factors in these ecosystems enabled better drainage of surface soils which favors higher microbial growth.

The diversity of soil in India equals those found in temperate regions, thus disqualifying assumptions and generalizations of diminished soil quality and nutrients. Indian soil was also classified according to the USDA soil classification system developed by the US Department of Agriculture. The varying presence of these 12 soil orders were important in understanding the past of soil formation, which then helped analyze predictive models for the future. Global warming and the climate change in the Quaternary have had an impact on the quality of the soil. It has been found that soil has become increasingly calcareous and sodic, which ultimately affects the physical and chemical properties of soil particles. Such changes can affect the crop yield and land productivity. $CaCO_3$ concentration in soil stipulates a decrease in the mean annual rainfall which results in alkaline and calcareous soil formation. Such examples show the significance of proper soil grouping and how it can help in decoding the change in climate pattern, using soil taxonomy as the key to accessing the vast memories stored in the soil.

2.2 Soil Classification Based on UNESCO Soil Map and USDA System

Another methodology used for soil classification is based on images as mentioned in [5]. This paper considered using soil identification based on image analysis techniques. The target of this study was the soil of Bangladesh, for analyzing the maximum use of land resources available for agricultural purposes. The samples collected for research were classified into eight different datasets – clay, compost, agronomy, EPI, loamy, silt, sandy, and SI. Image analysis is based on characteristics of soil like color and texture. The Munsell color chart has been used for identifying the type of soil and also helps in understanding the relationship between the color of the soil and organic carbon content. RGB image processing helped in the detection of iron and carbon in soil. CIElab, CMYK, and XYZ mode image processing were used to detect other soil factors like moisture content and other soil variables like nitrogen, clay, and sand to name a few. The study further presents and experiments with an algorithm that combines Q-HOG (quartile histogram-oriented gradients), φ-pixels, and a new selection method for classifying soil types.

2.3 Soil Characteristics Based on the Unified Soil Classification System (USCS) and AASHTO

The paper [6] surveyed the characteristics of soil in Indonesia to get data based on the suitability of soil for the construction of buildings. Samples were taken from Jababeka I and Lippo Cikarang to analyze the swelling potential of soil found in these samples. The aim of this study was to analyze the plasticity of soil. Soil plasticity refers to the ability of soil to adjust to changes in shape without forming cracks in the soil. Some soils like expansive soil are often labeled problematic in civil or geotechnical engineering, because of their ability to change (swell or shrink) with change in moisture content. Since the properties of soil are largely dependent on the composition and size of the grain, the classification system used in this study is Unified Soil Classification Systems (USCS). USCS describes the mechanical properties of soil – texture and grain size. The AASHTO soil classification system developed by the American Association of State Highway and Transportation is based on the quality of soil for planning construction purposes and was also used as a guide for classification.

2.4 Land Suitability Binary Classification System Based on FAO

The paper [7] describes in detail about FAO land suitability classification. This classification developed by the Food and Agriculture Organization (FAO) uses suitability, in other words, the fitness of land as criterion for soil classification. The two orders in this classification are S and N, where S denotes suitability and N denotes unsuitable land. These orders are further divided into moderately suitable, marginally suitable, unsuitable with suitability potential, and so on, based on varying degrees of suitability of lands. These depend on subclasses, which are limitations used to assess the suitability of land. Some examples of subclasses include climatic conditions (c), wetness limitation (w), topographic limitations (t), etc. These limitations help in providing a qualitative evaluation of land. This classification system also has land suitability units and the Capability Index which helps in understanding the relative importance of land development.

2.5 Land Suitability Assessment Based on Soil Vegetation Indices from Satellite Data

In this paper [8], the authors wished to create a land suitability system based on soil vegetation indicators derived from satellite remote sensing. The aim of this study was to evaluate land conditions for optimal planning in agricultural sectors. Samples were collected from Rangpur, Dinajpur, Kurigram, and Gaibandha districts of Rangpur Division, northern part of Bangladesh, where most of the inhabitants have agricultural livelihoods. Vegetation indices (Soil-Adjusted Vegetation Index [SAVI] and Atmospherically Resistant Vegetation Index [ARVI] to name a few), land surface temperature, slope, and elevation were factors which helped in assessing the land suitability of samples. Analysis was based on land suitability classification by the Food and Agriculture Organization (FAO). The aim of using FAO suitability classification was to assess per unit suitability of land for crop production and develop a yield map.

A tool for spatial planning, the land suitability rating system evaluates the adaptability of crops in Canada [9]. discusses the land suitability rating system (LSRS) and how it can be used to identify trends or changes in crop distribution. Based on climate, soil, and landscape potential, LSRS is used to determine a land's rate of appropriateness. Agriculture and Agri-Food Canada first released LSRS in 1995. The current version of LSRS, a site-specific tool, can be used with any soil map that contains data from the Canadian Soil Information Service. Classes and subclasses make up the LSRS system's basic organizational structure. According to how much a crop's potential to be produced on a certain piece of land is constrained, suitability classes range from 1 to 7.

2.6 Area-Specific Soil Classification

Many classification systems are area-specific, tuned according to the climatic or physiological conditions of that area. For example, the land suitability rating system is specific only to Canada. The need to assess and map soil conditions on a global scale brought about a different soil classification system, which does not depend on the climatic conditions of a single area but rather focuses on the physical and chemical properties of the soil. USDA soil classification system developed by the US Department of Agriculture is one such example. There are other soil classification systems which focus on the physical and behavioral properties of soil. These are useful for anthropogenic activities related to soil, such as construction and engineering purposes.

3 Crop Prediction

Crop recommendation, yield production, and measuring losses are some of the objectives related to crop growing on agricultural land. Numerous variables, including genotype, climate, and their interplay, establish crop prognostic qualities. A fundamental comprehension of the functional linkages between agriculture and interfering variables like genetics and climate is necessary for accurate crop prediction. It is important for decision-making at global, regional, and local levels. Forecasts for yield are based on variables related to soil, weather, environment, and plants. Decision support models are often used to extract important crop traits for prediction purposes. One can use remote sensing data to estimate the amount of light plants receive and predict yields. Remote sensing data is a powerful tool for estimating yields. It provides information about growing crops and their environment and can estimate crop yields. Predicting yields under different climatic conditions can help farmers and other partners provide basic guidance related to agronomic and product decisions. The model can be used to select the best crops and their yields for a region. This also increases the value and profit of agriculture.

3.1 Crop Monitoring Using Remote Sensing and Deep Learning

Numerous studies have influenced national and local agricultural policies, and remote sensing and statistical approaches have been widely employed for crop assessment at the regional, national, and global levels. The majority of the sensing systems have been satellite-based. But as UAVs have proliferated and sensing technology has improved, field-scale analysis has emerged as a new option. Deep learning architectures have emerged as a result of recent developments in data acquisition capabilities, computational platforms (especially utilizing massively parallel processing using high-performance GPU boards), and big data structures. These architectures are now capable of carrying out tasks that were previously impractical to carry out using traditional machine learning techniques. These architectures are characterized by their reliance on sizable heterogeneous datasets and their ability to learn directly from raw data without the need for a separate feature extraction stage.

Rice crop classification and yield estimation using multi-temporal Sentinel-2 data: a case study of terrain districts of Nepal in paper [10] stressed on how factors like climate change and increasing population are affecting agricultural production in terms of food security, crop productivity, and sustainability, particularly in developing countries. Agriculture occupies a large portion of the national economy of Nepal, a developing country. Despite the number of crop monitoring systems available on a regional to global scale, Nepal falls short due to the lack of timely, good quality, complete data. Using a deep learning approach, this study used Sentinel-2 (S2) imagery which is freely available, to classify and estimate the yield of crops. The target area for this study is the 20 districts of the Terai region in Southern Nepal, which is 49% of the total agricultural land of Nepal.

The main technique for rice crop classification and yield estimation was a deep neural network (DNN). For the yield estimation method, two CNN (2D and 3D convolutional neural network) designs were developed. These two architectures' performances were evaluated using the RMSE (root-mean-square error) metric. To evaluate these models' performance and determine their correctness when applied to the provided dataset, performance metrics like F1-score were also examined.

3.2 *Predicting Crop Losses with Remote Sensing and Machine Learning*

The authors in [11] proposed a method where the fusion of agricultural data with remote sensing data would help in addressing the problems caused by disasters. Ethiopia, the area chosen for this study, is often plagued by droughts which has immense adverse effects on the agricultural sector. Forecasts of agricultural outcomes or warnings of impending drought conditions can help government bodies, authorities, and communities to prepare in advance and, thus, minimize consequences.

The main objective of this study was to come up with scalable, machine learning models that can predict crop losses due to drought, relying only on data present from earlier growing seasons.

Five cereal crops were chosen for this study – wheat, barley, maize, teff, and sorghum. Data regarding previous crop losses in Ethiopia was sourced from the Central Statistical Agency Agricultural Sample Survey. Data for other variables, like precipitation and water availability, was also collected.

The machine learning models are trained on average crop losses reported by farmers from 2010 to 2015 collected by Ethiopia's Central Statistical Agency. This data combined with remote sensing data from satellites helped this study to come up with a model that predicts crop losses sooner in higher spatial resolution compared to other existing models.

4 Role of Remote Sensing and Machine Learning Techniques

4.1 Data Preparation

Data preparation is one of the most fundamental steps for machine learning. Machine learning tasks are mostly classification and regression. Appropriate feature selection represents the true data and is helpful for accurate prediction.

4.2 Data Acquisition

For soil classification and crop yield domain, data can be collected from various sources, which is the first fundamental step as shown in Fig. 1. One of the sources is remote sensing satellite data that can be collected in the form of images, from a geographical location (region of interest) which can be lithological data, climatic data, and agricultural data. The images are hyperspectral or multispectral. Another source of collecting soil features is from wet lab experiments compiled at national level and stored in regional datasets.

Data collection is in the form of comma-separated files (CSV), excel sheets, or tabular data. The GeoTIFF files are in the form of raster image file types that collect the satellite and aerial image data. The dataset can be generated in the form of TIFF and CSV files when extracted from the GeoTIFF files. The dataset generated is mixed data consisting of numerical or continuous values, categorical values, and image data.

For using machine learning models, data wrangling and preprocessing of data are required to have clean and suitable features. These processes include handling missing values and outlier detection.

Data transformation preprocessing steps include normalization, scaling, and feature engineering. The soil data, climate data, and crop-related data consist of mixed data consisting of categorical, numerical, and image data. The categorical and numerical data after preprocessing is made input to multilayer perceptron (MLP), and the image data is fed to convolutional neural network (CNN).

Datasets. Some of the datasets in demand for this type of work are as follows:

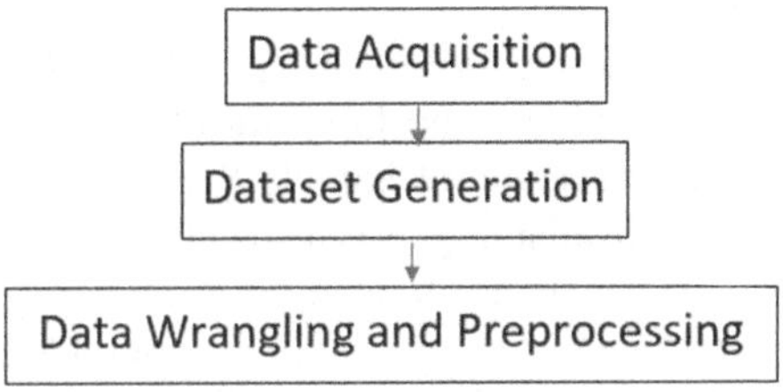

Fig. 1 Data preparation

- *Soil Dataset*: There are 207 soil datasets available on data.world. It consists of 16 attributes like PH, EC, OC, OM, N, P, K, Zn, Fe, Cu, Mn, sand, silt, clay, $CaCO_3$, and CEC.
- *Crop Dataset*: It consists of four attributes – temperature, humidity, PH, and rainfall.
- *Yield Dataset*: It consists of six attributes – nitrogen (N), phosphorous (P), potassium (K), organic care (Og), PH, and temperature.

Considering the crucial dimensions of space and time, it is important for machine learning algorithms to function on big datasets given the increasing proliferation of very large datasets. The data cannot be stored in the main memory if it is too huge. Similar to this, processing huge datasets would be impossible if learning time did not increase linearly with the quantity of training examples.

Consider the following when a dataset is too big for a certain learning algorithm:

- Use a limited subset rather than the entire dataset for training.
- Never evaluate performance using training data.
- Parallelization utilizing nearest neighbor approaches can be utilized to decrease learning's temporal complexity.

5 GIS and Remote Sensing

These methods take thermal emissions from soils into consideration for the soil series identification [12–16]. The information retrieved can be used by farmers as a base map to decide the type of crop to be cultivated, and the amount of fertilizers and pesticides to be applied. Researchers have used this technique for soil salinity mapping [17] too.

Geographic information systems (GIS): Similar to remote sensing, GIS helps in real-time analysis of soil, referred to as digital soil mapping (DSM) [18, 19].

Given a certain region or area of study, a ground survey should be carried out to gather soil data at various sites. Each soil sample would have its geolocation information tagged on it (soil sample number, date, latitude, longitude, and elevation). The nutrient content, composition (or type) (percentage of sand, silt, and clay), moisture content, texture, and other qualities, such as the acidity or pH level, will all be determined by laboratory examination of the soil sample that was taken. The GIS database should subsequently be populated with the mapped data from the soil survey.

6 Soil Classification Using Machine Learning Methods

The adoption of machine learning techniques to understand soil science research is referred to as pedometrics [20, 21] and has successfully been able to predict and classify soils and their distribution in space and time [22–24]. The soil classification systems such as the US Department of Agriculture (USDA) [25], Unified Soil Classification System (USCS) [26], OSHA Soil Classification System [27], and Indian standard soil systems [28] classify the soil series based on various soil attributes.

Below, we discuss some of the most commonly used techniques in soil mapping. In this section, we discuss some of the ML models and different variants of the algorithm that have been used in soil science mapping.

6.1 Machine Learning Implementation

Agriculture depends on predicting the best crop to grow, and in recent years, machine learning algorithms have become quite important in this process. In this age of data science and technology, the agriculture sector has a lot to gain from correctly applied methods. Two essential machine learning techniques are feature selection and classification. The goal of feature selection is to extract the most crucial dataset attributes.

Based on a benchmark that has been established, such as classification performance or class separability, which is essential in machine learning applications, it comprises choosing a part of relevant attributes from a larger collection of original attributes.

Datasets are collected from different resources and then divided into two sets, usually in the ratio of 80:20 as training vs. test dataset:

(i) Training Dataset: The data originally used to train the model as a first part of learning.
(ii) Testing Dataset: The dataset used to verify how good the model has learned the concept during training process.

Different supervised algorithms mentioned below are commonly used to create learning models. Their results are compared using appropriate performance measures.

6.1.1 Random Forest

Random forest integrates the soil inventory data with the available digital data using flexible statistical models and GIS tools, generating soil attributes [29]. Random forest is an ensemble learning technique that works on multiple decision trees during the training process and increases the prediction accuracy.

It finds a relationship between soil data and predictor data for prediction purposes. The model, in its simplest form, looks like the following:

$$y \sim f\left(x1 + x2 + x3 + \ldots\right) + \varepsilon,$$

where y is a response variable, xs are predictor variables, and ε is the associated error [30]. Random forests for soil characterization are widely in use [31] due to their resistance to overfitting.

Furthermore, if the considered dataset is very large, principal component analysis (PCA) can be used which will reduce the dimensions of the dataset without any information loss, using orthogonal linear transformations optimally describing the variance. Before proceeding with the analysis, the dataset is normalized. The result of PCA is a covariance matrix which is symmetric.

6.1.2 Support Vector Machines

This methodology builds a linear discriminant function that widely separates the critical border instances – referred to as support vectors – from each class.

This technique, which utilizes a linear kernel to categorize photographs of soil, is known to be effective in high-dimensional spaces for classifying soil texture [32]. A significantly faster and simpler similarity function is the linear kernel function. Cross validation is a method that can be used to gauge the accuracy of an SVM classifier.

The SVM model performs by providing the highest accuracy in soil classification, leading to prediction of soil properties, climatic factors, and crop production [33, 34]. Better precision accuracy can be achieved by fine-tuning the parameters in the training model.

By making it feasible to include nonlinear variables in the function, the instance-based method in this model overcomes the constraints of linear borders and enables the formation of quadratic, cubic, and higher-order decision boundaries.

PSO-SVM facilitates nonprobabilistic binary linear classification (particle swarm optimization-support vector machine). Using this method, one or more target classes may be found. Data is represented by a single dot (or point). It spreads as a result of the numerous distinctions among the various groupings.

According to where they fit within the gap, those new instances are divided into a number of target classes.

When the input datasets are unlabeled, nonlinear classification is feasible. Unsupervised learning is used to classify the data because there are no goal classes to assign to the instances. Following the construction of the function-based clusters, additional instances may be added.

6.1.3 Multiple Linear Regression

When the outcome class is numeric, this model is a staple method to consider and express the class as a linear combination of the attributes with predetermined weights. This model can be used to study relationships between multiple variables, where one is a continuous dependent variable and others are independent variables. It is represented as follows:

$$Y = m1^{*}\, x1 + m2^{*}\, x2 + m3^{*}\, x3 + \ldots\ldots\ldots + mn^{*}\, xn + c.$$

where Y is the predicted variable; $m1$, $m2$, and $m3$ are the regression coefficients; and e denotes the random error.

This model allows for the systematic description and comparison of soil properties and for the determination of the relationship between the selected soil properties [35].

6.1.4 Deep Learning and Convolutional Neural Networks (CNN)

Deep learning is a subset of machine learning, where automated feature engineering is attained using more than three layers (hence, it's called deep). Work carried out in [36] discusses the use of spectroscopy to classify soil types. The classification is based on the CNN model. This model is a kind of deep learning technique which takes a grid as input, such as an image [37]. The application of deep learning frameworks [38, 39] using TensorFlow and Keras is used for soil classification as discussed in [40, 41] with a dataset comprising 903 soil images with a good accuracy.

The paper [2] proposes soil classification using machine learning. The main objective of this study was to create a model for classifying different types of soil series, along with suggestions of suitable crops for growth. Soil series were taken from six upazilas in the Khulna District of Bangladesh as samples for this study. Database regarding soil was taken from Soil Resource Development Institute, Bangladesh, and a crop database was generated based on different types of soil found in each upazila. Three machine learning methods were used to find soil class – weighted K-NN (K-nearest neighbor), bagged tree, and Gaussian Kernel-based SVM (support vector machine). The authors in [42] proposed a method that would predict the estimate of crop yield in a specific area, based only on geographical and climatic data using machine learning. Karnataka, India, is the state of focus in this study. The factors used are rainfall, season, area, crop, and production. Regression algorithms like K-nearest neighbor (KNN), Gaussian process regression (GPR), decision tree (DT) regressor, and support vector regression (SVR) were used in this study followed by a feature scaling process for accuracy.

Images may be recognized using deep learning since feature extraction can be done without the use of artificial processing. In other words, even for photos where feature extraction is challenging, great accuracy can be attained. Deep learning is frequently employed in research because it is effective with substances like sand and clay, which are challenging for humans to characterize objectively. Compared to the traditional approach, the support vector machine approach has the benefit of requiring less training time because there are fewer parameters to be tuned.

Consequently, it is essential to carry out a procedure known as cross validation. Splitting the sampled data into training and validation and model training and validation is part of this process. The training and validation procedures must be repeated with a number of modifications in order to divide the data in such a way that as much of it is chosen as training or validation data.

6.1.5 Ensemble Learning

Most recent research focus on using this technique for soil classification [43]. This technique takes into consideration multiple learning algorithms resulting in better soil and crop predictions. Prominent schemes used in this regard are bagging, boosting, and stacking [44, 45].

The algorithm mentioned in Fig. 2 generates a diverse ensemble of classifiers obtaining excellent results.

Commonly used ensemble algorithms are random forest and gradient boosting machine [46, 47].

A big dataset can be divided into manageable chunks using the bagging-like scheme technique, and models can be learned separately for each one. The results can then be combined via voting or averaging. The benefit of boosting is that fresh chunks can be weighted according to the classifiers discovered from earlier chunks. The potential to derive performance guarantees led to the introduction of the concept of boosting machine learning research. It can be demonstrated that as the number of iterations rises, the combined classifier's error on the training set approaches zero relatively quickly. The boosting algorithm is shown in Fig. 3.

6.1.6 Rotation Forests

Rotation forest is an ensemble learning technique that tries to provide precise classifiers and has been widely utilized by academics to find precise land use maps [48, 49].

Model Generation Algorithm

Let n be the number of iterations in the soil training data.
For each of t iterations:
  Soil Sample n instances with replacement from training data.
  Apply the learning algorithm to the soil sample.
  Store the resulting model.

Classification Algorithm

For each of the t models:
Predict the class of soil instance using model
Return the class that has been predicted most often.

Fig. 2 Bagging algorithm

Model Generation Algorithm

Assign equal weight to each training instance

For each of t iterations:

Apply learning algorithm to weighted dataset;

Store the resulting model.

Compute error e of the model on weighted dataset

Store the error.

If e=0 OR e>=0.5;

Terminate model generation

For each instance in dataset:

If instance correctly classified by model;

Multiply weight of instance by e / (1- e),

Normalize weight of all instances.

Classification Algorithm

Assign weight of 0 to all classes;

For each of the t (or less) models:

Add –log (e / (1-e)) to weight of class predicted by model

Return class with highest weight

Fig. 3 Boosting algorithm

To build an ensemble of decsion trees, this method combines random subspace and bagging techniques with principal component feature creation.

The input properties are split into k distinct subsets in each cycle. Each subset is subjected to principal component analysis in order to produce linear combinations of the qualities that are rotations of the original axes. The values for the derived characteristics are calculated using the k sets of principal components. Each iteration's input to the tree learner is made up of these. This strategy can provide comparable performance to random forests with a lot fewer trees.

7 Conclusion

The significance of soil classification systems is that it helps researchers to predict the behavior of soil – which further helps in maximizing outputs in different sectors like agriculture, geotechnical engineering, construction, etc. The latest use of soil classification systems also includes finding out a sustainable way to use soil judiciously and predict climate pattern changes. Different classification systems have different criteria/factors which help group soils based on attributes. Therefore, there is no particular ideal soil classification system. Instead, it depends on the field of study and their relation to soil and its characteristics. The crop recommendation is based on many related factors, such as climate factors, soil health factors, and geographical location. Predictive analytics can help the agricultural community to plan crop plantation and production, provided the feature data captured from the sample collection give a true representation of the ecosystem belonging to the region of interest. In the future, predictive analytics and prescriptive analytics can be done using machine learning and deep learning technologies. The fields of geoscience and remote sensing provide advanced help in studying the present condition and progress of the land area, crop state, and its yield and soil health status.

Agribusiness crop yields may be increased by carefully selecting the optimum crops and putting in place supportive infrastructure. Several factors, including the weather, soil fertility, availability of water, water quality, crop prices, and others, are taken into consideration while developing agricultural projections. Since machine learning can anticipate crop productivity based on factors like location, weather, and season, it is crucial for predicting agricultural production. This helps farmers choose the crops they want to produce on their land in an informed manner.

References

1. Soil Science Division Staff, Soil survey manual, in *USDA Handbook 18*, ed. by C. Ditzler, K. Scheffe, H.C. Monger, (Government Printing Office, Washington, DC, 2017)
2. S.A.Z. Rahman, K. Chandra Mitra, S.M. Mohidul Islam, Soil classification using machine learning methods and crop suggestion based on soil series, in *2018 21st International Conference of Computer and Information Technology (ICCIT)*, (2018), pp. 1–4. https://doi.org/10.1109/ICCITECHN.2018.8631943
3. M. Velayutham, T. Bhattacharyya, D. Pal, Classification systems: Indian, in *Encyclopedia of Soil Science*, ed. by R. Lal, 3rd edn., (CRC Press, 2016), pp. 398–401. https://doi.org/10.1081/E-ESS3-120053879
4. T. Bhattacharyya, D.K. Pal, C. Mandal, P. Chandran, S.K. Ray, D. Sarkar, K. Velmourougane, A. Srivastava, G.S. Sidhu, R.S. Singh, A.K. Sahoo, D. Dutta, K.M. Nair, R. Srivastava, P. Tiwary, A.P. Nagar, S.S. Nimkhedkar, Soils of India: Historical perspective, classification and recent advances. Curr. Sci. **104**(10), 1308–1323 (2013). ISSN 0011-3891
5. M. Uddin, M. Hassan, A novel feature-based algorithm for soil type classification. Complex Intell. Syst. **8**, 1–17 (2022)
6. D. Daryati, Soil characteristics analysis based on the unified soil classification system. J. Phys. Conf. Ser. **1402**, 022028 (2019)

7. A. Mishra, Land suitability classification for different crops, in *Orissa Review*, (Orissa University of Agriculture & Technology, Bhubaneswar, 2007)
8. R. Mostafiz, R. Noguchi, T. Ahamed, Agricultural land suitability assessment using satellite remote sensing-derived soil-vegetation indices. Land **10**, 223 (2021). https://doi.org/10.3390/land10020223
9. M. Bock, P.Y. Gasser, W.W. Pettapiece, A.J. Brierley, A. Bootsma, P. Schut, C.S. Smith, The land suitability rating system is a spatial planning tool to assess crop suitability in Canada. Front. Environ. Sci. **6**, 77 (2018)
10. T. Baidar, *Rice Crop Classification and Yield Estimation Using Multi-temporal Sentinel-2 Data: A Case Study of Terai Districts of Nepal* (Universitat Jaume I, 2020)
11. M. Mann, J. Warner, A. Malik, Predicting high-magnitude, low-frequency crop losses using machine learning: An application to cereal crops in Ethiopia. Clim. Chang. **154**, 211 (2019). https://doi.org/10.1007/s10584-019-02432-7
12. K.W. Wong, M. Khoury, T. Thornburn, Automatic soil identification from remote sensing data. Photogramm. Eng. Remote. Sens. **43**(1), 73–80 (1977)
13. S. Grunwald, G.M. Vasques, R.G. Rivero, Fusion of soil and remote sensing data to model soil properties. Adv. Agron. **131**, 1–109 (2015)
14. X.X. Zhu, D. Tuia, L. Mou, G.S. Xia, L. Zhang, F. Xu, F. Fraundorfer, Deep learning in remote sensing: A comprehensive review and list of resources. IEEE Geosci. Remote Sens. Mag. **5**(4), 8–36 (2017)
15. X. Wang, F. Zhang, V.C. Johnson, New methods for improving the remote sensing estimation of soil organic matter content (SOMC) in the Ebinur Lake Wetland National Nature Reserve (ELWNNR) in northwest China. Remote Sens. Environ. **218**, 104–118 (2018)
16. Y. Zhang, B. Sui, H. Shen, L. Ouyang, Mapping stocks of soil total nitrogen using remote sensing data: A comparison of random forest models with different predictors. Comput. Electron. Agric. **160**, 23–30 (2019)
17. F.K. Khadim, H. Su, L. Xu, J. Tian, Soil salinity mapping in Everglades National Park using remote sensing techniques and vegetation salt tolerance. Phys. Chem. Earth. Parts A/B/C **110**, 31–50 (2019)
18. A.B. McBratney, M.M. Santos, B. Minasny, On digital soil mapping. Geoderma **117**(1–2), 3–52 (2003)
19. G.L. Zhang, L.I.U. Feng, X.D. Song, Recent progress and future prospect of digital soil mapping: A review. J. Integr. Agric. **16**(12), 2871–2885 (2017)
20. W. Ng, B. Minasny, M. Montazerolghaem, J. Padarian, R. Ferguson, S. Bailey, A.B. McBratney, Convolutional neural network for simultaneous prediction of several soil properties using visible/near-infrared, mid-infrared, and their combined spectra. Geoderma **352**, 251–267 (2019)
21. E. Mjolsness, D. Decoste, Machine learning for science: State of the art and future prospects. Science (New York, N.Y.) **293**, 2051–2055 (2001). https://doi.org/10.1126/science.293.5537.2051
22. C. Rudin, K.L. Wagstaff, Machine learning for science and society. Mach. Learn. **95**(1), 1–9 (2014). Version: Author's final manuscript
23. M. Appelbaum, H. Cooper, R.B. Kline, E. Mayo-Wilson, A.M. Nezu, S.M. Rao, Intent of reporting standards: Reply to Rossiter. Am. Psychol. **73**(7), 932 (2018). PMID: 30284892
24. L. Ma, Y. Liu, X. Zhang, Y. Ye, G. Yin, B.A. Johnson, Deep learning in remote sensing applications: A meta-analysis and review. ISPRS J. Photogramm. Remote Sens. **152**, 166–177 (2019)
25. U.S. Department of Agriculture, Agricultural Research Service. Food Data Central (2019). fdc.nal.usda.gov
26. A.K. Howard, *Soil Classification Handbook: Unified Soil Classification System* (Geotechnical Branch, Division of Research and Laboratory Services, Engineering and Research Centre, Bureau of Reclamation, Denver, 1986)
27. Occupational Safety and Health Administration, *Occupational Safety and Health Standards: Occupational Health and Environmental Control* (1970)

28. B.B. Mishra, Indian system of soil classification: A way forward. Agric. Res. Technol. Open Access J. **3**, ARTOAJ.MS.ID.555606 (2016). https://doi.org/10.19080/ARTOAJ.2016.03.555606
29. T. Hengl, G.B. Heuvelink, B. Kempen, J.G. Leenaars, M.G. Walsh, K.D. Shepherd, J.E. Tondoh, Mapping soil properties of Africa at 250 m resolution: Random forests significantly improve current predictions. PLoS One **10**(6), e0125814 (2015)
30. M.N. Wright, A. Ziegler, Ranger: A fast implementation of random forests for high dimensional data in C++ and R. arXiv preprint arXiv:1508.04409 (2015)
31. K. Were, D.T. Bui, Ø.B. Dick, B.R. Singh, A comparative assessment of support vector regression, artificial neural networks, and random forests for predicting and mapping soil organic carbon stocks across an Afromontane landscape. Ecol. Indic. **52**, 394–403 (2015). ISSN 1470-160X
32. W. Wei, A.-D. Li, X.-H. He, R. Ma, H.-B. Liu, J.-K. Lv, A comparison of support vector machines, artificial neural network and classification tree for identifying soil texture classes in southwest China. Comput. Electron. Agric. **144**, 86–93 (2018)., ISSN 0168-1699
33. E. Harefa, W. Zhou, Performing sequential forward selection and variational autoencoder techniques in soil classification based on laser-induced breakdown spectroscopy. Anal. Methods **13**(41), 4926–4933 (2021)
34. P.A. Harlianto, T.B. Adji, N.A. Setiawan, Comparison of machine learning algorithms for soil type classification, in *2017 3rd International Conference on Science and Technology – Computer (ICST)*, (IEEE, 2017), pp. 7–10
35. J. Ren, X. Li, S. Li, H. Zhu, K. Zhao, Quantitative analysis of spectral response to soda saline-alkali soil after cracking process: A laboratory procedure to improve soil property estimation. Remote Sens. **11**, 1406 (2019). https://doi.org/10.3390/rs11121406
36. X. Li, P. Fan, Z. Li, G. Chen, H. Qiu, G. Hou, Soil classification based on deep learning algorithm and visible near-infrared spectroscopy. J. Spectrosc. **2021**, Article ID 1508267, 11 pages (2021). https://doi.org/10.1155/2021/1508267
37. J. Anguera et al. (eds.), 2017 Microelectronics, Electromagnetics and Telecommunications, Lecture Notes in Electrical Engineering, vol 471. Springer. ISBN: 978-981-10-7328-1
38. T. Goswami, Machine learning behind classification tasks in various engineering and science domains, in *Cognitive Informatics, Computer Modelling, and Cognitive Science: Theory, Case Studies, and Applications*, (Academic, 2020). https://doi.org/10.1016/B978-0-12-819443-0.00016-7
39. T. Goswami, Impact of deep learning in image processing and computer vision, in *Microelectronics, Electromagnetics and Telecommunications*, Lecture Notes in Electrical Engineering, ed. by J. Anguera, S. Satapathy, V. Bhateja, K. Sunitha, vol. 471, (Springer, Singapore, 2018). https://doi.org/10.1007/978-981-10-7329-8_48
40. P. Srivastava, A. Shukla, A. Bansal, A comprehensive review on soil classification using deep learning and computer vision techniques. Multimed. Tools Appl. **80**, 14887 (2021). https://doi.org/10.1007/s11042-021-10544-5
41. M.G. Lanjewar, O.L. Gurav, Convolutional neural networks based classifications of soil images. Multimed. Tools Appl. **81**(7), 10313–10336 (2022). https://doi.org/10.1007/s11042-022-12200-y
42. V. Sandhya, A. Padyana, Machine learning based crop yield prediction on geographical and climatic data, in *2021 Sixth International Conference on Image Information Processing (ICIIP)*, (2021), pp. 186–191. https://doi.org/10.1109/ICIIP53038.2021.9702556
43. T.T. Nguyen, T.D. Pham, C. Nguyen, J. Delfos, R. Archibald, K.B. Dang, N. Hoang, W. Guo, H. Ngo, A novel intelligence approach based active and ensemble learning for agricultural soil organic carbon prediction using multispectral and SAR data fusion. Sci. Total Environ. **804**, 150187 (2021). https://doi.org/10.1016/j.scitotenv.2021.150187
44. B. Das, P. Rathore, D. Roy, D. Chakraborty, R.S. Jatav, D. Sethi, P. Kumar, Comparison of bagging, boosting and stacking algorithms for surface soil moisture mapping using optical-thermal-microwave remote sensing synergies. Catena **217**, 106485 (2022)

45. Y. Zhang, J. Liu, W. Shen, A review of ensemble learning algorithms used in remote sensing applications. Appl. Sci. **12**(17), 8654 (2022)
46. V.F. Rodriguez-Galiano, M. Chica-Olmo, F. Abarca-Hernandez, P.M. Atkinson, C. Jeganathan, Random forest classification of Mediterranean land cover using multi-seasonal imagery and multi-seasonal texture. Remote Sens. Environ. **121**, 93–107 (2012)
47. M. Powell, G. Hodgins, T. Danaher, J. Ling, M. Hughes, L. Wen, Mapping wetland types in semiarid floodplains: A statistical learning approach. Remote Sens. **11**, 609 (2019)
48. Ö. Akar, Mapping land use with using rotation forest algorithm from UAV images. Eur. J. Remote Sens. **50**(1), 269–279 (2017)
49. K. Archana, K.G. Saranya, Crop yield prediction, forecasting and fertilizer recommendation using voting based ensemble classifier. SSRG Int. J. Comput. Sci. Eng. **7**, 1–4 (2020)

Precision Agriculture: A Novel Approach on AI-Driven Farming

Elakkiya Elango, AhamedLebbe Hanees, Balasubramanian Shanmuganathan, and Mohamed Imran Kareem Basha

1 Introduction

Nearly $150 billion is currently the overall market value of drone technology utilized in all sectors of the economy, including commercial and government applications, and this value is likely to keep rising in the near future. Among the many other areas, agriculture is one where the use of drones will continue to grow. Drones in agriculture can aid in enabling better agricultural methods to handle difficulties in the future. Farming drones are viewed as the newest cutting-edge technology that will be essential to feed the expanding population, which is expected to reach nine billion people by the year 2050, with a corresponding increase in the need for food of more than 70%. Furthermore, the employment of drones in agriculture will be crucial in the battle over extreme climate conditions, which continuously trouble farmers and imperil nutrition security in several regions around the world. Figure 1 enlightens the commercial uses of drone technology in farming and agriculture.

E. Elango (✉)
Department of Computer Science, Government Arts College for Women, Sivagangai, Tamil Nadu, India

A. Hanees
Department of Mathematical Sciences, Faculty of Applied Sciences, South Eastern University of Sri Lanka, Sammanthurai, Sri Lanka

B. Shanmuganathan
Department of Computer Science, DDE, Alagappa University, Karaikudi, Tamil Nadu, India

M. I. Kareem Basha
Department of Computer Application, Merit Haji Ismail Sahib, Arts and Science College, Vellore, Tamil Nadu, India

S. Balasubaramanian et al. (eds.), *Intelligent Robots and Drones for Precision Agriculture*, Signals and Communication Technology,
https://doi.org/10.1007/978-3-031-51195-0_7

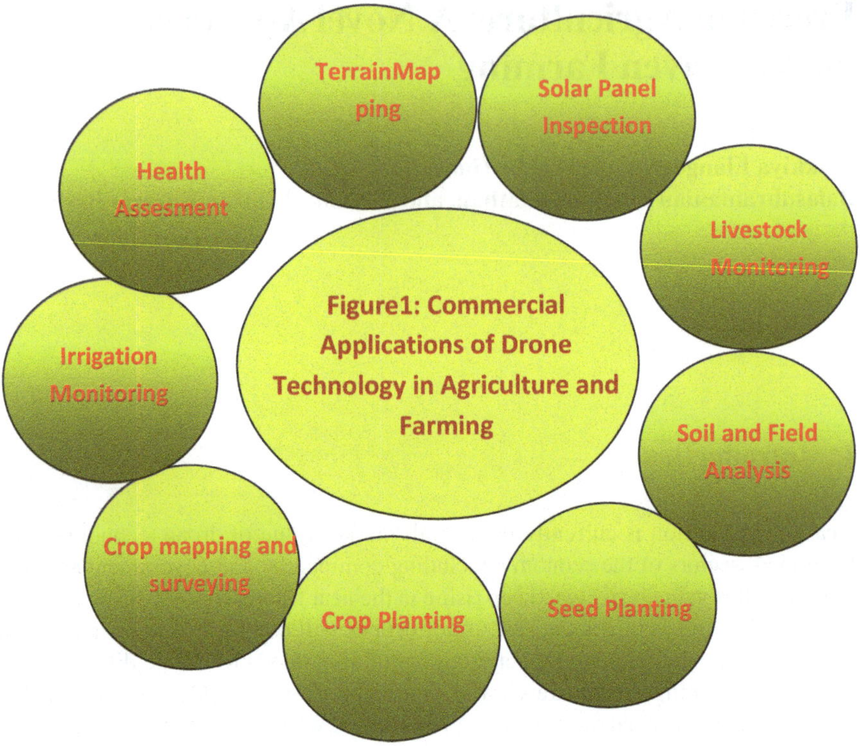

Fig. 1 A few commercial uses of drone technology in farming and agriculture [1]

2 Terrain Mapping

Drones are the ideal equipment for performing terrain mapping because of their natural capacity to survey large areas of land. AI drones are already being produced by businesses like Folio3 [2] that can be programmed to carry out a variety of activities, such as terrain mapping for better management of the agricultural sector. Farmers employ contemporary drones fitted with LiDAR and other sensors to map the landscape across large areas of land as a scouting tool for land preparations. Drones with advanced computer vision skills can provide farmers with real-time feedback while correctly mapping the terrain for effective field planning and management. This process would be flawless if automated terrain mapping with drones were used, which is desperately needed in the field of civil engineering. Folio3 has created AI-based solutions for this. Drones make sure that the region being studied is being recorded live using 3D cameras. Through a variety of CAD systems, this stream can be utilized to gather data for the creation of 3D models. The drone is deployed over the field or region under consideration and equipped with a 3D camera and a LiDAR detector.

By strengthening a layering of a terrain image utilizing the collection of multispectral of the terrain that used an autonomous 4 × 4 copter UAV (unmanned aerial vehicle) armed with a multispectral camera on board, our goal is to provide farmers with an integrated tool for determining and assessing real-time greenery.

2.1 *Working Techniques*

Figure 2 shows the suggested architecture. Our goal is to use visual mosaicing methods to integrate spatial data from the nearby infrared (NIR) image as well as the abovementioned sensor data into a spatial multispectral image of both harvest terrains. This study is an ongoing research project that combines methods for mosaicing a landscape with NDVI calculations used on rice crop fields. The main components of the system, as shown in Fig. 2, are a base station, an advancing technology pelican quadcopter, and a connected multispectral camera. Table 1 provides a concise summary of the system's key parameters. The pelican is provided with such a GPS waypoint navigation system with only a 1-m precision. Data from the sensors is sent to the ground station using the 2.4-GHz XBee network. Tetracam ADC Lite multispectral camera was intended to capture apparent light wavelengths greater than 520 nm and near-infrared frequencies up to 920 nm.

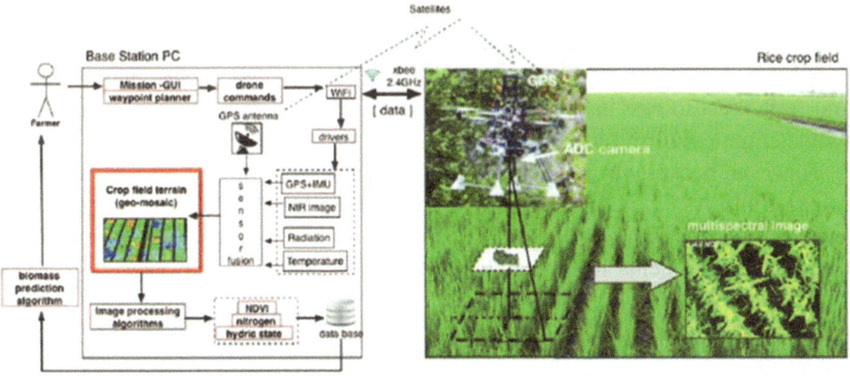

Fig. 2 Unmanned UAV-based aerial surveillance and sensing structure

Table 1 Pelican quadrotor specifications

Parameter	Description
Amount of rotors	4
Maximum weight for launching	1650 g
Payload	650 g
Flight time (with payload)	10 min
Maximum manual speed	15 ms^{-1}
Maximum autonomous speed	3 ms^{-1}
Recommended fighting area	1 km^2
Battery	LiPo 6100 mAh
Processor	MastermindIntelCoreTMi7

The primary use of this product is for light frequencies scattered and absorbed by plants and crop coverings. The weight is 200 gf (making it suitable for UAV applications). Because the images this camera produces are in raw format, we had to convert them using the Pixel Wrench 2 program to JPEG, TIFF, etc. With a field of view of 100 × 75 m and a height above ground of 122 m, this camera can capture images with a ground resolution of 48.8 mm per pixel.

3 Solar Mapping

Drones and artificial intelligence are used by raptor maps [3] to detect and forecast solar farm outages as followed in Fig. 3.

In rural locations where the availability of electricity is intermittent, solar energy is a preferable option to conventional electricity [4]. Solar cells are used in the current generation of solar energy converters to convert energy. The main disadvantages of these silicon-based solar cells are their low conversion efficiency and higher cost as a result of their larger size. Utilizing nanomaterials, which have greater efficiency of convergence and are smaller than silicon solar cells in size, is a more recent trend in the conversion of solar energy. The lower solar irradiation captured by these solar energy converters is another factor contributing to the poorer conversion. The solar radiation captured by devices using sun-tracking mechanisms can only be slightly improved. By tracing the Sun path from east to west in the morning and from west to east in the evening, these traditional solar-tracking technologies will attempt to gather a little extra energy.

However, because the route and angle of the Sun vary with the seasons and hemispheres, this system is unable to harness its full potential. The most energy can be gathered whenever the devices understand and follow the Sun's movement across the year. The systems in typical smart farming controllers, however, need to be

Fig. 3 Raptor maps uses drones and AI to detect and forecast solar farm outages

trained using machine learning algorithms. Only when a solar panel made of nanomaterials can follow the Sun's direction will the maximum amount of energy be harvested. To do this, a machine learning technique is used.

Reinforced machine learning uses feedback to learn from the environment. An appropriate response will be rewarded, and an improper response will be penalized. Although the Sun route varies depending on the season, this approach is acceptable for our goal of continuously tracking the Sun path throughout the year. The reinforcement algorithm picks up knowledge from its surroundings.

For utility-scale solar farms, this entails either visiting hundreds of acres and meticulously evaluating thousands of panels by hand or, more frequently, inspecting just a representative sample of panels in an effort to spot systematic problems. In some circumstances, expensive examinations by a small aircraft may be used.

3.1 Working Techniques

In the past year, conducted research, contrasting its 100% IR drone inspections with pertinent manual inspection scenarios across four sites in terms of time, expense, and outcomes [5].

- *Site 1*: 74 MW in Sumrall will undergo a 100% IR drone inspection for maintenance as opposed testing to clamp and 20% IV curve tracing testing.
- *Site 2*: A 30-MW maintenance project in Sprague, Connecticut, will use 100% IR drone inspection as opposed to eye inspection and Voc/I scat the combined box testing.
- *Site 3*: For the commissioning of a 21 MW in Rincon, Georgia, a 100% IR drone inspections were assessed to a 100% custom IR scanning procedure and 15% IV curve mapping analysis.
- *Site 4*: 100% IR drone inspection compared to 100% IV curve tracing testing for the commissioning of 12.5 MW in Herald, California.

According to relevant data from the four sites, manual inspections increased inspection efficiency by 97% as compared to drone inspections. The turnaround time is remarkably rapid given the volume of data handled during a normal examination (about 800 pictures per MW). The analysis and delivery of each of these inspections took less than 5 business days.

4 Livestock Monitoring

In today's livestock operations, it's critical to manage animal health and spot problems early. Wearable devices and thermal imaging cameras might be used to track livestock's anxiety levels, eating habits, and movement patterns in addition to vital indications like temperature. There may be opposition to implanted and orally ingested sensors if society becomes more compassionate and aware of animal health issues.

Wearable technology have the distinct advantage of being the least intrusive method of making farming "wiser," especially whenever tied to other technologies like decision support systems. Governments uphold norms for the well-being of cattle, which the agricultural industry is required to abide by. Keeping track of the livestock's head count is necessary for the procedure, but it can be difficult when herds are dispersed across large open fields. When physical cattle-herding is done on-site, it takes a lot of time and effort. It could be expensive because farmers may need to traverse grassland, bushes, wind, and snow in order to track and hunt for them. In this job, counting people too close to the cows, or cattle, frequently leads to accidents, especially when the cows are stressed. Since cattle are heavier than humans and are capable of moving extremely fast, cattle frequently injure people by kicking or crushing them. Agitated cattle pose a special risk, and handling them properly requires training. Furthermore, because cows move, hide behind one other, or stand behind obstructions like large hay bales or trees, hand head counts can require several tries to get it right. Even when the animals are lying down, counting can be challenging.

Farming and livestock monitoring are two positive applications where AI has a big potential to improve human life. The production of the farm is increased, and operational costs are decreased by automatically tracking the cattle [6]. The use of drones to monitor livestock has attracted a lot of interest from farmers. The number and behavior of the livestock may be ascertained from the drone's livestock photographs, which are then relayed in real time to the monitoring equipment so that farmers can take appropriate action. This strategy is quite effective, particularly when it comes to keeping an eye on the cattle on a sizable agricultural property and locating human poachers.

Due to the similarities in some species' physical characteristics, this application sometimes misclassifies animals, which results in inaccurate information being given to farmers. False information can also be effectively conveyed through the use of lighting and background effects. The use of drones in livestock monitoring has been the subject of numerous research as shown in Fig. 4. The usage of a drone with a quadcopter-like layout for sheep livestock surveillance was proven by Al-Thani et al. [7]. The drone's components include a Raspberry Pi Module V2 with both a machine learning algorithm to measure the sheep and monitor its location, an ArduPilot Mega (APM) flight controller, and a 4G connection.

It can be expensive because farmers may need to traverse grassland, bushes, wind, and snow in order to track and hunt for them. It was found that when compared to offline processing, online processing gave findings that were more accurate and looked more promising. Li and Xing [8] investigated how many drones would be needed to follow and monitor the livestock with the least amount of equipment. Each animal was given a GPS collar so the drone's receiver could receive information about its location and health. The study also used the density-based clustering technique DBSCAN to address the minimal average drone-animal distance. This approach yielded more cattle covered than the traditional K-means clustering technique, with a smaller average drone-animal distance.

Fig. 4 Livestock monitoring

The convolution neural network (CNN) technique was implemented by Rivas et al. [9] to examine the images produced by the GoPro Hero 5 camera that was mounted on the drone to detect cattle in real time. This detecting module has incorporated a Raspberry Pi to power the algorithm, allowing for instantaneous livestock counting without the need to first analyze the photographs on a distant computer. Xu [2] used a Mask R-CNN algorithm to categorize cattle using photos from a DJI Mavic Pro drone, and the results of an experiment showed 96% classification accuracy and 92% counting accuracy.

4.1 Working Techniques

The pipeline of the processing is shown in Fig. 5. We use a drone to record videos of the pasture before identifying and counting the cows. As described in the subsections below, the pipeline is made up of various stages such as data fetching, cow detection, and data visualization.

4.1.1 Data Fetching

The machine learning model's limitation enables just one image to be used for a given input. As a result, we had to separate the recorded footage into a series of frames, identify cows, and note where they were in each frame. In plenty of other words, we created for each individual frame; a counting and a listing of cow locations are provided.

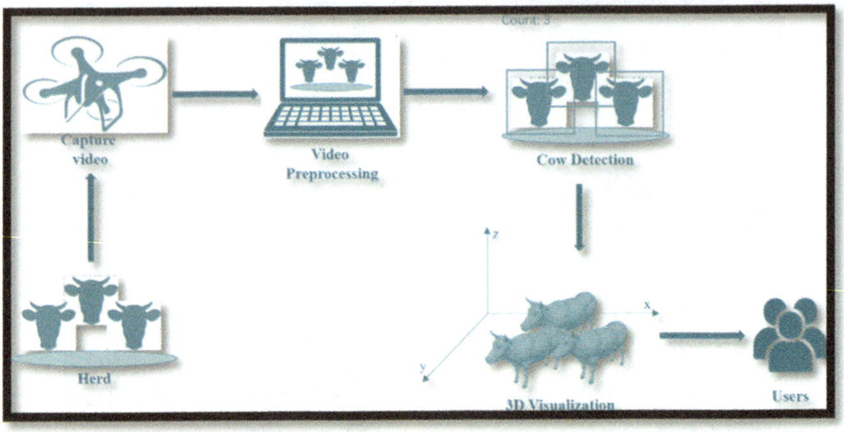

Fig. 5 Processing pipeline

4.1.2 Cow Detection

FeiYang [5] investigated the accuracy performance of two previously trained models for the detection of cows. Among them, YOLO-v3 was just one [10]. We set up MXNet from the Gluon CV Toolkit package. We prepared each image into one of the three grids, measuring 8*8, 16*16, and 32*32, in accordance with the input structure of the model. We discovered that YOLO (You Only Look Once) is a decent compromise between accuracy and speed even if it does not have the maximum accuracy. To integrate YOLO-v1 and YOLO-v2 data, Redmon et al. [11] found YOLO-v3 was an improved version [12]. Particularly when identifying small objects, although maintaining the performance gain of many other YOLO approaches, it improves the detection rate. The image is divided into several sections of various sizes using the YOLO-v3 model.

The item parameters are then classified based on the computed probability. YOLO-v3 applies the Darknet-53 modeling approach that has 53 convolution operations. This convolution layer is followed by a batch normalization layer and a faulty ReLU layer that serve as the kernel function. The discovered outcome is just not optimal since the model's frame rate, which requires about 10 s, is too stagnant to enable a real-world application. Additionally, the precision is not high enough, particularly whenever the image's cows stack on top of one another and resulting in extra counting.

4.1.3 Data Visualization

It investigated a different pre-trained Mask R-CNN model for the aforementioned reasons. We integrated R50-FPN-3x and installed the Detectron2 package that is given access via Facebook. Just Mac or Linux systems running Google Colab can be utilized to evaluate this model. Instance segmentation is carried out by Mask R-CNN, which requires accurate segmentation in addition to proper localization of the identified objects. Thus, instance segmentation integrates semantic segmentation with object detection. Faster R-CNN (FCN) is enhanced by Mask R-CNN. Each containing a proposal candidate box is semantically divided.

The instance segmentation Mask R-CNN framework consists of three steps:

(i) *Target detection* – It entails drawing a bounding box around the observed object.
(ii) *Target classification* – It entails identifying the target's associated class to identify if it belongs to a person, an automobile, or another category.
(iii) *Pixel-level target segmentation* – The foreground and backdrop must be segregated for each goal. These model's outcomes are the bounding box position, its classified label, and also the label probabilities. The quantity of cows is used to determine the bounding box coordinates for every frame.

5 Soil and Field Analysis

The foundation of a productive farm is a healthy soil. One of the most crucial farm resources is the soil. It serves as a vital reservoir of water and nutrients for crops. Each kind of soil contains unique characteristics. Farmers could always just enhance the soil's quality by regulating nitrogen levels and soil pH even when soil texture can indeed be altered. Periodic soil analysis serves as one of the key elements in regulating soil health [13].

A significant farming technique is soil analysis, which establishes the precise quantity of crop nutrients that exist in the soil. Furthermore, it offers a clear overview of numerous chemical, physical, and biological aspects of the soil. The following are some of the most fundamental but important micronutrient measurements such as (i) measuring the amounts of calcium, magnesium, potassium, phosphorus, and nitrogen, (ii) analysis of soil pH, and (iii) measuring the amount of organic matter, accessible lime, and humus.

Albeit with today's modern advanced farming techniques, it can be challenging to keep the appropriate nitrogen balance within the soil as shown in Fig. 6. Nitrogen deficiency will cause crops to suffer, but if chemical-intensive agricultural methods are not used correctly, topsoil might suffer long-term harm. Fortunately, current revelations are providing farmers and researchers with a seeing glimpse of the problem. To prevent losses from such dangers, artificial intelligence in farming is playing an important role in soil analysis prior to planting, crop health or maturity level monitoring, pesticide management, wind forecasting, and other natural disaster prediction.

Different algorithms are used for agricultural analysis in artificial intelligence (AI) and machine learning (ML) systems that track the quality of the soil and fertility. Utilizing both supervised and unsupervised techniques, machine learning applications enhance data methodologies and produce adequate data for statistical

Fig. 6 Successful farm production starts with healthy soil

solutions to issues needing various methods. Farmers can identify potential nutrient deficits in soil quality using artificial intelligence technology, especially electronic technologies for deep learning.

5.1 *Soil and Field Analysis Using Artificial Intelligence Model*

We develop AI-based techniques that enable self-driving vehicles, robotics, drones, weed-controlling machinery, and certain other comparable equipment to enter the agricultural sector. All of these technologies will aid in boosting agricultural productivity on a massive scale.

5.2 *Latest Developments in Soil and Crop Remote Sensing*

5.2.1 Platforms and Sensors

Working Techniques

While there are already many technologies used for remote sensing [14] including satellites, aircraft, balloons, helicopters, unmanned aircraft, and towers, the much more potential platforms for smart farming are microsatellites and multirotor drones. Figure 7 explicates the types of remote sensing of soils and crops.

- *Optical Domain* – A number of sensors can be utilized for remote sensing and monitoring soils and crops. Inside the optical signals, data can be gathered that are multispectral and hyperspectral (visible, near-infrared, and shortwave-infrared wavelength ranges: 400–2500 nm). While hyperspectral sensors often realize maximum (1–10 nm) and continuous spectra, multispectral sensors frequently capture reflectance profiles in just few broad (10–100 nm) spectral bands. The bulk of satellite sensors have at least four spectral bands in the visible to near-infrared range, even if hyperspectral satellite sensors aren't yet practi-

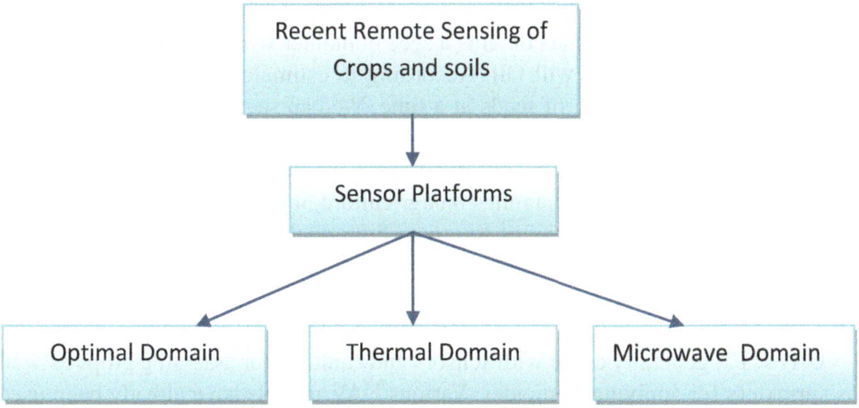

Fig. 7 Recent remote sensing of soil and crops

cally viable. Ongoing (fairly frequent) monitoring of the planet's surface is made possible by the functioning of a constellation or collection of connected satellites. This spatial resolution can be utilized in large agricultural areas all over the world (1–10 m).

- *Thermal Domain* – In catching the thermal emission from a range (3–15 m), thermal sensors like thermographs and infrared thermometers could be utilized to determine the target surfaces' intensity, warmth, and reflectivity. Thermal sensors are borne in just a few satellites, such as Landsat 8; however, their spatial resolution and revisit frequency are inadequate for most smart farming applications. A basic tool determining the outer temperature of soil, leaves, or canopies is a portable infrared thermometer. Recently, a portable thermograph has been used to evaluate plant covering in order to detect water shortages and infections. Compact and lightweight thermal imaging cameras nowadays are available for unmanned aerial vehicles.
- *Microwave Domain* – Microwave satellite sensors (SAR (synthetic aperture radar)) seem to be very attractive for precision farming, especially in monsoon regions because they are hardly affected by clouds. Electromagnetic fingerprints including such scattering coefficient, radiation, and polar metric fingerprints within the microwave domain can be utilized to detect changes in crop cultivation or soil condition. In fact, Inoue et al. [15] used an exploratory study to show how rice development causes unique seasonal variations in the polar metric backscattering signature of a wide broad range of frequency bands.

6 Seed Planting

The agricultural sector is currently suffering in rural areas for this single reason. No laborers are available for planting, and even if they were, they would demand more money. However, the crucial thing to remember is that farming is what the former depends on [16]. The whole planting equipment is attached to the bottom of the drone, and the planting drone contains a seed container where seeds are stored and a funnel into which seeds will fall. According to estimates, the planting drone is capable of carrying 1–5 kg of seeds at a time. We can seed approximately 10,000 plants in a single day with the aid of two operators and ten drones; if the number of operators and drones rises, they will plant approximately 35,000 plants in a single day. Despite how heavily India relies on agriculture, it nevertheless is well short of integrating the latest technologies to create high-quality farms. Unmanned aerial vehicle (UAV) applications in remote sensing, photogrammetry, and precision farming has indeed started in industrialised countries. It is incredibly efficient and might alleviate a farmer's workload.

UAVs [17] generally come with sensors and cameras for assessing crops as well as sprinklers for applying pesticides. Various UAV models have already been used for both military and commercial applications. Yamaha creates the very first UAV model for use in agriculture. For use in agricultural pest management and crop monitoring. Crop monitoring using the Yamaha RMAX, an unmanned aircraft. In precision agriculture, inductive reasoning to UAVs evaluates their viability for activities involving precision farming, crop height measurement, and tree planting, among others.

6.1 How Reforestation Works

Figure 8 explains how reforestation works. In blend with our custom pods and accelerated seed planting, we use an assortment of advanced unmanned aerial vehicles [18].

Fig. 8 Reforestation working process

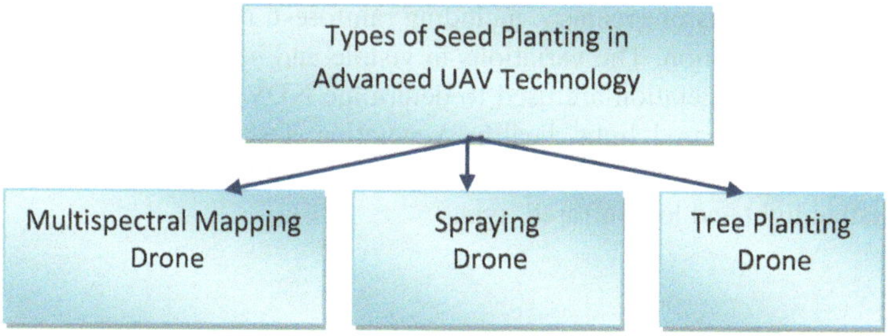

Fig. 9 Types of seed planting (advanced UAV)

Figure 9 illustrates three methods of seed sowing utilising sophisticated UAVs (unmanned aerial vehicles):

(i) Multispectral Mapping Drone: We gather data to determine the best microsites to plant each tree and our growing seedlings.
(ii) Spraying Drone: We give our seedlings nutrients and continuous support for the first crucial years.
(iii) Tree Planting Drones: We mount drones with pneumatic firing devices and use our software to fire pods into selected sites.

7 Crop Planting

Drone technology is utilized to monitor crops, assess crop growth, locate severe damage to crops, and identify areas that need more irrigation and fertilization. It offers 2D and 3D visuals, enabling farmers to gauge plant development, crop yields, and sturdiness. Analyzing the 2D/3D imagery could provide insightful information about crop growth and highlight manufacturing inefficiencies [19]. Near-infrared (NIR) light is monitored and sensed by drones using sensors. Sturdy plants fluoresce while scattering NIR, but inferior plants absorb more NIR while reflecting more light energy, making NIR useful. The NIR approach provided farmers with detailed maps of information about plant nutrition and far more closely monitors interventions. Drone crop surveillance provides significant crop vegetation index surveillance using spectral analysis of high-resolution radar data for various places and crops, allowing for the observation of both positive and gloomy crop growth dynamics.

7.1 Working Techniques

Agronomists can then employ spectroscopic analysis to examine crop health and calculate yield utilizing UAV imagery. The Normalized Difference Vegetation Index (NDVI), which is already employed via satellite surveillance to evaluate the quality of substantial regions of greenery, including rainforest, represents the most crucial statistic in this situation. The variations in visible and near-infrared light reflected and absorbed by vegetation are used to determine NDVI. While reflecting a large amount of near-infrared light, healthy vegetation absorbs the majority of light waves. In comparison to healthy vegetation, unhealthy vegetation radiates both more visible light and less infrared light.

7.2 Crop Yield

When it comes to concerned crop conditions, NDVI only offers a superficial study of vegetation coverage. Crop health and many other traits can be thoroughly analyzed by computer vision algorithms that have been taught with customizable data labels. Phenotyping plants require a lot of time and effort, and it might be difficult to phenotype crops over a wide geographic range. Crop phenotyping [20] is being scaled up significantly by drones outfitted with sophisticated sensors and cameras, enabling farmers to examine their plants over vast and varied geographical regions. A younger group of scalable crop phenotyping is exemplified by drone-enabled increased crop phenotyping (HTPP).

With over 98% reliability, the model classified and graded the variety of individual lettuces. The system would assess the size of each lettuce and determine how much of a crop was covered. In order to focus on trouble spots using fertilizer or irrigation, it gives growers information regarding how well lettuces are developing in certain areas within a possibly broad territory. Several efforts had trained machine learning algorithms with non-NDVI typical RGB camera pictures using polygon identification, picture annotation, binary classification, and bounding box. A capability of such approaches has only been constrained by the amount of information provided by the pictures; machine learning algorithms are better suited for making use of the enormous amounts of information which high-resolution photographs could provide, given that perhaps the algorithms are properly trained and evaluated.

8 Crop Mapping and Surveying

In a drone research, an unmanned aerial vehicle (UAV) is used to gather airborne data utilizing downward-facing sensors such as RGB or multispectral cameras and LiDAR payloads. Throughout the study, the area was shot by a drone equipped with an RGB camera. Each image is labeled with given parameters. These maps can also be required to obtain data, including such quantitative measures or high precision distances. Drones have the potential to operate at considerably lower heights than manned aircraft or satellite imagery, which allows for such rapid, inexpensive, and autonomous collection of high-resolution, high-accuracy data.

8.1 Using Drones for Surveying in AI

Figure 10 explains surveying drones in AI.

- *Land Surveying/Cartography* – In locations with poor quality, obsolete, or even no information, surveying drones produce high-resolution orthomosaics and intricate 3D models. These make it possible to swiftly and efficiently construct very accurate cadastral mapping, especially in challenging or difficult circumstances. Using the imagery, surveyors also can identify elements like signs, hedges, road markings, sprinkler systems, and sewers. These very same photographs could create extremely intricate contour models, contour lines, and divisiveness after being post-processed with photogrammetric software, in addition to 3D restorations of ground areas or constructions.
- *Development and Land Management* – Drone-captured aerial imagery significantly speeds up and simplifies topographical investigations for land planning and maintenance. This appears to be accurate for site selection, allotted design, and actual implementation of streets, structures, and facilities.
- *Precise Measurements* – Researchers can make extremely precise ground and distance assessments because of high-resolution orthophotos.
- *Stock Pile Volumetric Measurements* – Volumetric measurements can also be taken from identical imagery using 3D mapping software. For inventory or performance evaluation, this quick and affordable technique of fluid column is especially helpful for calculating stockpiles in crushers and quarries additionally; they may accomplish this far more safely than if they were to personally gather the information by climbing upward and downward a stack. Drones are

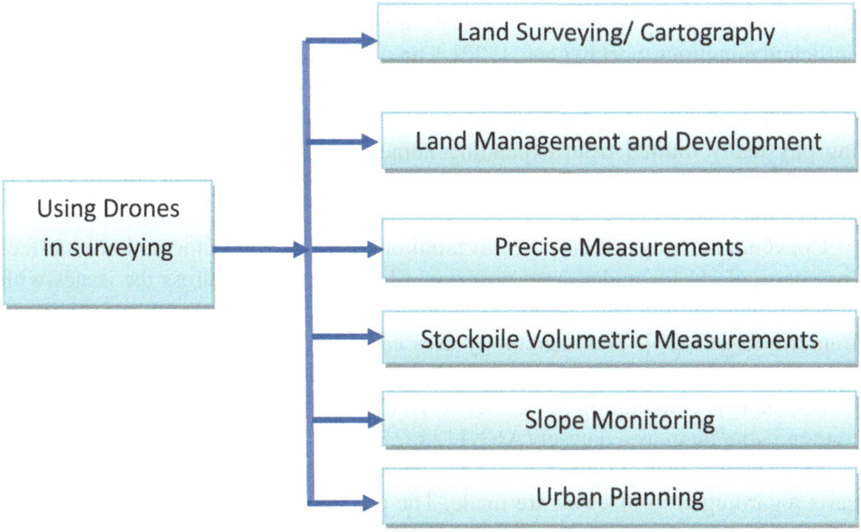

Fig. 10 Using drones in surveying

collecting information from above, so site procedures won't be hampered. The quick sampling frequency makes it possible to take a picture of the website at a particular moment in time.

- *Slope Monitoring* – From the DTMs and DSMs produced using drone images, gradient data can be extracted using computerized GIS technique. The locations can be categorized and utilized for gradient surveillance reasons, particularly landslide hazard mitigation and prevention, depending on how precipitous the ground's surface is. The Earth's motion can be monitored as well as its speed measured using orthomosaics collected at various intervals. This information can be used to anticipate catastrophes and guard against possible harm to bridges, railroads, and roadways. Drones offer more detailed statistical collection as compared to conventional methodologies, in which sensors are mounted on specific sites. Since all these places are frequently difficult to find or perhaps even hazardous, drones with PPK capabilities that don't necessitate the setting out of several GCPs are ideal for this scenario.
- *Urban Planning* – Urban areas are becoming more congested and complicated, which necessitates meticulous planning and hence time-consuming and expensive information gathering. Drones allow urban planners to collect a large amount of current data rapidly and with far less personnel. Planners can assess the locations' current social and ecological characteristics and take into account the effects of various scenarios to the content captured in this way.

9 Irrigation Monitoring

Because of the growing population, there is a steady demand for nourishing feed, which puts impacts on food producers and suppliers to provide high-quality food in sufficient quantities worldwide [21, 22]. Three main elements affect crop yields: the accessibility of nutrients (fertilizer and insecticides), the amount of water available (irrigated agriculture), and the weather conditions. Water and nutrients are becoming physically limited from expanding numerous needs for water as well as the rapid depletion of fossil fuels as just a result of global warming and climate change at the same time.

Conventional ways of crop health monitoring take a long time and are ineffective, particularly for wider areas covering a lot of land. To address the issues with conventional farming, numerous academics have created a variety of AI-based frameworks [17, 23]. The technology uses aerial photographs or unmanned aerial vehicles (UAVs) with HD cameras to collect cropped imagery. Artificial neural networks, such as convolution neural networks (CNNs), are trained just on image histogram using them as a dataset (ANN) [24, 25].

From photos that are given to CNNs, valuable features are extracted, and forecasts regarding crop diseases are made. The process delivers superior answers for determining the number of herbicides to be applied and the ideal time to apply a herbicide since it is efficient at continuously monitoring the condition of plants. Researchers can use weather prediction in the automated irrigation system. Why do we say that, exactly? A device which warns that the soil is parched but we don't need to water since it will rain in a few hours seems like a fascinating used scenario. Nowadays, every nation around the globe is required to do using freshwater incredibly efficiently. Over one-third of the worldwide people will experience total water shortages by 2025, according to current research on the global water crisis. For this reason, drip irrigation, which is utilized like an automated irrigation system for small farms, is just one of the irrigation systems being used to enhance usage of water.

10 Health Assessment

Plant health monitoring with drones is successful. Drones having thermal and infrared sensors is frequently utilized to collect real-time footage for precision agriculture analysis. Bespoke crop health monitoring software is available from companies such as Folio3, which can be utilized to rapidly determine its chlorophyll concentration and calculate soil quality using crop pictures captured in the thermal and infrared spectra. The amount of visual and near-infrared radiation that plants reflect changes according to their state of well-being and degree of stress [26]. Drones having sensors that can examine crops in visible and near-infrared light can be used to monitor crop's overall health over time.

11 Conclusion

AI and smart farming are really the agricultural sector's destiny. They will enhance farming by assisting in the early detection of plant pathogens and illnesses while raising the overall caliber of something like the harvest. AI-based crop production forecasting that is accurate will assist nations in achieving food security. As we will see, artificial intelligence has significant advantages for the agricultural sector. It enables better interaction, greater effectiveness, and cheaper manufacturing. Although innovation is the new big thing for agriculture, it cannot function independently. However, there are a number of obstacles to using AI, such as a shortage of varied samples and a steep learning curve. Issues over privacy and security and a lack of digital literacy are further issues. Farmers now need to produce more food to support a rising population as the world's population rises, and the advent of robotics and a digital workforce can provide robotic aid.

12 AI's Future in Agriculture

As drones are really being designed to function as robotic pollinators and to integrate smart applications, the future of drones in farming is also bright. Drones are really potent and reasonably priced technology that could help deal with the issue of rising food scarcity. Customers can acquire delicious fresh food year-round thanks to genetically manipulated nutrients and packaged foods, which imply farms must rely on data to design longer seasons, wider farms, or various growing periods. Given that the majority of cutting-edge innovations are only utilized sizable, well-connected fields, the AI's future in farming will need to place a strong emphasis on universal coverage. The future of machine learning automated agricultural goods and data science in agriculture would be secured by expanding accessibility and reach even to tiny farms in rural countries around the world.

References

1. https://www.folio3.ai/blog/drones-in-agriculture/
2. https://www.folio3.ai/ai-drones/
3. https://greentownlabs.com/raptor-maps-uses-drones-and-artificial-intelligence-to-identify-and-predictsolar-farm-outages/
4. R. Vatti et al., Solar energy harvesting for smart farming using nanomaterial and machine learning. IOP Conf. Ser.: Mater. Sci. Eng. **981**(3), art. no. 032009 (2020). https://doi.org/10.1088/1757-899X/981/3/032009
5. https://www.timeanddate.com/calendar/june-solstice.html
6. https://wingtra.com/drone-mapping-applications/surveying-gis/
7. X. Li, L. Xing, Reactive deployment of autonomous drones for livestock monitoring based on density-based clustering, in *2019 IEEE International Conference on Robotics and Biometrics (ROBIO)*, (IEEE Press, 2019), pp. 2421–2426. https://doi.org/10.1109/ROBIO49542.2019.8961763
8. A. Rivas, P. Chamoso, A. González-Briones, J.M. Corchado, Detection of cattle using drones and convolutional neural networks. Sensors (Basel) **18**(7), 2048 (2018). https://doi.org/10.3390/s18072048. PMID: 29954080; PMCID: PMC6068661
9. C.A. Mücher, S. Los, G.J. Franke, C. Kamphuis, Detection, identification and posture recognition of cattle with satellites, aerial photography and UAVs using deep learning techniques. Int. J. Remote Sens. **43**(7), 2377–2392 (2022)
10. J. Redmon, A. Farhadi, YOLOv3: An incremental improvement, **1804**, 02767 (2018). https://doi.org/arXiv:1804.02767v1
11. J. Redmon, S. Divvala, R. Girshick, A. Farhadi, You only look once: Unified, real-time object detection, 779–788 (2016). https://doi.org/10.1109/CVPR.2016.91
12. J. Redmon, A. Farhadi, YOLO9000: better, faster, stronger. In *Proceedings of the IEEE conference on computer vision and patternrecognition* (2017) (pp. 7263–7271)
13. https://www.ayadata.ai/blog-posts/how-drones-and-computer-vision-are-used-to-enhance-crop-yield
14. https://www.tandfonline.com/doi/full/10.1080/00380768.2020.1738899
15. I. Inoue, F. Namiki, T. Tsuge, Plant colonization by the vascular wilt fungus Fusarium oxysporum requires FOW1, a gene encoding a mitochondrial protein. The Plant Cell **14**(8), 1869–1883 (2002). https://doi.org/10.1105/tpc.002576

16. K.V. Malini, K. Prashantha, K.R. Rakshith Kumar Naik, M.G. Vijay Kumar, Manjunatha, Seed planting robotic drone. Int. J. Emerg. Technol. Innov. Res. (www.jetir.org) **6**(5), 639–641 (2019). ISSN: 2349-5162
17. D. Sinwar, V.S. Dhaka, M.K. Sharma, G. Rani, AI-based yield prediction and smart irrigation, in *Internet of Things and Analytics for Agriculture, Volume 2. Studies in Big Data*, ed. by P. Pattnaik, R. Kumar, S. Pal, vol. 67, (Springer, Singapore, 2020). https://doi.org/10.1007/978-981-15-0663-5_8
18. https://interestingengineering.com/innovation/these-drones-will-plant-1-billion-trees-in-just-8-years
19. V. Rana, Mahima, Impact of drone technology in agriculture. Int. J. Curr. Microbiol. Appl. Sci. **9**(1), 1613–1619 (2020). https://doi.org/10.20546/ijcmas.2020.901.177
20. N. Gnanasankaran, E. Ramaraj, T. Manikumar, An intelligent framework for rice yield prediction using machine learning based models. Int. J. Sci. Eng. Res. **12**, 422–431 (2021)
21. P. Shah, M. Manohar, Analyzing implementation of AI, robotics, and UAV technologies in agriculture to boost irrigation and solve crop monitoring challenges. Int. J. Mech. Eng. **7**(4) (2022). ISSN: 0974-5823
22. Special Issue, *Unmanned Aerial Vehicle for Optimize Irrigation and Crop Monitoring* (Springer, 2022)
23. S. Sengupta, W.S. Lee, Identification and determination of the number of immature green citrus fruit in a canopy under different ambient light conditions. Biosyst. Eng. **117**, 51–61 (2014). https://doi.org/10.1016/j.biosystemseng.2013.07.007
24. H. Pathak, G.A.K. Kumar, S.D. Mohapatra, B.B. Gaikwad, J. Rane, *Use of Drones in Agriculture: Potentials, Problems and Policy Needs*, Publication no. 300 (ICAR-NIASM, 2020), p. 13+iv
25. N. Poompavai, E. Elakkiya, Feed the globe utilizing IOT-driven precision agriculture. Adv. Comput. Sci. Technol. **15**(1), 11–20 (2022)
26. B. Shanmuganathan, E. Elango, *Exploring Recent Advances of IOT in Ambient Intelligence (AMI) & USE Case, Applications of IOT in Science and Technology* (Innovation Online Training Academy (IOTA) Publishers, 2023/9), pp. 159–167

Embracing IoT and Precision Agriculture for Sustainable Crop Yields

P. Geetha and R. Karthikeyan

1 Introduction

Farmers today are not getting a proper yield in agriculture because of several challenges encountered in various situations beginning with sowing, harvesting, and selling products. Farmers are also confronting unexpected natural disasters, such as crop infection from mice, rats, insects, birds, and diseases in leaves, among other things. To combat this, farmers require a different methodology from the beginning to the finish of each cultivation.

Pervasive automation is a technology that decreases human labor, often known as sophisticated smart farming technology. Farmers employ numerous technologies in diverse scenarios to achieve good yields in their crops. This study highlights the potential of 5G technology to bring in a new era of smart and efficient farming, with applications ranging from real-time monitoring to autonomous farming operations. It promises to greatly improve the agricultural sector's sustainability, productivity, and responsiveness [1]. India's agricultural sector is the largest producer of pulses, rice, wheat, spices, and spice products in world, and 8% of India's gross domestic product (GDP) provides employment to 50% of the country's workforce. Illiteracy is the big drawback of farmers in agriculture field, since most of the farmers are aged and don't have the knowledge about current trends and technology. They are unaware of updates arriving in agriculture field. Most of the people in India depend on agriculture, but water scarcity really affects farmers and people. Farmers are

P. Geetha (✉)
PG Department of Computer Science, Dr. Umayal Ramanathan College for Women, Karaikudi, India

R. Karthikeyan
Department of Computer Science, Alagappa University, Karaikudi, India
e-mail: karthikeyancs@srmasc.ac.in

S. Balasubaramanian et al. (eds.), *Intelligent Robots and Drones for Precision Agriculture*, Signals and Communication Technology,
https://doi.org/10.1007/978-3-031-51195-0_8

unable to handle the recent updates like GPS, mobile apps, Facebook, and Twitter, so the recent trends in agriculture are not aware of them. This leads to contingency expenses in agriculture due to natural calamities, disease/fungal infection in crops, improper storage mechanism of crops, lack of market demand products, broker's intervention, and lack of money investment by farmers. Most farmers are economically weak so they can't get the past rainfall history data, new disease information, and statistics about fertilizer growth. Karunathilake et al. [8] provides insights into the rapidly evolving field of precision agriculture, emphasizing its potential to transform existing farming practices. It discusses recent advances, ongoing problems, and precision agriculture's optimistic trajectory and significance in the future of agriculture and sustainable food production.

This research aims to apply profitable agriculture to those farmers living in rural area to get more yields on their cultivation with the support of pervasive automation. The progression of agricultural practices from traditional approaches to the modern concept of Agriculture 5.0 emphasizes the need for technological advancement to meet the growing demand for high-quality food in a world with an ever-increasing population:

Agriculture 1.0: The era of old farming practices and manual labor with low productivity. Agriculture 2.0: The introduction of technology during the first Industrial Revolution, which resulted in greater food production and decreased physical labor.

Agriculture 3.0: Increased automation through embedded systems, software development, and communication technologies, as well as the use of green renewable energies and information technology.

Agriculture 4.0: The rise of precision agriculture fueled by data collection and analysis and the use of Industry 4.0 technologies.

Agriculture 5.0: The next step in agricultural evolution, characterized by autonomous robotic farming, integrated artificial intelligence (AI), and unmanned operations. This concept aims to address issues such as population increase, urbanization, resource scarcity, and environmental protection, ushering in a new era of farm productivity and profitability. Environmental protection ushers in a new era of agricultural productivity and profitability.

Precision agriculture focuses solely on field-level precision and variability, whereas Agriculture 4.0 embraces a larger range of technology and practices that extend beyond field-level precision to farm-level management and integration. Agriculture 5.0 marks the peak of this evolution, including AI-driven autonomous systems and decision assistance for sustainable and highly efficient farming practices. Jha et al. [7] highlight the significance of automation, particularly AI-driven solutions, in addressing the challenges faced by agriculture. It discusses various technologies and their potential to enhance agricultural practices, improve productivity, and mitigate environmental impacts.

Farmers initially relied on manual labor and draughts for farming, as well as minimal irrigation systems based on soil gravity and conventional crop selection and agricultural practices with limited use of fertilizers and pesticides. Later, the advent of revolutions such as steam engines and mechanical reapers for improved plowing and planting techniques coincided with the birth of scientific agriculture through the use of early chemical fertilizers. When the green revolution began, high-yield crop types and hybrid seeds, as well as the use of synthetic fertilizers and pesticides, led to advances in irrigation technology and the expansion of farm machinery. The end of the twentieth century focuses on sustainability and organic farming, with the adoption of biotechnology, particularly genetically modified crops. Precision agriculture and remote sensing improved agricultural storage and transportation. Agricultural techniques have evolved over generations since the beginning of the twenty-first century due to the integration of digital technology and IoT. Monitoring is carried out using autonomous machinery and drones. Advanced devices are used in climate-smart agriculture and sustainability practices. Block chain and traceability systems were introduced to measure crop productivity in advance.

2 Review of Literature

The agriculture industry is at the forefront of a technological transformation, with smart agriculture, powered by data analytics, machine learning, and IoT, changing farming practices. In this collection of chapters and titles, we look at the many facets of smart agriculture, putting light on its transformational potential.

This study goes into the world of smart farming, a paradigm change facilitated by information and communication technology that includes machinery, equipment, and sensors networked inside network-based, high-tech agricultural systems. It investigates how modern technologies like as the Internet of Things (IoT) and cloud computing are set to alter agriculture, paving the way for the incorporation of robotics and artificial intelligence into farming practices. While these developments carry enormous promise, they also pose a new set of obstacles [2].

The primary goal of this research [3] is to evaluate the effectiveness of deep learning in agriculture by comparing it to other established artificial intelligence models typically utilized in the agricultural sector.

Ghobadpour et al. [4] investigate the developing environment of green energy-powered off-road electric cars and autonomous robots in the agricultural sector. It addresses the significant issues confronting agriculture, such as population expansion, rising energy demands, labor constraints, and environmental problems such as global warming. The study emphasizes the significance of switching to renewable energy sources and electric agricultural vehicles in order to realize smart farming in Agriculture 5.0.

Aide et al. [6] emphasize the crucial role of soil in helping to mitigate climate change through carbon sequestration. It examines numerous tactics and factors for improving soil health and maximizing carbon storage in various habitats, particularly deciduous forests.

Kumar [9] proposes an IoT-driven method to safeguard crops from wild animal attacks, emphasizing the relevance of technology in modern agriculture and the potential benefits it provides to crop security and farming practices.

Andrew et al. [10] investigates the potential of precision agriculture to improve productivity and resource utilization when compared to traditional agricultural practices, resulting in cost savings and greater yields. It focuses on two feasibility studies incorporating IoT solutions: automated irrigation for crop farming and automated animal monitoring for livestock farming. The studies demonstrate the utility of various IoT technologies in everyday life.

Mondal and Rehena [11] intend to boost agricultural productivity and adapt to changing climatic conditions by utilizing IoT-based smart farming approaches, ultimately contributing to sustainable and efficient farming practices.

Suzuki et al. [12] describe an agricultural cloud support system that makes use of support vector machine (SVM) technology. Using sensor data, this SVM-based smart irrigation system changes the amount of water applied to crops automatically. By doing so, it enables persons without agricultural knowledge to properly and efficiently regulate irrigation in greenhouse horticulture, potentially enhancing crop yields and sustainability.

This research describes a revolutionary low-power and cost-effective IoT network intended exclusively for smart agriculture. The network employs a custom-developed sensor to detect soil moisture levels. This breakthrough offers great promise for improving the sustainability and affordability of smart agricultural practices [14].

Expert system technology has the potential to be extremely useful in the agricultural sector. This study is meant to present several features of agricultural expert systems. It also suggests a structural vision of an agricultural expert system that will be extremely beneficial to farmers in increasing productivity, conserving nature, and producing less pesticide-contaminated food [15].

Machine learning is vital in crop yield prediction and agricultural decision-making. This study [19] conducted a systematic literature review (SLR) to assess the algorithms and features typically employed in agricultural yield prediction studies. Fifty of the 567 retrieved studies were chosen for in-depth study. Temperature, rainfall, and soil type were the most commonly used features, with artificial neural networks serving as the leading algorithm. Additionally, the study ran an extra search to discover deep learning-based studies, indicating Convolution Neural Networks (CNN) as the most prominent deep learning algorithm in crop production prediction research, followed by long short-term memory (LSTM) and deep neural networks (DNN). This detailed evaluation provides insights for future research in this topic. These works highlight the transformative impact of technology in agriculture and provide useful insights for future research in this dynamic subject.

3 Methodology

3.1 Rule-Based Agriculture System (RBAS)

The rule-based agriculture system (RBAS) methodology starts with smart agriculture and precision agriculture which are the two main aspects carried out in this work. This all-encompassing approach to modern smart agriculture includes a number of tactics targeted at optimizing agricultural practices in response to changing environmental conditions. The procedure starts with a comprehensive examination of environmental elements such as climate, soil quality, and weather patterns. This study informs crop selection by utilizing innovative technology and data-driven insights. Once crop selection has been chosen, the tactics apply to all stages of the agricultural lifecycle, from planting and culture to harvest and crop sale. Smart technology and data-driven decision-making are critical in optimizing each of these stages, guaranteeing effective resource allocation, precision irrigation, and timely pest management. Furthermore, the techniques emphasize maximizing profitability throughout the crop cycle. This entails using real-time market data and demand estimates to make informed judgments about crop sale timing and pricing. Overall, these adaptation tactics reflect a comprehensive and technology-driven approach to modern agriculture, with the goal of increasing productivity, sustainability, and profitability in the face of changing environmental conditions. This research has three phases, namely, (i) survey, (ii) growth, and (iii) warehouse:

- The first phase is survey, that is, before the seed or crop process, which is various kinds of tests performed in water, soil, and suitable fertilizers for the land to improve better productivity and check for climate and rainfall record of the current area. It starts with data collection in various forms like GPS-based yield monitors, ground-related soil test, structure, and chemical properties like nitrogen levels on each field which seed suits most on that field, Mobile alerts for weather-related information monitor the field conditions and soil moistures.
- The second step is to select high-quality seeds or crops that improve early crop growth in tropical or cold regions. Spacing between crops or seeds is also important to promote growth. Additional fertilizers, such as herbicides and pesticides, are required to obtain adequate yields. Fencing is a technique used to protect crops from birds, insects, and animals. Predicted decisions are informed to farmers regarding how, when, and where to sow, water and fertilizer, pesticide, and harvest and how to manage the data in real time to improve their product.
- The third phase is warehouse, which includes harvest, storage, procurement, and trading/marketing in a timely manner. Large amount of farming data are stored in database, and farmers are to make a timely decision to improve the yield in agriculture.

Plants and ecosystems must be studied further if the Earth is to survive and thrive in the face of human and animal population increases [5]. The four different types of landforms are mountain or hill, forest, plain field, and seashore represented as

Kurinji, Mullai, Marutham, and Neithal. Target beneficiaries are farmers living in rural areas with illiteracy background, where they cannot get the recent updates in agricultural technology. The research objectives are stated below:

- To use pervasive automation in agricultural sector to reduce human workload.
- To find the best profitable agriculture for farmers in rural areas.
- To find the problems faced by the farmers and eradicate them.
- To extend the support for farmers to get maximum yield in crop cultivation.
- To educate farmers using various devices like mobile devices and drone.
- To provide historical data of market demand products in agriculture.
- To provide weather information and rainfall data to farmers.
- To practice farmers getting better yield even in difficult situations.
- To provide training in marketing aspects for farmers.

The three phases of rule-based agriculture system (RBAS) starts with land survey, plant growth, and value addition of agricultural products with the help of warehousing (Fig. 1).

The RBAS is further subdivided into various stages starting from soil analysis to technology integration. Several regulations and best practices in agriculture must be followed in order to produce higher crop yields. Furthermore, the integration of IoT (Internet of Things), robotics, and drones can considerably boost a country's agricultural sector. Here are some agricultural production guidelines, followed by an explanation of how these modern technologies can help: Farmers can improve crop output and promote agricultural sustainability by taking the following 12 steps:

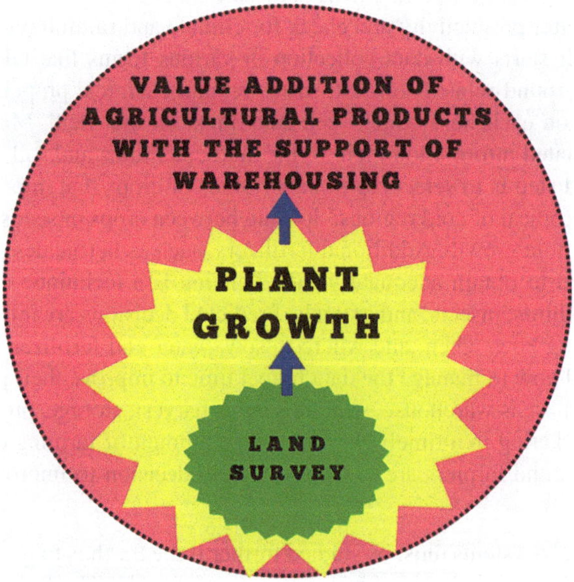

Fig. 1 Three phases of rule-based agriculture system (RBAS)

- *Soil analysis and preparation* Begin with soil testing and apply the essential amendments for optimal soil health.
- *Crop selection and planning* Choose appropriate crops and plan rotations to avoid soil depletion.
- *Seed selection and preparation* Use high-quality seeds that have been pest- and disease-treated.
- *Planting and early crop care* Use precision planting, efficient watering, and integrated pest management.
- *Crop monitoring and maintenance* Keep an eye on nutrients, pests, diseases, and watering, and prune as needed.
- *Harvesting and postharvest handling* Harvest at the suitable stage, treat with care, and store properly.
- *Record keeping and data analysis* Maintain records and analyze data to make educated decisions.
- *Continuous learning and improvement* Stay current on agricultural breakthroughs and adjust practices.
- *Sustainable practices* Implement sustainable strategies to safeguard the environment and increase productivity.
- *Market access and marketing* Establish market connections and look for value-added opportunities.
- *Risk management* Create ways to mitigate weather, pest, and market risks.
- *Technology integration* Utilize IoT, drones, and precise equipment for efficient crop management. Adapting these processes to specific crops and local conditions can lead to increased agricultural productivity and resilience in agriculture. Smart technologies such as IoT (Internet of Things) help in soil monitoring, weather forecasting, and livestock monitoring; robotics in precision planting, weeding and pest control, and harvesting; and drones in aerial imaging, crop spraying, and livestock management (Fig. 2).

Fig. 2 The 12 stages of extended rule-based agriculture system (RBAS)

Marques and Pitarma [13] emphasize the importance of monitoring and managing environmental parameters in agricultural settings to improve energy efficiency and productivity. Real-time monitoring allows for the early detection and rectification of unfavorable conditions, optimizing resource usage, and protecting crops from disease. The study describes an autonomous method for monitoring agricultural surroundings that makes use of the Internet of Things (IoT). The field of smart agriculture is undergoing a fundamental transition as a result of technological breakthroughs, data analytics, and machine learning. This work investigates several aspects of smart agriculture, ranging from the categorization of IoT farm sensor data using machine learning and web semantics to the integration of AI and ML for improving crop yields and resource efficiency. Precision agriculture, in which technology-driven practices are used for effective farming, and the importance of IoT in linking the world through smart gadgets are both highlighted. Furthermore, the collection provides a thorough examination of genetic engineering's merits and downsides in agriculture. These chapters and titles, taken together, provide insights into the changing terrain of smart agriculture and its potential to revolutionize the farming business.

4 Strategies for Adaptation

Market-driven crop farming that takes into account environmental conditions is a comprehensive method that balances profitability and sustainability. It optimizes crop selection and production practices to fulfill market demands while protecting the local ecology, resulting in a win-win situation for farmers, consumers, and the environment. The study of Mariammal et al. [16] focuses on predicting the suitability of crops for cultivation based on soil and environmental variables, with machine learning techniques used to assist farmers in crop choices. The development of a unique feature selection approach called modified recursive feature elimination (MRFE), aimed at discovering critical features for crop prediction, is a fundamental element of this research. To determine the effectiveness of the MRFE approach, it is compared to other classifiers such as kNN, NB, DT, SVM, RF, and bagging. The results show that the MRFE strategy, especially when combined with the bagging classifier, beats other feature selection methods in terms of accuracy (ACC), with a phenomenal 95% ACC. The study emphasizes the significance of precise crop prediction in agriculture, as well as the potential benefits of MRFE in supporting farmers with crop selection. Bhadouria et al. [17] emphasize the substantial concerns that changing climatic circumstances, particularly global warming and climate variability, bring to agriculture. It emphasizes that climate change has become a worldwide concern, necessitating urgent policy formulation at both national and international levels. One of the worries raised is the melting of Himalayan glaciers, which could dramatically affect temperature regimes critical for the growth of numerous crop species. To solve these difficulties, the current generation must devise methods to offset the negative effects of environmental changes on agricultural products. Proposed options include biodiversity conservation, agricultural practice adaptation, and the search for plant species that can tolerate varied environmental pressures, both abiotic and biotic. These initiatives are regarded as necessary steps in combating the detrimental effects of climate change.

4.1 Crop and Technology Awareness for the Formers in Three Different Phases

The term smart agriculture refers here to a wide range of information technology in agriculture field to support farmers in agriculture activities. This research has three phases; each phase has information technology involvement for better productivity. By improving the pervasive automation in every phase, farmers can handle any situation which affects the crop in a different manner.

In first-phase survey, farmers must know the following before they start cultivating:

- He should have a working knowledge of GPS mobile-supported device with Internet connection.
- He should have the working knowledge of camera drones.
- He should have the rainfall database for a period of year in rural area.
- He should do water/soil test for minerals available and chemicals required for land.
- He should know the demand in market-rich product in every season.
- He should get alerts regarding weather conditions.
- He should have the knowledge of disease management for crops in seasonal timings.
- He should have the three-phase fully automatic mobile starter [work from home by calling motor start/stop].
- His own ideas and implementation knowledge about the agriculture climate change.
- Source of water [well water/bore well water/river canal irrigation/rain water-based].

For second-phase crop, farmers must know the following:

- Previous database of crop growth-related information.
- When, where, and how to sow and irrigate.
- Record every week snapshot of grown crop to cross validate from the first week of crop/seed stimulates.
- Managing drought by drop water cultivation.
- Analyzing the crop images regularly through drone-based images.
- Adding proper fertilizers to boost up crops through historical data.

- Using pesticides and herbicides for insect infection in crop through disease management and rectifying the diseases.
- Restricting birds and animals from crops till harvesting.

For third-phase harvest, after harvesting, farmers must know the following:

- Store matured seed/crops in underground or surface level to maintain its nature.
- Take a snapshot of seed/crop and cross verify the image with database.
- Check the quality and quantity of seed/crop.
- Analyze the demand in market with current requirement of crops/seed.
- Analyze market strategy trading/marketing/processing/retailing that best suits for better profit.

Human-machine integration for smart agriculture (Fig. 3) explores the study of the dynamic and exciting topic of smart agriculture, addressing future innovation, difficulties, and opportunities. These innovations show the transformative potential of technology in determining the future of agriculture. The future of smart agriculture is in the hands of technologies such as IoT (Internet of Things). Automation of skills includes data-driven farming, chat bots, and drone technologies, such as soil and field analysis, planting, crop spraying, crop monitoring, irrigation, and crop health assessment.

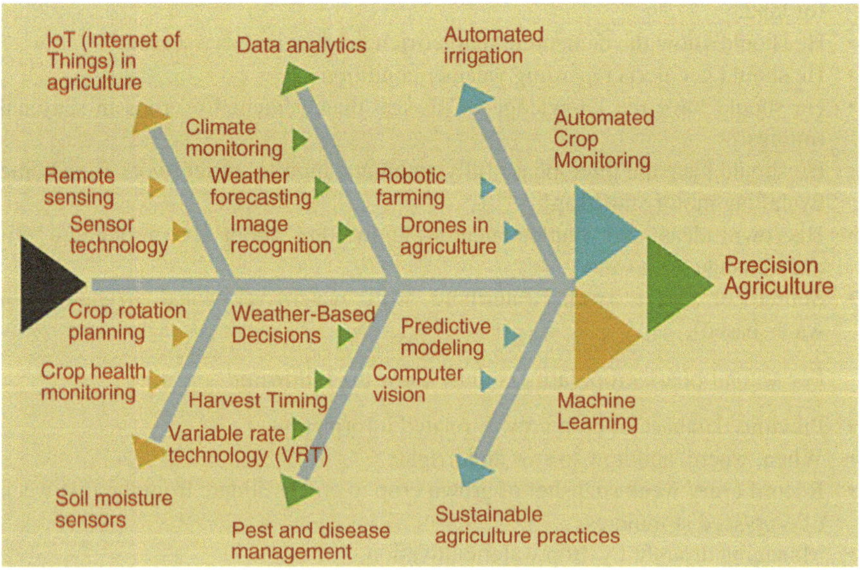

Fig. 3 Human-machine integration for smart agriculture

4.2 Analysis of Environmental Conditions and Crop Suggestion for Namakkal District

Improving crop yields and increasing precision agriculture need extensive monitoring and analysis of many environmental conditions. Temperature, pressure, wind speed, humidity, rainfall, rainfall days, average daylight, sunshine length, UV index, and cloud cover are all aspects to consider. Farmers and agricultural experts can make informed judgments and implement precise strategies to maximize crop growth and yield by painstakingly examining these elements. This data-driven method enables agriculture to adapt to changing weather conditions, optimize resource allocation, and ultimately boost agricultural yields while lowering environmental impact. Namakkal District-based environmental conditions and crop suggestions were provided based on the district graph details retrieved from [20]. Figure 4 shows the environmental factors for crop cultivation and the importance of the plan. Figs. 5, 6, 7, 8, 9, 10, 11, 12, 13, 14 and 15 show the environmental conditions of Namakkal District one by one.

Figure 16 shows the spatial information of Namakkal District (red in color), Tamil Nadu, India.

Here is a summary of the weather statements provided for Namakkal, India [21]:

- The driest month in Namakkal is January, with 3.8 days of rain and 6 mm of precipitation.

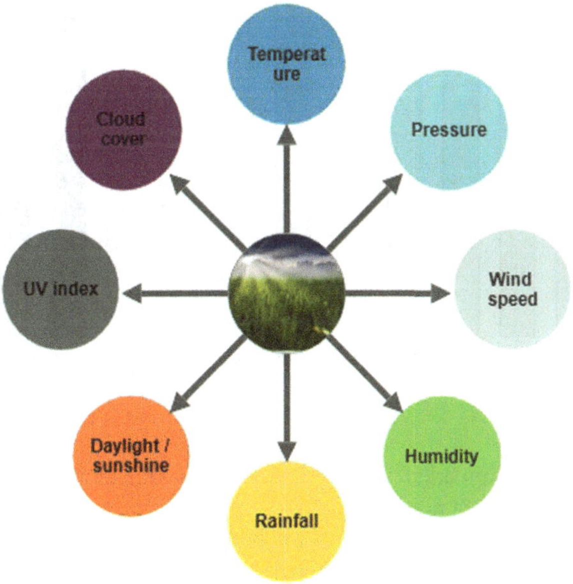

Fig. 4 Environmental factors for crop cultivation

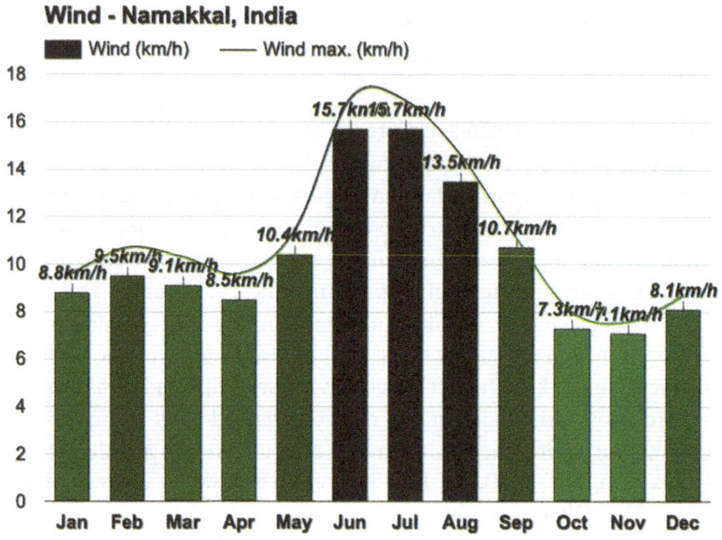

Fig. 5 Wind

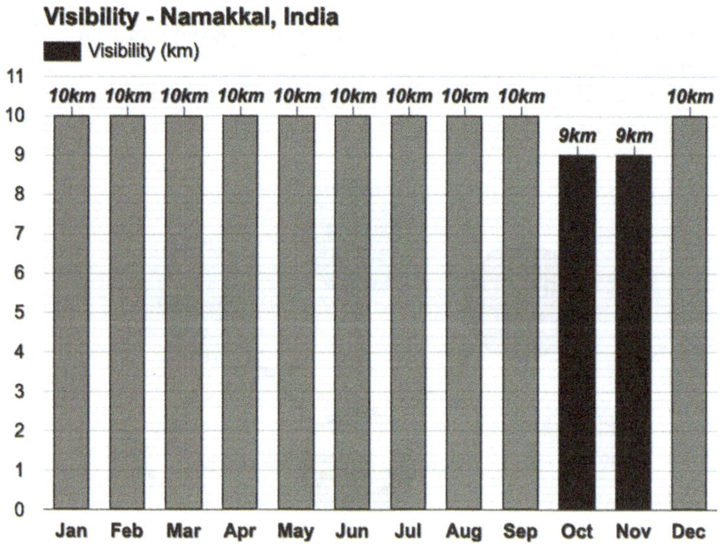

Fig. 6 Visibility

- Annual rainfall: Namakkal receives rain on approximately 180.4 days out of the year, for a total of 698 mm.
- January, October, and December have the lowest UV index, with an average maximum UV index of 6. The months from March through May have the greatest UV index, with an average maximum UV index of 8.

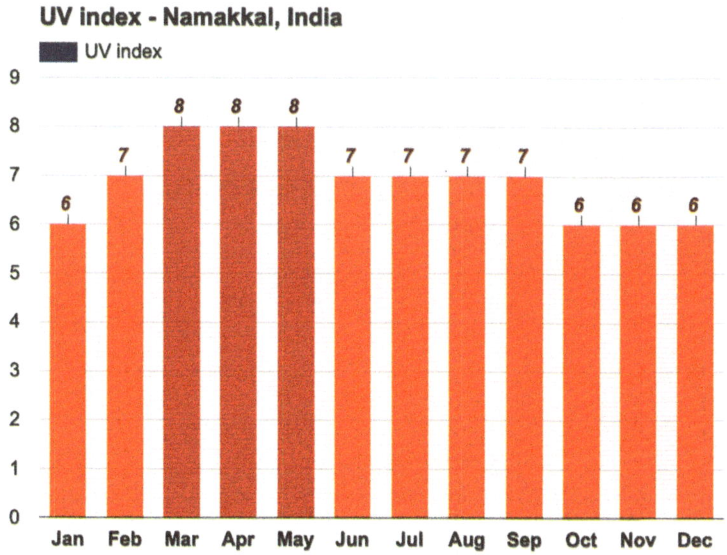

Fig. 7 UV index

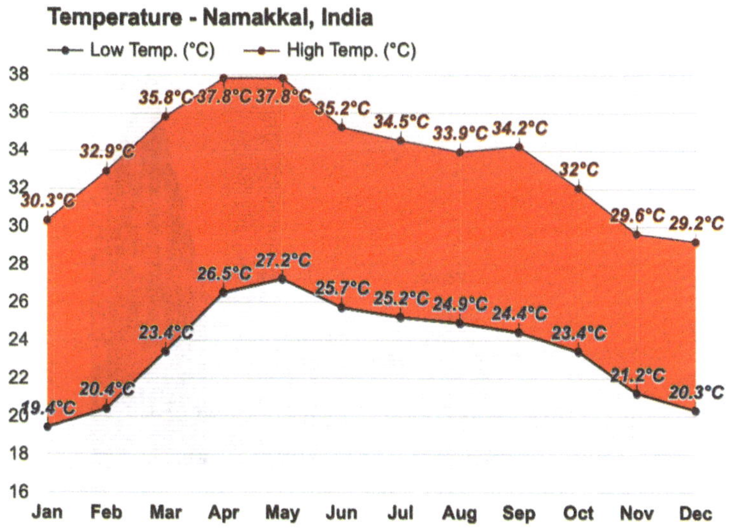

Fig. 8 Temperature

- The hottest months in Namakkal are April and May, with an average high temperature of 37.8 °C.
- April has the lowest relative humidity, with an average relative humidity of 49%.
- May is the sunniest month in Namakkal, with an average of 11.7 h of sunshine per day.

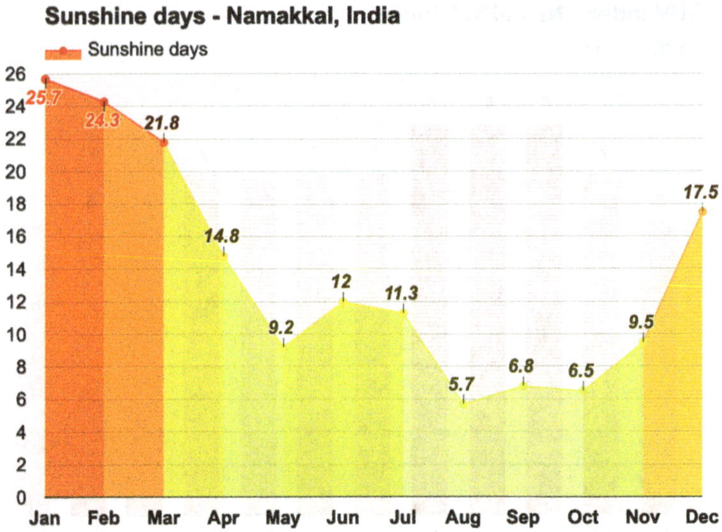

Fig. 9 Sunshine days

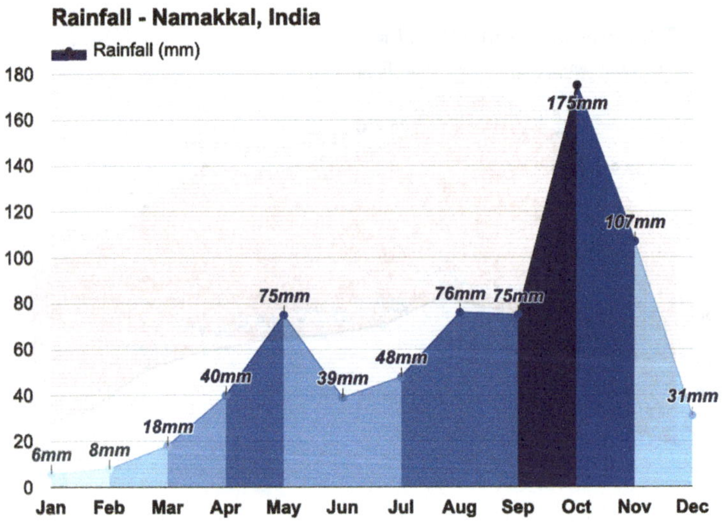

Fig. 10 Rainfall

- June offers the year's longest days, with an average of 12 h and 48 min of daylight. October is the wettest month, with 22.9 days of rain and 175 mm of precipitation.
- The most humid month is November, with an average relative humidity of 78%.

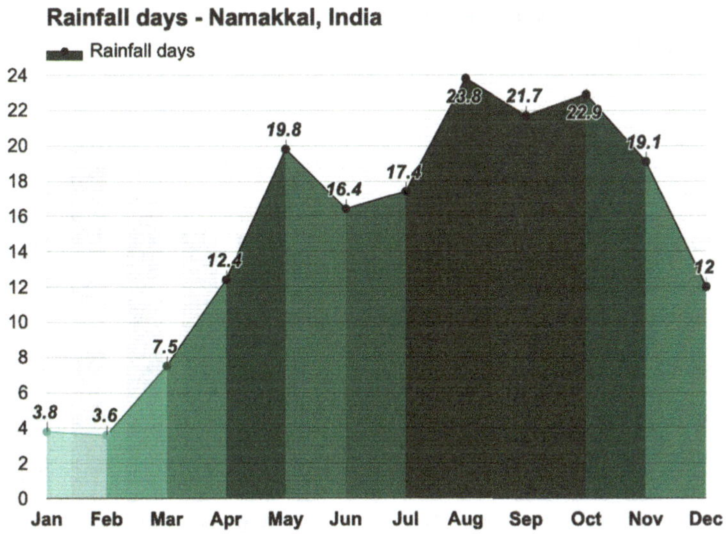

Fig. 11 Rainfall days

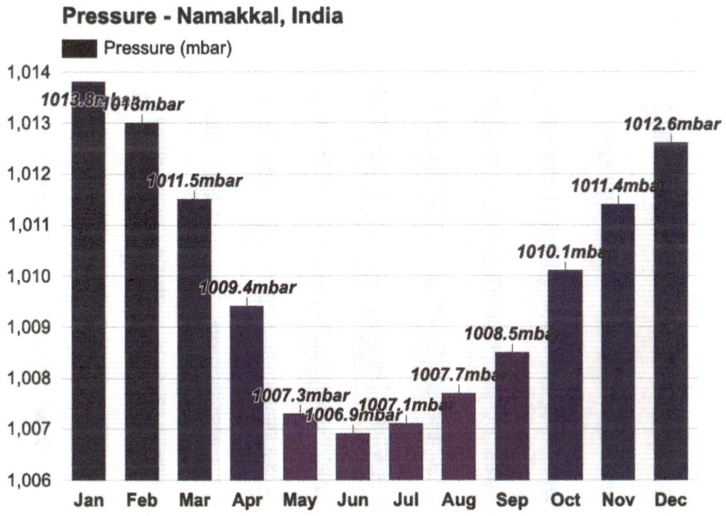

Fig. 12 Pressure

- November has the least amount of sunshine, with an average of 5.9 h.
- The coldest month is December, with an average high of 29.2 °C and a low of 20.3 °C. The shortest days are in December, with an average of 11 h and 30 min of daylight.

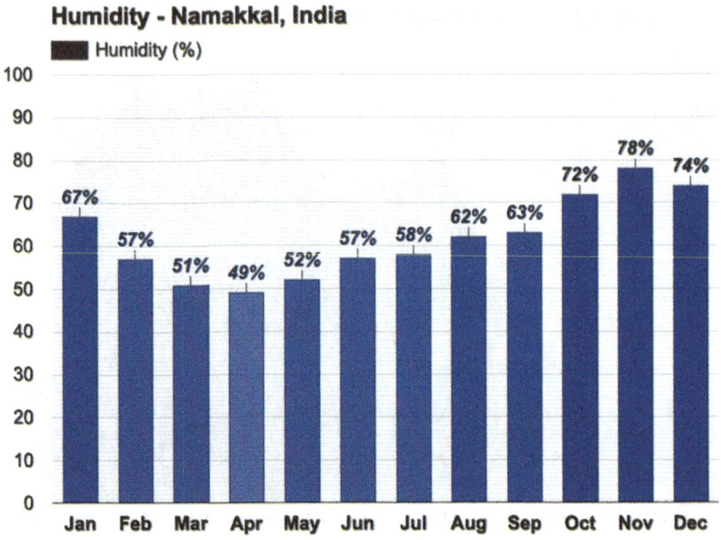

Fig. 13 Humidity

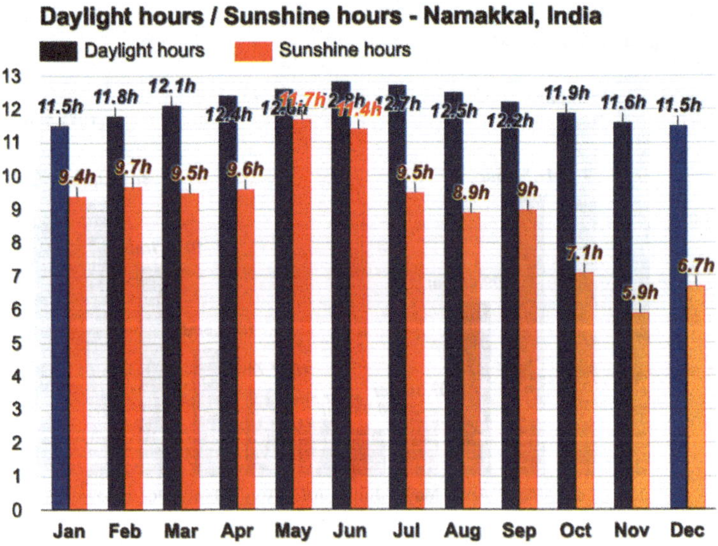

Fig. 14 Daylight/sunshine hours

Based on meteorological observations in Namakkal, India, the following conclusions can be taken about acceptable crops for cultivation:

1. *Dry Conditions* Namakkal has its driest month in January, with little precipitation and only 3.8 days of rain. This means that drought-tolerant crops or those with a short growing cycle may be more suitable during this time.

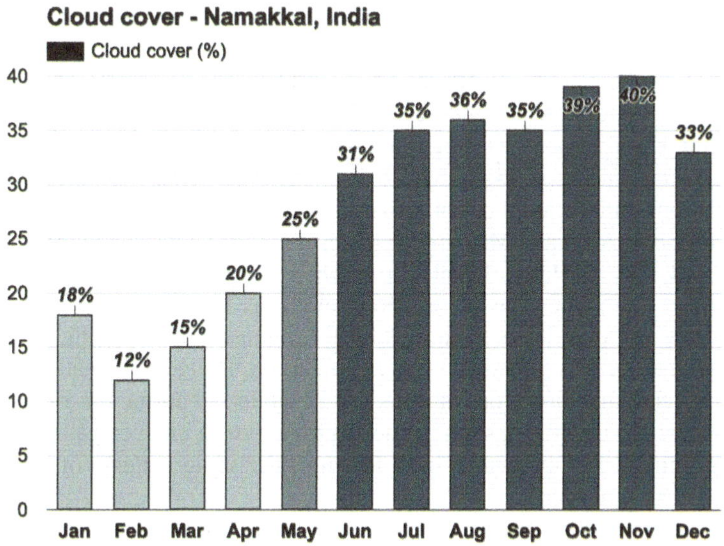

Fig. 15 Cloud cover

Fig. 16 Spatial information of Namakkal (red color), Tamil Nadu, India

2. *Annual Rainfall* With around 180.4 days of rain per year and an annual precipitation of 698 mm, crops that can thrive in locations with moderate rainfall and can also tolerate occasional heavy rains, such as rice or some varieties of millets, may be feasible possibilities.
3. *High UV Index* The months of March through May have the greatest UV index, suggesting intense sunlight. Sun-loving vegetables (e.g., tomatoes, eggplants) and crops utilized in oil production (e.g., sunflowers) may perform well.
4. *Hot Months* April and May are the hottest months, with average high temperatures of 37.8 °C. Heat-tolerant crops such as sorghum, pearl millet, or drought-resistant pulse types may be suitable alternatives during these months.
5. *Low Humidity* April has the lowest relative humidity at 49%, which may favor crops that are less sensitive to fungal infections. However, irrigation solutions for water-demanding crops may need to be addressed during this time period.
6. *Sunniest Month* May has the greatest sunshine, with an average of 11.7 h per day. This time of the year is ideal for crops that require plenty of sunlight for growth and fruiting.
7. *Longest Days* June has the longest days with 12 h and 48 min of daylight, allowing crops to thrive for longer periods of time.
8. *Wettest Month* October is the wettest month, with substantial rainfall (22.9 days of rain and 175 mm of precipitation). Crops that can endure severe rainfall and waterlogging, such as paddy rice or some vegetable kinds, may thrive during this season.
9. *High Humidity* November is the wettest month, with an average relative humidity of 78%. This period may be ideal for crops that benefit from increasing moisture levels.
10. *Shorter Days and Cooler Temperatures* December features the shortest days and the coldest temperatures. Crops that flourish in milder circumstances or have a longer growing season may be more appropriate during this time.

The crop selection in Namakkal should take into account the changing weather patterns throughout the year. Diversifying crops to reflect seasonal climate variations can help optimize yield and reduce risks associated with extreme weather occurrences. Additionally, good irrigation and water management practices are critical for successful agriculture in this region. Crop yield prediction is an important part of agricultural planning and decision-making. Statistical models have traditionally been utilized for this purpose, but they are time-consuming and labor-intensive. This research [18] investigates the use of deep learning models, specifically recurrent neural network (RNN) and long short-term memory (LSTM), to forecast wheat crop yields in northern India. The RNN-LSTM model surpasses typical machine learning models in terms of accuracy by exploiting deep learning's capacity to extract features from huge datasets. The results show that the RNN-LSTM model delivers more accurate predictions and is a potential strategy for agricultural yield prediction in the region.

5 Conclusion

This study proposes a structured strategy to agricultural management divided into three distinct stages, survey, growth, and warehousing, with the help of rule-based agriculture system (RBAS). The crop suggestions based on Namakkal District can be extended to various other areas to obtain a greater yield based on several environmental parameters. The overall theme of automation, precision agriculture, and the use of IoT-based technology binds these stages together. This human-machine collaboration permits thorough data analysis, allowing us to obtain insights into historical data and present trends and future predictions in crucial areas such as soil quality, weather conditions, fertilizer consumption, and sophisticated agricultural practices. In the field of machine learning in agriculture, we explore crop production guidelines, alternative cropping systems, agricultural finance, and cooperative development. The future of agriculture depends on our ability to accept these breakthroughs, adapt to changing environments, and build a brighter, more sustainable future for future generations.

References

1. P. Majumdar, D. Bhattacharya, S. Mitra, Utilities of 5G communication technologies for promoting advancement in agriculture 4.0: Recent trends, research issues and review of literature, in *5G and Beyond, Springer Tracts in Electrical and Electronics Engineering*, (2023), pp. 111–125
2. M. Dhanaraju, P. Chenniappan, K. Ramalingam, S. Pazhanivelan, R. Kaliaperumal, Smart farming: Internet of things (IoT)-based sustainable agriculture. Agriculture **12**, 1745 (2022). https://doi.org/10.3390/agriculture12101745
3. V.S. Magomadov, Deep learning and its role in smart agriculture. J. Phys. Conf. Ser. **1399**, 044109 (2019)
4. A. Ghobadpour, G. Monsalve, A. Cardenas, H. Mousazadeh, Off-road electric vehicles and autonomous robots in agricultural sector: Trends, challenges, and opportunities. Vehicles **4**, 843–864 (2022). https://doi.org/10.3390/vehicles4030047
5. B. Bahadur et al. (eds.), *Plant Biology and Biotechnology: Volume I: Plant Diversity, Organization, Function and Improvement* (Springer, 2015). https://doi.org/10.1007/978-81-322-2286-6_1
6. M. Aide, I. Braden, S. Nakasagga, S. Svenson, Improving forest soil health and ecosystem services to minimize the impact of climate change. Agric. Sci. **14**, 1153–1168 (2023). https://doi.org/10.4236/as.2023.149077
7. K. Jha, A. Doshi, P. Patel, M. Shah, A comprehensive review on automation in agriculture using artificial intelligence. Artif Intell Agric. **2**, 1–12 (2019)
8. E.M.B.M. Karunathilake, A.T. Le, S. Heo, Y.S. Chung, S. Mansoor, The path to smart farming: Innovations and opportunities in precision agriculture. Agriculture **13**, 1593 (2023). https://doi.org/10.3390/agriculture13081593
9. J. Kumar, Agricultural field protection from wild animal. J. Emerg. Technol. Innov. Res. **5**(10), 206–208 (2018)
10. R.C. Andrew, R. Malekian, D.C. Bogatinoska, IoT solutions for precision agriculture, in *International Convention on Information and Communication Technology, Electronics and Microelectronics (MIPRO)*, (Opatija Croatia, May 21–25, 2018), pp. 345–349

11. M.A. Mondal, Z. Rehena, IoT based intelligent agriculture field monitoring system, in *8th International Conference on Cloud Computing, Data Science & Engineering (Confluence)*, (2018), pp. 625–629
12. Y. Suzuki, K. Nakamatsu, H. Mineno, A proposal for an agricultural irrigation control system based on support vector machine, in *Second IIAI International Conference on Advanced Applied Informatics*, (2013), pp. 104–107
13. G. Marques, R. Pitarma, Agricultural environment monitoring system using wireless sensor networks and IoT, in *13th Iberian Conference on Information Systems and Technologies (CISTI)*, (2018), pp. 1–6
14. S. Heble, A. Kumar, K.D. Prasad, S. Samirana, P. Rajalakshmi, A low power IoT network for smart agriculture, in *Proceedings of the 2018 IEEE 4th World Forum on Internet of Things (WF-IoT)*, (2018), pp. 609–614
15. B. Maurya, M.R. Beg, S. Mukherjee, Expert system design and architecture for farming sector, in *IEEE Conference on Information & Communication Technologies*, (2013)
16. G. Mariammal, A. Suruliandi, S.P. Raja, E. Poongothai, Prediction of land suitability for crop cultivation based on soil and environmental characteristics using modified recursive feature elimination technique with various classifiers. IEEE Trans. Comput. Soc. Syst. **8**, 1132 (2021)
17. R. Bhadouria, R. Singh, V.K. Singh, A. Borthakur, A. Ahamad, G. Kumar, P. Singh, Chapter 1 – Agriculture in the era of climate change: Consequences and effects, in *Climate Change and Agricultural Ecosystems*, ed. by K.K. Choudhary, A. Kumar, A.K. Singh, (Woodhead Publishing, 2019), pp. 1–23. https://doi.org/10.1016/B978-0-12-816483-9.00001-3. ISBN 9780128164839
18. N. Bali, A. Singla, Deep learning based wheat crop yield prediction model in Punjab Region of North India. Appl. Artif. Intell. **35**(15), 1304–1328 (2021). https://doi.org/10.1080/08839514.2021.1976091
19. T. van Klompenburg, A. Kassahun, C. Catal, Crop yield prediction using machine learning: A systematic literature review. Computers and Electronics in Agriculture **177**, 105709 (2020). https://doi.org/10.1016/j.compag.2020.105709
20. https://www.weather-atlas.com/en/india/namakkal-climate
21. https://commons.wikimedia.org/wiki/File:Namakkal_district_Tamil_Nadu.png

Internet of Things-Based Smart Agriculture Advisory System

Mahalakshmi Jeyabalu, Akil Shabbir Ghodi, Sundaravadivazhagan Balasubramanian, Balakrishnan Chinnayan, and Jayapriya Jayapal

1 Introduction

The Internet era provides a lot of automation tools for data analysis, and it is the need of the hour to develop new analytical tools to manage the big data. For task automation, machine learning and expert systems are of primary importance to study the behavior of computer thinking to involve computers in sensible work, known as "computational intelligence." In India, agriculture is the major backbone, and it is an immediate requirement to deploy technology to agriculture. There is a need for the integrated framework to focus on the development of new tools for crop management, especially the plants that contain various important medicinal components. This research work involves a multidimensional approach to the data, and the main aim is to create a secured automation tool, to help the stakeholders in the crop management.

Smart agriculture is a beneficial use case in data analytics and IoT, as they bring precision farming to agriculture to maximize the yield per unit of land fit for farming by using futuristic farming methods. In digital India, the technology enables the systems to support the agriculturalists in many ways including sustainable growth.

IoT sensors provide information to farmers about crop type, rainfall acquired, pest infestation, and soil nutrition that are required for the prediction of yield. The main aim is to create a smart environment using enabling technologies for sensing

M. Jeyabalu (✉) · A. S. Ghodi · B. Chinnayan · J. Jayapal
Department of Computer Science, CHRIST (Deemed to be University), Bangalore Yeshwanthpur Campus, Karnataka, India
e-mail: mahalakshmi.j@christuniversity.in

S. Balasubramanian
Department of Information Technology, University of Technology and Applied Sciences, AL Mussanah, Oman

S. Balasubaramanian et al. (eds.), *Intelligent Robots and Drones for Precision Agriculture*, Signals and Communication Technology,
https://doi.org/10.1007/978-3-031-51195-0_9

soil moisture and nutrients, early pest and disease identification and control, and precision farming. The Specific focus towards the research work is to design and develop a secure agro-advisoru framework and smart environment for farmers through IoT, supporting agriculturalist in their activities to the elite outcome of the crops and getting up-to-date information on soil, livestock and climatic conditions with the help of sensor devices.

The tremendously scalable and availability feature, in addition, offers a shared pool of resources between a large number of services, making cloud computing a most notable computing pattern. This article focused on multidimensional data analytics as an automation tool, which too provides insights to agriculturalists and practitioners. Hence, protection of data is considered.

In connection with this, a cryptographic service, one of the most notable features for data encryption, is offered as a service from the cloud paradigm, which provides the privilege for the users to convert the data to unintelligible format, when received from the sensors used to measure various metrics, by means of the application models deployed. Generally, data storage takes multiple patterns in the sensors, used depending on the type of the metrics imposed. It is then transferred by the embedded software applications, for evaluation, storage, and sharing, using software such as Arduino, eclipse Internet of Things, Kinoma, open Internet of Things device, etc. This phase gives a wide variety of objective functions for specific problems and has a lot of security vulnerabilities for the data being transferred from the open-source embedded sensors.

This research article focuses specifically on the data from IoT sensors in one phase and the prediction of crop management for smart agriculture in another phase. The received data from sensors is then passed to the analytics tool, since it is the most required tool for predicting as well as assisting the agriculturalists and the practitioners to make use of the digitally available software and embedded applications to enhance the plant yields, finally leading to smart farming.

Sensors provide the data that may be an image file (the leaves/roots) or may be the soil optimum pH values (decimal values including nutrients values) or may be simple CSV files (inclusive of entire data) and image dataset for finding the disease symptom of the leaves. The proposed application model for data encryption is the first phase of this research work, which will alter the file, and decryption takes place only upon the privileged authenticated access.

The article proceeds as follows: Sect. 2 states the primal information on the Internet of Things, data security services available on the cloud, and prediction techniques found in the literature. The primary notations are briefly explained as mathematical preliminaries in the preceding sections. The meta-heuristic simulated annealing algorithm, which is one of the optimization techniques for the key generation of the cryptographic process, is clearly stated separately. Then, the article covers the multilayer perceptron, artificial neural network algorithm for the classification of crops acquired from the sensor data/satellite images, etc. Followed by the classification technique, the prediction of disease symptoms in various parameters was discussed. The experimental result section then comprised of the analysis between the trained the tested dataset for five sample different dataset and the accuracy rate of prediction. Finally, the conclusion section comprised the result yield from the given algorithm that clearly states the pros of using the deep learning algorithms for the smart agriculture.

2 Related Research Work

Ayaz et al. [1] elaborated on the wireless sensor network communication techniques with the Internet of Things, which is in association with smart farming methods. Every individual process holds a different application, developed and deployed for a specific purpose, which was briefly explained by the authors of this work. The authors explored the specific use of the application for agricultural practitioners in varied ways right from the sowing until the harvesting period.

Channe et al. [2] elaborated data on multidisciplinary application models for smart agricultural systems. The advantage explained by the author's works involves IoT sensors, cloud computing, mobile computing, and big data analysis for the improved farming system.

Gnanasankaran and Ramaraj [3], in their research work, discussed about the effective yield of paddy cultivation in specific geographic regions, using smart farming. Crop monitoring using machine learning techniques is detailed in their research model.

Pendyala et al. [4] discussed temperature monitoring sensors for smart agriculture using the NodeMCU. The authors reported how the connected sensors did automatic watering while the threshold range is met and automatically noted the change of degrees in the soil field.

High-volume data analytics involves precision farming to provide farmers with crop yields along with the individual component's values, pest control infestation, and soil nutrition for individual types [5].

The authors reported on the protection against various open medium threats and vulnerabilities and the way cloud encryption service offers various features to its users [6]. Tremendous scalability suggests cloud is the best way for computing services [7].

Patil and Khairnar [8], in their detailed research work, discussed IoT-based smart farming systems. The authors of the work reported on the devices used for sensing plants on various parameters without manual intervention. The thus collected parameters were evaluated and processed for further data processing.

Ray [9] elaborately discussed the Internet of Things to overcome real-life problems by power fact-IoT notion exploration. The author explained the various frameworks available for industrial needs accordingly.

Sheetal et al. [10] discussed the dampness and soil nutrient values of the grains during the rainfall duration. Arduino has been widely used by the authors to control the contact process.

Sivachandran et al. [11] reported on the integrated soil analyzer, which includes the evaluation of soil nutrients and other metrics. The system, explained by the authors, is an integrated model, which holds a signal control unit for finding the threshold values, a process control for finding the difference in the pH values, and an LED display to show the extracted results from the sensors embedded. The methods stated by the authors of the work discussed on the nutrient's evaluation such as potassium, nitrogen, phosphorus, pH, and temperature change (Fig. 1).

Srinivasan et al. [12] explored other ways for crop maintenance using the WSN. The author stated various wireless sensor applications, threats, and issues found in the model, and the solution enhanced smart cultivation. The authors also explained various computational methods that are in practice for agriculture modernization with the aid of the cloud and its enhancement in decision-making for agricultural practitioners. Also, the authors explored the way how wireless sensors are deployed for smart farming.

The author reported on numerous wireless applications, and the issues and threats in their data security. Along with that, in a determined way, various cloud techniques available for the smart agriculture solution for stakeholders are also provided by the authors of the work. Encryption as a service from cloud offers an application to inarticulate the thus received files [13]. A new computational archetype, cloud in the technology era, offers resources such as infrastructure, platform, software, security, database, etc. as a shared service (pay-as-you-use) whenever and wherever on demand [14].

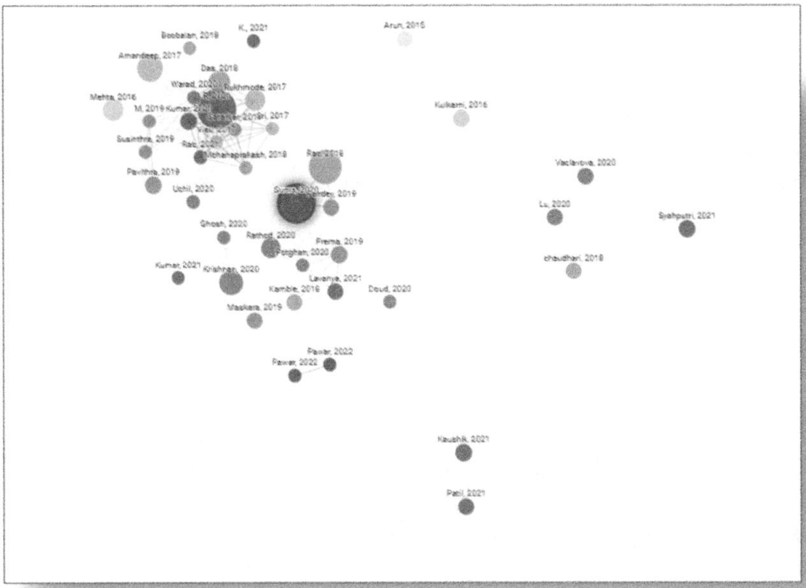

Fig. 1 Related research work by various authors between 2015 and 2022

3 Preliminaries

3.1 *Importance of the Proposed Research Work in the Context of the Current Status*

The designed and deployed research work objective is to implement a new automated software tool that helps the agriculturalist and other stakeholders such as the medical practitioners. This work helps to extract the potential information (nutrition and medicinal component information) for the specific crops and plants for specific geographical location using computational intelligence. Nowadays, there exist a lot of automated tools and websites to list information about crops and their uses. The proposed work differs in a way that it works based on the geographical location and the crop variety suitable for the location and their data management.

As an integrated tool, it also focuses on maintaining the nutrition information of the crops and any relevant medicinal values of the crop using IoT. The IoT has its scope in the varied arenas that include the things for the connection, the software solutions used for embedded connections, the platform on which it will take place, the applications that are to be included, and the geographic regions, where it is going to be implemented. IoT enables the ease of access and connection between things and their associated applications making it easier to share the data between the devices easier.

To enhance customer experience and for further evaluation, various smart devices are available in the market such as smart phones, smart wearable gadgets, etc., to collect and formulate the data from the users based on their requirements and usage. This drastic improvement requires proper data analytics to predict the near future. The Internet of Things and Industry 4.0 discuss the above-said statement and work on how easier access can be done to create a technology-driven environment.

3.2 *Encryption Service for Secure IoT Data Storage in Cloud*

Cloud, a new computing technology, renders varied service to the users with enhanced scalability and reliability. One among them is the encryption services, to ensure security and authenticity of the data, which is the most important feature taken into consideration in this research work.

Data encryption is basically converting the cosigner information into an inarticulate format to the consignee using cryptography algorithms. In this given research article, we proposed the combination of traditional block cipher methods along with the optimization technique, as an encryption service to the users. The interesting fact is that the data is now received from various sensors in the fields and preprocessing is also done.

The data thus received from the sensors may be reported in a multidimensional way, since it includes temperature data, humidity data, water range sensing, nutrient values, etc. Hence, preprocessing is a must for data encryption. Symmetric key generation is considered for this work implementation since varied data blocks are included. For stronger key generation, an improved cipher blockchain symmetric key algorithm is proposed, along with the hybridization of the traditional meta-heuristic algorithm to suboptimal key generation.

The proposed technique promises that the key generated as a result of the hybridization results in the minimum execution time and generates keys on varied bit sizes, like 64 bits–512 bits. The wireless sensor devices are used at the end of the system and the tools, which observe data from crops, transfer the collected data, through any open medium on the cloud. The given research work as a first phase proposes a novel methodology for smart farming by including a smart sensing system for data security through cloud paradigm.

3.3 Simulated Annealing: A Meta-Heuristic Optimization Algorithm

In ease of way, varied data-type files can be encrypted using the meta-heuristic approach for better results. For ease of access, the fundamental notation to generate the key, as well as blocks of data, matrix method is followed. The elements of the matrix are represented in zero and one binary format. Consider M as a matrix and then M_{xy} as a square matrix, for which x indicates the row and y indicates the column (Fig. 2). For data encryption, the data received from the sensors after preprocessing is broken down into several words such as w_1, w_2, etc. assigned with the binary string.

For better evaluation results, cipher blocks are considered for the formation of the matrix structure. Since it is a predefined and constant structure, the data can be encrypted and decrypted in a fast rate and with minimal execution time. The square matrix may be of any size, say, 4 * 4 and 8 * 8 are precisely fixed depending on the data received from the sensors, and the number of rounds for the cryptographic process is fixed. In the case of decryption, after key generation forms the improved block ciphers and pseudorandom number generators, a strong key value is constructed. The cryptographic method proposed in this article stands against various

Fig. 2 8 × 8 Matrix structure (Ref. [1])

$$Z_{xy} = \begin{bmatrix} 10101101 \\ 10010101 \\ 01011101 \\ 10010010 \\ 10101011 \\ 01010101 \\ 11001011 \\ 10011010 \end{bmatrix}$$

vulnerable threats in the open channel. The blockchain structure of the matrix is the self-synchronizing mode, which leads to less error propagation.

3.4 Optimization Technique and Objective Function Formulation

For fortified key generation, various traditional optimization techniques can be combined with the cipher blocks. A suboptimal key from the optimization technique will improve the authenticity of the data being encrypted. The objective function is developed to find either the suboptimum or local minimum feasible solution. Optimization algorithms were evolved to find the objective function to be either the maximum or the minimum for the given problem. Based on the requirement, a feasible solution may be attained.

For the key generation (Fig. 3), the finest meta-heuristic simulated annealing algorithms are considered. The simulated annealing (SA) algorithm is one of the meta-heuristic algorithms, also probabilistic. It is used for specific objective

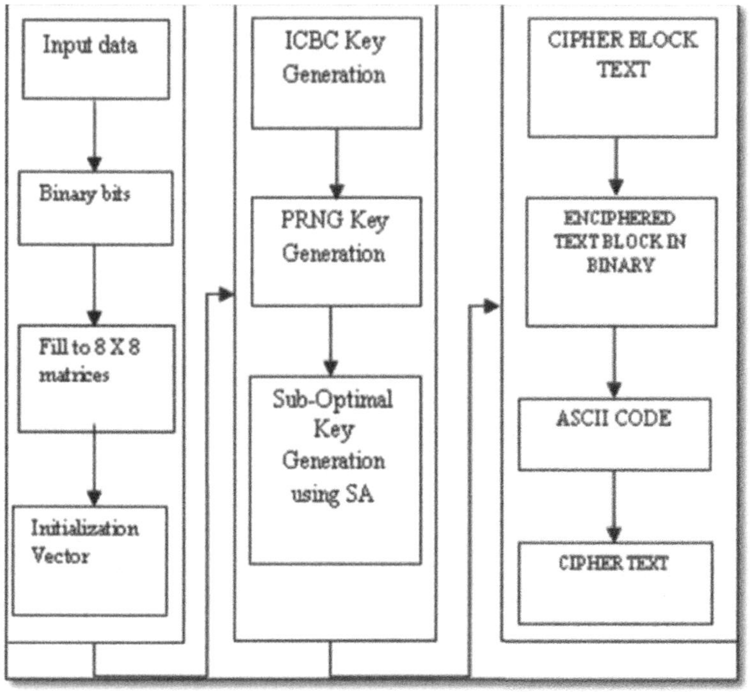

Fig. 3 Block diagram of encryption and decryption process. (Source: https://doi.org/10.1007/978-981-13-7968-0_4)

functions, in discrete to find the global optimum solution in a large search space. It is also considered as an alternate technique to that of a gradient descent algorithm.

4 Proposed Methodology

The given research work focuses on multidimensional data analytics, for crops and plants, that especially classifies the disease from the deep learning models. This research work involves a multidimensional approach to the data, and the main aim is to create an automation tool, to help the stakeholders in the crop management using machine learning algorithms (Fig. 4). Agriculture crop management is implemented through big data (crops and plants) analytics to help farmers in making informed decisions for enhancement of cultivation and information on the disease symptoms and remedies.

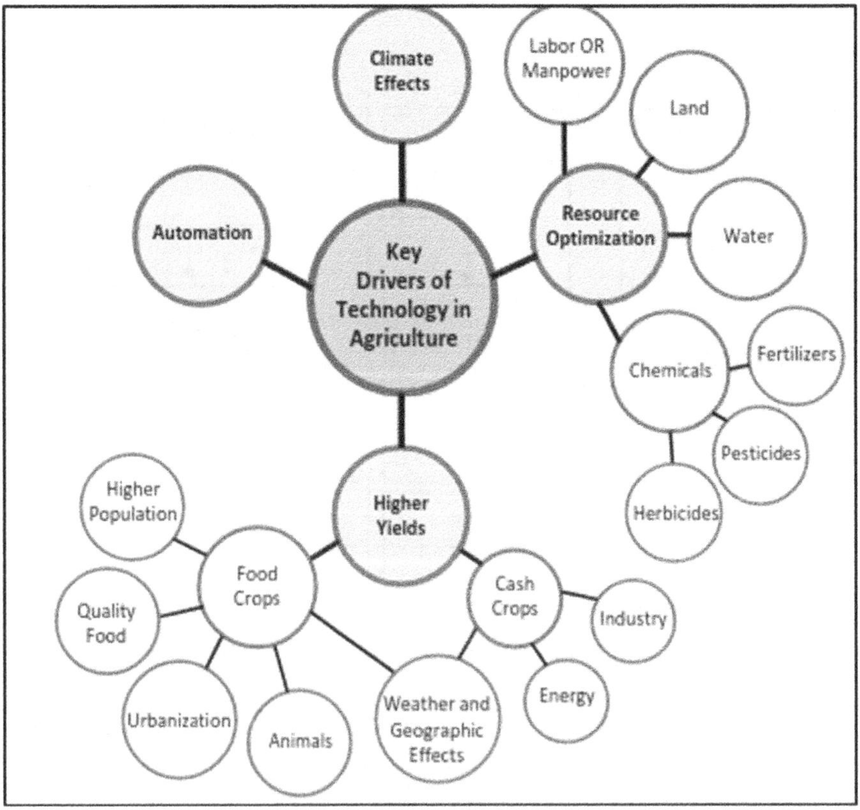

Fig. 4 Factors for smart agriculture – key drivers

The proposed research work involves the analysis of satellite images or images captured with sensors implanted. The research work will identify crop management in chosen areas. This will provide timely information to farmers in crop management for better agriculture processes and further management support.

4.1 Precision Agriculture

The proposed analytical tool aims to develop a prediction system to identify disease symptoms at earlier stage and provide a solution for agricultural crop nutrients that helps the agricultural practitioners. This system obtains better precision based on sensors using IoT. The model attempts to provide a complete, integrated solution for the decision-making support for the agriculturalists focusing on crop protection and providing them with balanced nutrient management strategies. Many factors such as weather, soil testing, sampling at regular intervals, soil fertility of the land, supply of water requirements, bio and organic manures, lack of macronutrients and micronutrients, medicinal component function, and deficiency symptoms are taken into consideration to reach an enhanced result.

4.2 Prediction Algorithm

The proposed research work extracts the input from the temperature sensor, leaf and stem sensor to know its components, and humidity sensor to monitor the climatic changes. Soil sampling is done as the primary process. The following is the procedure and algorithm for prediction.

Process Involved

- Collection of data from the temperature (DHT11) and humidity sensor
- Soil sampling
- Data collection from the leaf extracts to find out the disease components
- Data preprocessing
- Data security
- Prediction on yield and cultivation

In the conventional agricultural process, the attributes include soil value and its nutrient optimum pH values, temperature range between certain intervals, rainfall values, and climate changes, along with manual intervention. In the proposed method, the strategy deployed works as a multidimesional data model, and detect the capability to predict the exact cultivation, yield as well as individual components uses. In traditional strategies, based on the previous past outcomes, pest control and precision are done, whereas, on the other hand, this research work focuses on integrated tool. A machine learning algorithm with a suboptimal method is used to predict. The proposed research work involves the MLP-NN (multilayer perceptron neural network) to deeply analyze the components of crops.

4.3 MLP-NN (Multilayer Perceptron Neural Network)

Neural networks are created as in the form of interlinked nodes/neurons through tight coupled connections between them. Artificial neural network is one among the trained interlinked connections, used as estimation models for data analytics. ANN is formed by three different layers named as input, hidden, and output. The input layer is a compilation of several nodes that holds the input values, while a hidden layer is located as the middle layer, in which the function puts weights to the inputs and channels them through an activation function as the output. Finally, the output layer produces the output for a given objective function.

Perceptron falls under the supervised learning algorithm, intended to do binary classification on the data provided. Perceptron belongs to a linear classification of the data and breaks the given problem into simpler terms. Basically, a vector is considered as input, weights, and channels are included for the training of the data received from the sensors. MLP-NN sets better results for both linear and nonlinear data, including all the parameters in the three-layer setup. Perceptron algorithm provides better MSE (mean square error) value, regression analysis can be done easier, and feature selection for the training data can be done easier.

The below given is feed-forward neural network model (Fig. 5), which comes under linear classification, for which the classification of the data or the prediction of the given data is done by the output layer streaming. In multilayer perceptron,

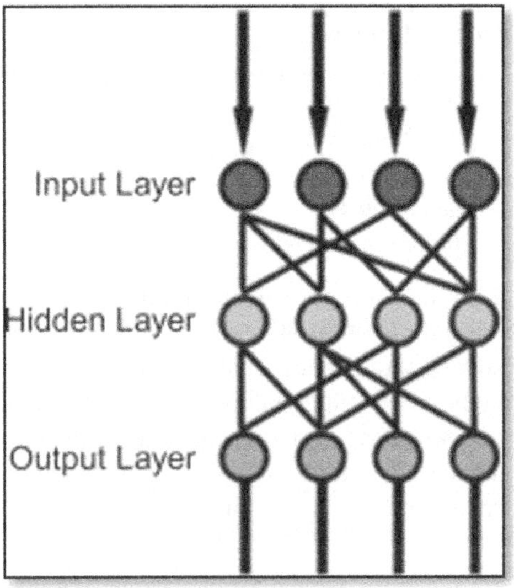

Fig. 5 Block diagram depiction of MLP with single layer

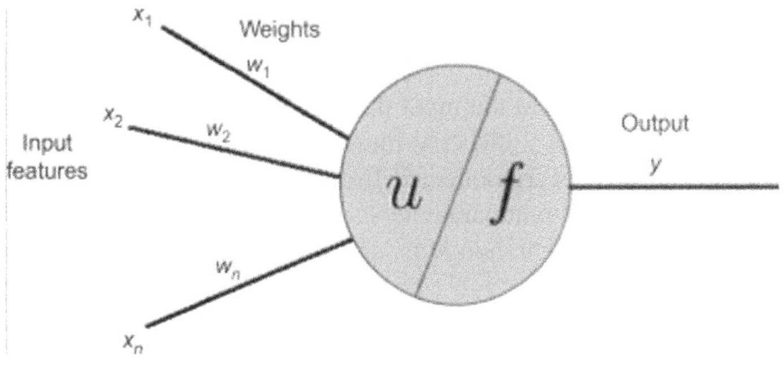

Fig. 6 Perceptron with n input

similar to the feed-forward network, the input data moves in the forward direction from input to output (Fig. 6). As per the traditional way, the neurons in the multi-layer perception are trained with the back-propagation learning algorithm for given data prediction. MLPs are designed to give suboptimum to any continuous function and can solve problems that are nonlinear.

5 Experimental Setup Discussions

The stated proposed work is implemented using Python (Keras and TensorFlow) and deployed in the private cloud for data access management and security. The specific objective is to implement the deep learning technique MLN-ANN, for the classification of the crop images (leaf images for sample five crops), for smart agriculture. Using various sensors, the flawless monitoring of the crops and plants are achieved. The extracted data, on other hand, safely imported to cloud as encrypted data for further process.

In the data collection procedure, a total of ten different crop images are included for the experimental work. The dataset thus comprised entirely the image dataset and all the preprocessing, and the prediction is done using the AWS SAGEMAKER – GPU(G2Boost) for better results. The crop leaves considered for the prediction of the disease symptoms and rectifications are as follows: grapes, apple, strawberry, cotton, tomato, ragi, and paddy.

5.1 Data Preprocessing

The preliminary stage is the preprocessing of the dataset to remove the duplicated or null valued data. As in the case of the image dataset, the unclear data are removed using Python Keras. The dimension of the image is considered as the parameter so that all the images can be equally preprocessed.

5.2 Classification of the Leaves

Figures 7, 8, 9, 10, and 11 are the input dataset (image captured from the various sensors), sequentially apple, strawberry, tomato, potato, and cotton between regular time intervals. The images are collected from the sensors imbibed to it and also acquire the data under different parameters such as geographic locations, temperature, soil nutrient management, humidity, and water level.

5.3 Algorithm

The following is the procedure to obtain the images from the sensors and the process flow to train and test the data. The classification of the images under different factors and prediction on the tested dataset for various disease symptoms is clearly stated stepwise in the given algorithm:

Step 1: Define the MLP-ANN model.

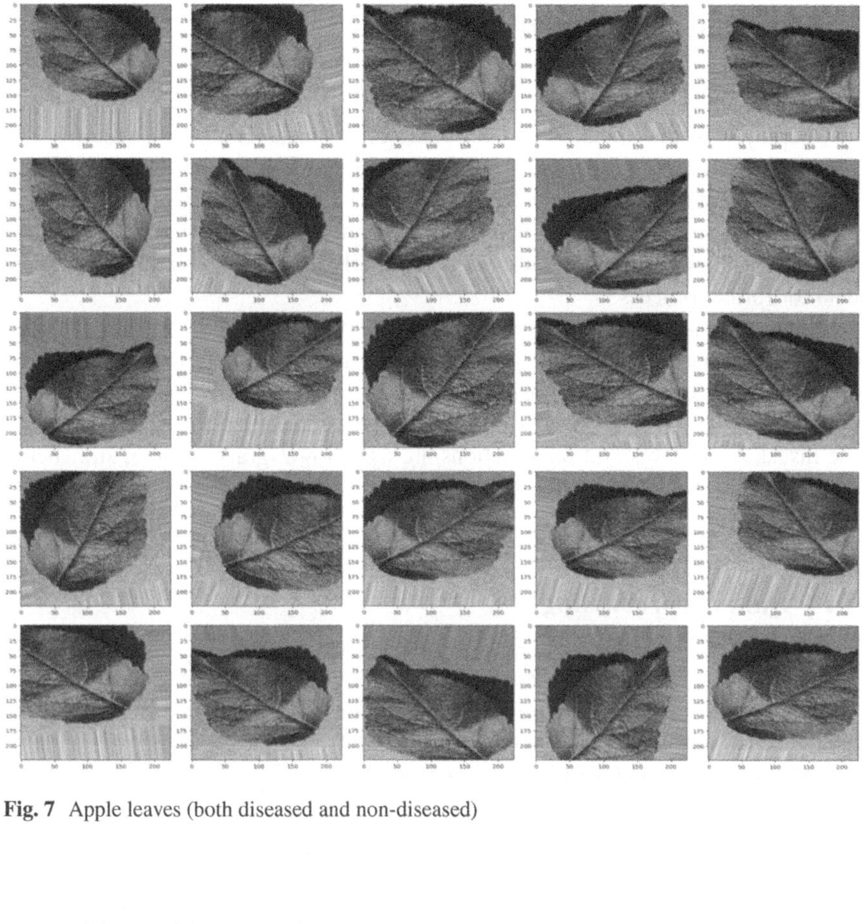

Fig. 7 Apple leaves (both diseased and non-diseased)

Fig. 8 Strawberry leaves (both diseased and non-diseased)

Fig. 9 Tomato leaves (both diseased and non-diseased)

Step 2: Define the models (sequential models).
Step 3: Extract the layers of the images acquired.
Step 4: Convert the acquired images to 2D.

Fig. 10 Potato leaves (both diseased and non-diseased)

Fig. 11 Cotton leaves (both diseased and non-diseased)

```
model = models.Sequential
([
    Layers converted to 2D layers.Conv2D(32, (3, 3), activation='
relu', input_shape=(224, 224, 3)),
    layers.MaxPooling2D((2, 2)),
```

```
    layers.Conv2D(64, (3, 3), activation='relu'),
    layers.MaxPooling2D((2, 2)),
    layers.Conv2D(128, (3, 3), activation='relu'),
    layers.MaxPooling2D((2, 2)),
    layers.Flatten(),
    layers.Dense(128, activation='relu'),
    layers.Dense(4, activation='softmax')
])
```

Step 5: Compile the model.

```
model.compile(optimizer=object',
              loss=finding entropy of data
              metrics=['accuracy'])
model.summary()
```

Step 6: Load and preprocess the training and testing data.

```
Train the datagenerator form the keras library # tf.keras.prepro-
cessing.image.ImageDataGenerator(rescale=1./255)
Test the datagenerator form the keras library
 #tesdatagen from keras.preprocessing.image.ImageDataGenerator(re
scale=1./255)
train_datagenaratorflow_from_directory using data path, size
and mode
    train_data_path,
    target_size=(224, 224),
    batch_size=32,
    class_mode='categorical'
)
```

Step 7: Train the data.

```
object = model.fit(
    train_generator,
    steps_per_epoch=len(train_generator),
    epochs=1 - n,
    validation_data=test_generator,
    validation_steps=len(test_generator),
)
```

Step 8: Evaluate the model.

Model: mode/type

Layer (type)	Output shape	Param #
conv2d_6 (Conv2D)	(None, 222, 222, 32)	896
max_pooling2d_6 (MaxPooling2D)	(None, 111, 111, 32)	0
conv2d_7 (Conv2D)	(None, 109, 109, 64)	18496
max_pooling2d_7 (MaxPooling2D)	(None, 54, 54, 64)	0
conv2d_8 (Conv2D)	(None, 52, 52, 128)	73856
max_pooling2d_8 (MaxPooling2D)	(None, 26, 26, 128)	0
flatten_2 (Flatten)	(None, 86528)	0
dense_4 (Dense)	(None, 128)	11075712
dense_5 (Dense)	(None, 4)	516
Include trainable and nontrainable parameters using the checkpoints.		

5.4 *Result Analysis*

The experimental results acquired as a result of the deep learning classification algorithm are presented below. The implementation is done with the Python language (with required libraries), under the configuration of Windows 10 operating system with Core i7 and 8 GB RAM. The image optimization is done with AWS SageMaker. The obtained results are encrypted with varied key sizes and block sizes from 64 bits, 128 bits, and 256 bits, and the results based on various parameters are analyzed. Various types of leafs are experimented and tested. The below given graphs (Figs. 12, 13, and 14) show the results obtained from both trained and tested data by using the MLP-ANN algorithm.

Reviews investigate a lot of algorithms with own pros and cons on the way of parameters being analyzed. By exploring various parameters, the efficiency and strength of the algorithm are investigated. Time complexity and computational complexity are the two main constraints for any algorithm being experimented.

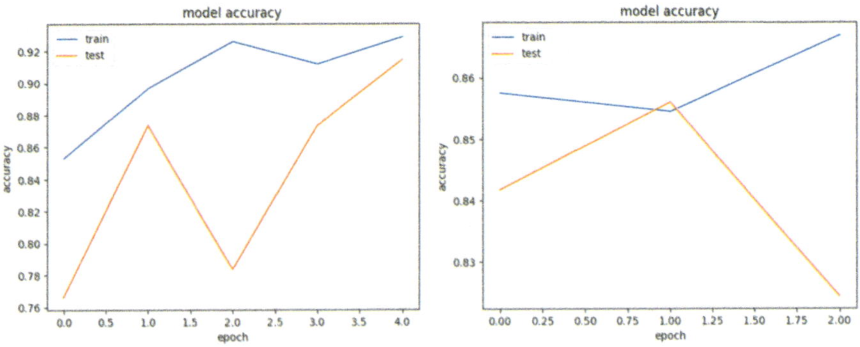

Fig. 12 Training and test results of the apple and tomato leaves

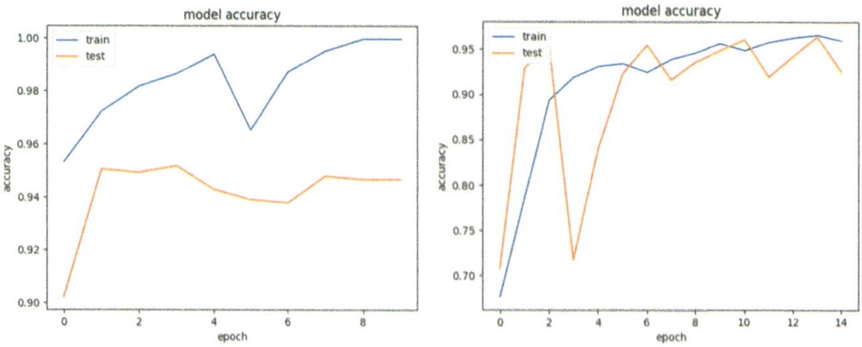

Fig. 13 Training and test results of the cotton and strawberry leaves

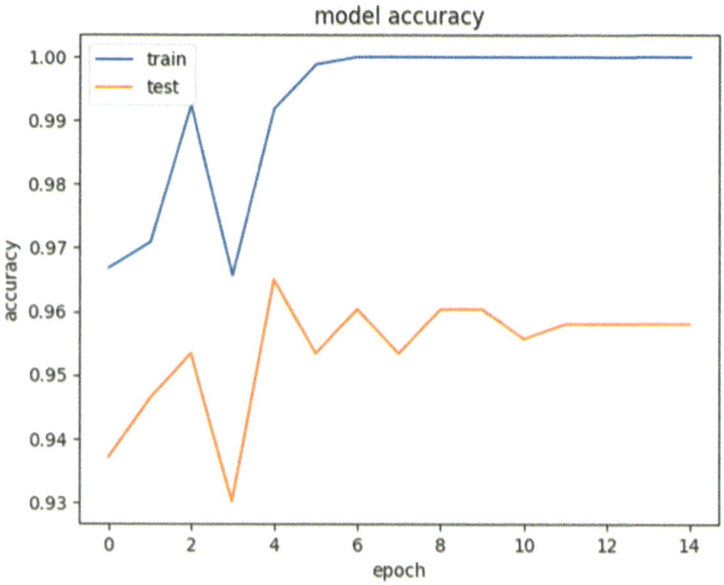

Fig. 14 Training and test results of the potato leaves

Considering this note, the experimented MLP-ANN algorithm ends with less time complexity and in computational complexity. The input in the image dataset, which is converted to 2D data after preprocessing, ends in faster computations.

After examining all the parameters from the obtained results, the multilayer perceptron algorithm is a better choice for the image dataset for less computational complexity and time complexity with higher accuracy rate. As the graph indicates, the X-axis scale denotes the accuracy rate, and the Y-axis scale indicates the total number of epochs for the system. From the listed graphs, it is clear that the accuracy rate of disease prediction in the crops and plants is close nearly to 97.1%.

Table 1 Various parameter value comparison between existing and proposed algorithm (Ref. [15])

Algorithm	Parameter (accuracy rate)
Random forest	0.928
SVM	0.965
VGG-19	0.962
Inception-v3	0.97
Multilayer perceptron ANN	0.971

5.5 *Comparative Study*

A comparative analysis is performed for the examined algorithm with the state-of-the-art algorithms in various parameters. Also, the proposed algorithm achieves the specific objective with high speed and lower computational complexity. The high voluminous data is being stored and accessed from the cloud, with data security. Optimized encryption algorithm is used for the cryptographic process (Table 1).

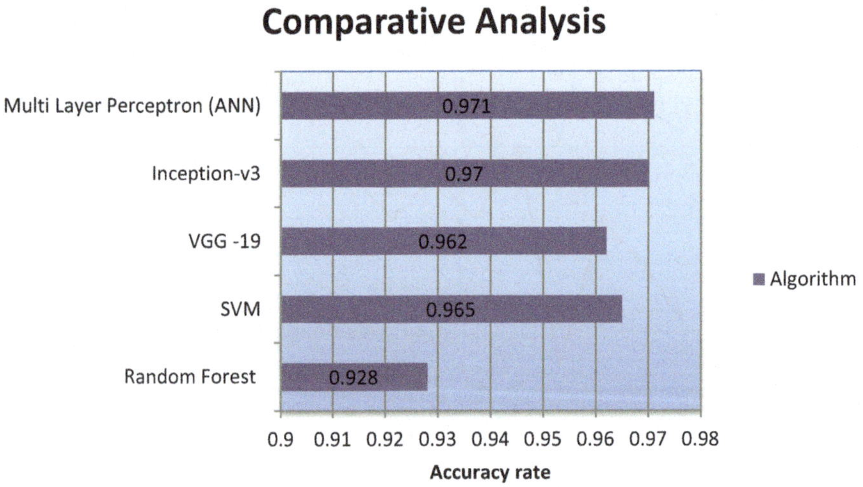

6 Conclusion

The proposed research work looks for better and more efficient crop cultivation. The integration of new techniques will improve the prediction of each and every component of plants and crops and thus helps agriculturalists in decision-making. This proposed research work considered multidimensional aspects of the data to make the results more specific. To achieve this, various smart sensors for data collection and cloud for data security and sharing are also involved. As a result from various reviews and studies, the multilayer perceptron ANN algorithm shows better results close to the deep learning inception V3 algorithm. In the future, this study extends to find different nutrient components in the crop leaves for the ease of use for medical practitioners.

References

1. M. Ayaz et al., Internet-of-Things (IoT) based smart agriculture: Towards making the fields talk. IEEE Access **7**, 129551 (2019)
2. H. Channe, S. Kothari, D. Kadam, Multidisciplinary model for smart agriculture using internet-of-things (IoT), sensors, cloud-computing, mobile-computing & big-data analysis. Int. J. Comput. Technol. Appl **6**(3), 374–382 (2015)
3. N. Gnanasankaran, Ramaraj, The effective yield of paddy crop in Sivaganga District: An initiative for smart farming. Int. J. Sci. Technol. Res. **9**(2), ISSN: 2277-8616, 6553–6556 (2020)
4. H. Pendyala et al., IoT based smart agriculture monitoring system. Int. J. Sci. Eng. Res., ISSN (Online): 2347–3878 **9**(7), 31 (2021)
5. X. Li, W. Li, D. Shi, Enterprise private cloud file encryption system based on tripartite secret key protocol, in *International Industrial Informatics and Computer Engineering Conference*, (Atlantis Press, 2015), pp. 166–169
6. J. Liu, H. Wang, M. Xian, H. Rong, K. Huang, Reliable and confidential cloud storage with efficient data forwarding functionality. IET Commun. **10**(6), 661–668 (2016)
7. J. Mahalakshmi, K. Kuppusamy, IoT sensor-based smart agricultural system, in *Emerging Technologies for Agriculture and Environment, Lecture Notes on Multidisciplinary Industrial Engineering Book Series*, (2020), pp. 39–52. ISSN: 978-981-13-7967-3, https://link.springer.com/book/10.1007/978-981-13-7968-0
8. N. Patil, V.D. Khairnar, IoT based smart farming system. Int. J. Adv. Res. Ideas Innov. Technol **7**(1), ISSN: 2454-132X (2021)
9. P.P. Ray, A survey on internet of things architectures. J. King Saud Uni.-Comput. Info. Sci. **30**(3), 291–319 (2018)
10. V. Sheetal, A. Bakshi, T. Tanvi, Green house by using IoT and cloud computing, in *IEEE International Conference on Recent Trends in Electronics, Information & Communication Technology*, (2016, May)
11. S. Sivachandran, K. Balakrishnan, K. Navin, Real time embedded based soil analyser. Int. Res. J. Eng. Technol **3**(3), 1497–499 (2014)
12. G. Srinivasan, N. Vishnu Kumar, Y. Shafeer Ahamed, S. Jagadeesan, Providing smart agricultural solution to farmers for better yielding using IoT. Int. J. Adv. Sci. Eng. Res. **2**(1), 40–43 (2017)
13. M. Thamizhselvan, R. Raghuraman, S.G. Manoj, P.V. Paul, Data security model for Cloud Computing using V-GRT methodology, in *Intelligent System and Control ISCO 8th IEEE Conference*, (2014), pp. 224–228
14. Q.H. Vu, M. Colombo, R. Asal, A. Sajjad, F.A. El-Moussa, T. Dimitrakos, Secure cloud storage: A framework for data protection as a service in the multi-cloud environment, in *2015IEEE Conference on Communications and Network Security (CNS)*, (2015), pp. 1–6. https://doi.org/10.1109/cns.2015.7346879
15. C. Jackulin, S. Murugavalli, A comprehensive review on detection of plant disease using machine learning and deep learning approaches, 100441, ISSN 2665-9174. Measure. Sensors **24** (2022). https://doi.org/10.1016/j.measen.2022.100441

Machine Learning (ML) Algorithms on IoT and Drone Data for Smart Farming

Meganathan Elumalai, Terrance Frederick Fernandez, and Mahmoud Ragab

1 Introduction

Many countries' economies rely heavily on the agriculture industry. Food costs are on the rise as producers try to keep up with demand from a growing global population. India's agriculture is well-known. Most people are farmers. Plant diseases reduce agricultural productivity, nullifying any gains. Identifying and treating plant diseases early can save a farm from destruction. Many illnesses can swiftly spread and damage a crop due to microorganisms in its surroundings. Viruses, fungi, bacteria, and other microbes cause these disorders [1]. Crop diseases directly affect production and quality [2]. At the outset, illness detection is essential to hinder massive expenses and reduce pesticide excessive utilization. In less developed nations and on smaller ranches, growers diagnose crop ailments by watching physical symptoms. To meet expanding customer needs and reduce the environmental impact of chemical inputs, researchers have devised accurate, rapid, and reliable methods for early diagnosis of crop diseases [3, 4]. Several solutions have been proposed to automate illness identification. Direct and indirect techniques identify crop diseases automatically [5–7].

Direct approaches include molecular and serological techniques, which allow precise and direct detection of disease-causing pathogens but take a long time to

M. Elumalai (✉) · T. F. Fernandez
Institute of Computer Science and Engineering, Saveetha School of Engineering, Saveetha Institute of Medical and Technical Sciences, Chennai, India
e-mail: meganathane1036.sse@saveetha.com; frederick@pec.edu

M. Ragab
Information Technology Department, Faculty of Computing and Information Technology, King Abdulaziz University, Jeddah, Saudi Arabia
e-mail: mragab@kau.edu.sa

S. Balasubaramanian et al. (eds.), *Intelligent Robots and Drones for Precision Agriculture*, Signals and Communication Technology,
https://doi.org/10.1007/978-3-031-51195-0_10

collect, handle, and analyze samples. Morphological change and transpiration rate can be utilized to diagnose illnesses and predict crop health. Indirect disease detection uses fluorescence and hyperspectral imaging [8]. Low-income farmers generally have trouble obtaining hyperspectral equipment due to its expensive cost, inconvenient size, and restricted availability, despite the fact that hyperspectral photographs provide more data than conventional shots [9]. Electronics stores sell cheap digital cameras.

Farmers must analyze every crop frequently to detect illnesses, which is complex and time-consuming. Drones will accomplish this duty, removing the need for humans and saving time. Drones are also called DRONEs. Drones/UAVs can be remotely or computer controlled. Drones are replacing satellites in agricultural applications. Early drones were used for combat and surveillance. UAVs may capture high-quality images affordably. They can take sharp photographs at low altitudes. Farming drones are gaining popularity quickly. This study employs machine learning and image processing to identification illness in plant leaves [10]. We use the drone's Raspberry Pi-connected camera to capture plant leaves. All plant leaves can detect disease early. Viruses are also undetectable by the human eye, a light-sensitive microscopy, or both. A skilled eye can see the mosaic leaf pattern, yellowing, or crinkling of virus-infected leaves. Fungal diseases cause leaf spots, leaf yellowing, and bird-eye-shaped fruit spots. Bacterial diseases can be identified by reddish-brown patches, bacterial oozing after cutting a leaf, etc. Leaf analysis requires picture processing [11, 12].

Preprocessing removes image noise after collection. Histogram equalization improves high-intensity image quality. After that, each pixel and leaf part is split for feature extraction. Image segmentation uses clustering and genetics. This review focuses on approaches and algorithms for automatic crop disease identification using image processing and spectroscopy. COVID-19 affects food production and distribution [13, 14]. Many farmers lacked access to labor, seeds, fertilizer, and pesticides, resulting in fewer crops [15]. This is due to basic agricultural, know-how, limited electricity, and inexperienced farmers. Agriculture supports 73% of India's population. India is affected. Farmers continue to apply insecticides and fertilizers after the seeds sprout. Spraying insecticides and fertilizer with the old-fashioned approach takes longer and has lower yields [7]. Due to COVID-19, conventional farmers had trouble tracking when and where they sprayed crops, fertilizer, and pesticides [16]. Drones can help with agriculture problems [17].

Farmers utilize drones to see their crops from above. Water, soil, pests, and fungus infections are all included. Infrared and visible drone crop shots show detail. Insights concerning plant health can be extracted from these photographs. This system can track monthly or hourly yields. Farmers can manage their crops more successfully with greater crop data [18, 19]. The drone's equipment can inform precision agriculture research. "Payload" is a drone's carrying capacity. This research focuses on crop health monitoring and herbicide use. We looked at how pesticide-spraying drones have progressed and how far we've gone in producing an accurate drone.

2 Background

Manually identifying agricultural diseases is labor-intensive and error-prone, making it impracticable for most farms. Sickness detection using automation is faster and more reliable [20]. We'll review the outcomes of each study [21, 22] in turn.

2.1 Classification of Diseases of Crops and Its Signs

Bacteria, fungi, viruses, and depletion can damage crops. Pathogens are separated into autotrophs, which feed on living tissue, and saprophytes, which eat disintegrating materials [23]. The disease affects crop growth and development clearly. Strange-colored leaves are an early indicator of plant illness [24]. The form and texture of leaves assist diagnose numerous illnesses. Leaf images can identify mildew, rust, and powdery mildew [25, 26]. Figure 1 and Table 1 list the three most common plant diseases.

2.1.1 Virus Diseases

The symptoms of infectious plant diseases are the most challenging to recognize and diagnose, and they are often misunderstood as indications of nutritional shortage or injury because there is no reliable signal that can be tracked over time. Insects such as whiteflies, leafhoppers, aphids, and those that crawl on cucumbers frequently carry viruses.

2.1.2 Fungal Diseases

Downy mildew, anthracnose, and powdery mildew are all fungus-caused foliar diseases. Old, lower leaves that are spotted with grayish green or are drenched in water are the first to show symptoms. These areas become darker and fungal growth occurs as the infection develops.

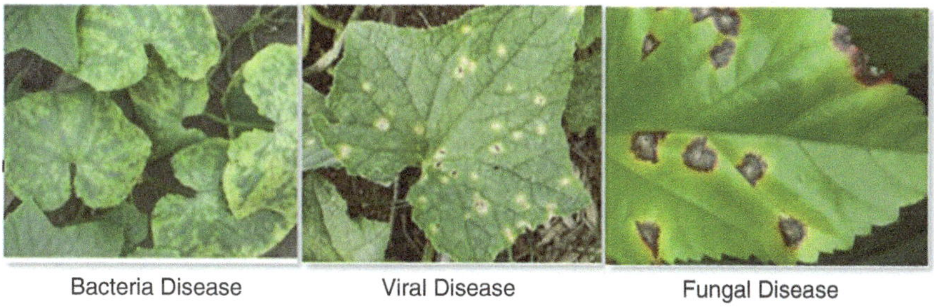

Fig. 1 Pathogens: viruses, fungi, and bacteria

Table 1 Symptomatology and categorization of a few diseases of the leaves

Plant leaf	Diseases	Symptoms	Pathogen category
Rice	Brown spot/*Bipolaris oryzae* Blast leaf/*Pyricularia oryzae* Cavara	Whitish-gray center An irregular dark brown	Fungi
Cotton	Foliar leaf/*Stemphylium solani* Areolate mildew/*Cercospora* Leaf spot/*Alternaria* spot Bacterial blight/*Xanthomonas campestris*	Spot of light yellow color with dark brown margins Tanned brown spot Circular dark brown leaf spots to black Halo yellowish green	Fungi, bacterial, virus Fungi Fungi Bacterial
Tomato	Early blight/*Alternaria tomatophila* Late blight/*Phytophthora infestans* Powdery mildew/*Leveillula taurica* Yellow curl/ infectious tomato chlorosis virus	Dark ring spot around it yellow The dark spot is growing rapidly Curly and yellowish leaf Soaked in the water ringed by a yellow halo	Fungi Fungi Fungi Virus
Wheat	Rust/*Puccinia triticina* Eriks. Powdery mildew/*Blumeria graminis* Bacterial blight/*Pseudomonas syringae*	Pale leaves spots White gray or brown spot Halo yellowish green	Fungi Fungi Bacterial

2.1.3 Bacterial Diseases

Vegetables are particularly vulnerable to the virulence of certain diseases. They gain access to the crop not via direct contact but rather via preexisting openings or wounds. Pests, insects, and even picking and pruning tools can cause damage to crops.

3 Machine Learning and Image Processing in Disease Identification

The investigation and diagnosis of leaf diseases rely heavily on image processing [27]. Figure 2 provides a glimpse into the various methods used by the authors to identify the sickness in the leaves through image processing and artificial intelligence.

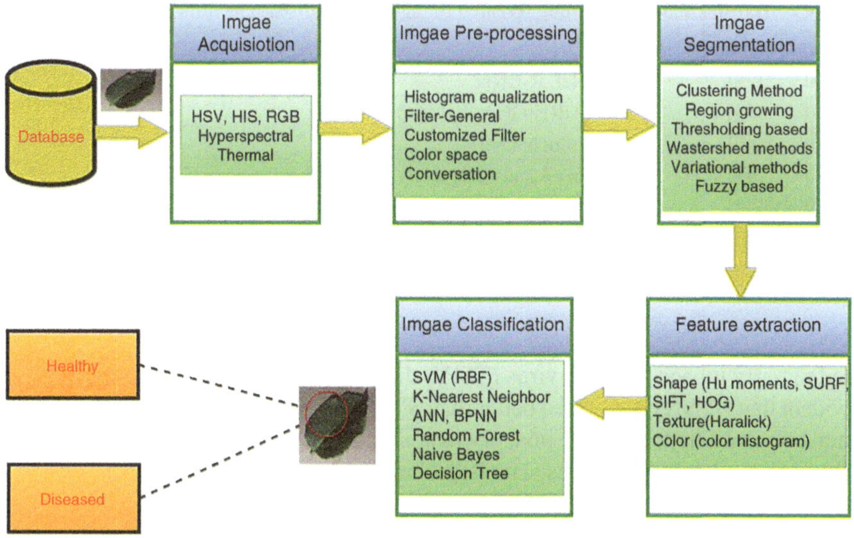

Fig. 2 Various techniques for identifying leaf diseases

3.1 Deep and Transfer Learning in Disease Identification

In the last 10 years, the agricultural sector has benefited greatly from the implementation of deep learning [28, 29] and transfer learning. Fine-tuning outperforms a freshly trained CNN model, according to Mohanty et al. Neural networks (NN) are often used to analyze hyperspectral data for early disease diagnosis [21, 30]. Photographs are the first step in sickness diagnosis [27]. Digital cameras or imaging systems can usually retrieve images. Noise in raw photographs must be removed. The second stage, picture preprocessing, removes unwanted distortions and boosts contrast to make image features more apparent and legible. Gaussian functions reduce visual noise by gently blurring images. The third process, picture segmentation [31], divides the region of interest (ROI) to emphasize significant characteristics.

Extraction of features is the fourth phase, and it reveals the hidden data and specifics in an image [32]. The characteristics of the leaves, such as their form, texture, and color, are commonly used in crop diagnosis. The selected features become a feature vector that is supplied to the classifier. You can tell one type of object from another by using this vector. Classification is the last process [33]. Keep in mind that not every classifier will work for every situation. The feature vector created in the fourth phase is then used by the classifier to identify the photos by placing them into one of the several categories. The categorization job serves this aim with its two stages, training and testing. Its ability to combine training sets improves diagnostic accuracy over rival machine learning systems, i.e., hyperspectral imaging to detect

Table 2 Transfer and deep learning research for agricultural disease identification

Year	Authors	Model	Accuracy (%)
2020	[38]	MobileNet, R-CNN	70.53
2020	[39]	DNN, SURF, GOA	98.28
2019	[40]	CNN-Multichannel	93.67
2019	[41]	InceptionV3 and CNN using a hierarchical approach	97.74
2019	[42]	Nine-layer deep CNN	96.46
2018	[43]	CNN, Faster R-CNN	91.67
2018	[28]	OverFeat, VGG16, AlexNet	99.53

and classify tobacco mosaic virus (TMV) disease pre-symptomatically [34]. It was evident that BPNNs performed better than SVMs, RFs, LDAs, ELMs, and LS-SVMs and examined hyperspectral imagery as a noninvasive method for detecting TMV illness in the beginning stages. The BPNN model was 95% accurate, while the chemometric models were 80% [35].

Table 2 shows that the number of researchers engaged in deep learning has risen considerably, notably in 2018. CNNs, AEs, RNNs, and limited Boltzmann machines are common crop image classification models. So many articles have been written about utilizing deep learning to diagnose agricultural diseases. Ma et al. developed a DCNN model that could recognize more than four cucumber diseases [36]. In contrast with more standard techniques like the SVM, naive Bayes, and AlexNet, the DCNN identified cucumber illnesses with 93.4% accuracy and deployed a dense YOLOV3 model to avoid overfitting. Despite wavering lights, tangled fruit, and intricate backgrounds, they applied their method to apple orchards [37].

3.2 Hyperspectral Imaging (HIS) Used to Identify Disease

Hyperspectral photography has made gains in detecting abiotic and biotic plant stresses [44, 45]. For collecting harmonic and temporal information, hyperspectral imagery incorporates visualization and spectroscopy. Hyperspectral reflectance is utilized to distinguish healthy and diseased TSWV tobacco leaves. Zhu and others employed the best EW wavelength for hyperspectral imaging with SPA to detect early TMV infection. Hyperspectral data is multicollinear due to associated spectral values. EWs simplify the hyperspectral analysis, maximize data use, and speed processing. The subsequent projection technique (SPA) [46], partial least squares regression models [46], and genetic algorithms (GAs) [47] have all been used to address multicollinearity.

3.2.1 Internet of Things' Use in Leaf Disease Detection

Table 3 shows in the agricultural sector the IoT has led to significant improvements. Farmers can use IoT applications to monitor their fields from anywhere and be informed about the state of their crops and the weather at any time. Farmers can better prepare for the upcoming harvest by using IoT technologies [48]. As an added bonus, they can safeguard their harvest by identifying crop illnesses in their earliest stages and thereby preventing further spread. There's no denying the importance of agricultural IoT apps in boosting agricultural output and lowering crop losses due to illnesses.

Finally, the crop disease identification and classification models were assessed using a variety of measures, such as sensitivity, precision (*P*), recall (*R*), quality measure (QM), and F1 score, which are model-specific. Here, we present a set of statistical evaluation measures that can be utilized to conduct a quantitative analysis of the efficacy of models for detecting agricultural diseases that make use of deep and transfer learning:

$$\text{Precision} = \frac{\text{TP}}{(\text{TP} + \text{FP})}$$

Precision (*P*) is the fraction of true positives (TP) to the sum of TP and false positives (FP). *P* is averaged across classes for multiclass categorization:

$$\text{Sensitivity} = \frac{\text{TP}}{(\text{TP} + \text{FN})}$$

A test's sensitivity/recall (*R*) measures both false positives and false negatives (FN). *R* calculates the mean of many classes:

Table 3 Analysis of the existing research on IoT infrastructure

Researchers	Detection techniques and algorithms	Parameter evaluation
[49]	The RideNN Cycling Neural Network, based on the SCA, is used in an IoT-based model for monitoring and detecting plant diseases	The SCA-based RideNN model was 91.56% accurate
[50]	IoT models utilizing GLCM, RFC, and k-means clustering	RFC-GLCM-based disease detection and classification were 99.99% accurate
[51]	Utilizing support vector machines and K-means clustering in an Internet of Things	The farmer receives instantaneous text message warning
[48]	RiceTalk is an IoT platform that employs an AI model	Net prediction accuracy was 89.4%
[52]	The Internet of Things system uses deep learning and transfer learning to remotely monitor rice crops	Constant tracking of environmental conditions

$$\text{Specificity} = \frac{\text{TN}}{(\text{TN} + \text{FP})}$$

Specificity is measured by the ratio of negative to healthy samples. This metric evaluates a model's false-negative predictions:

$$\text{Accuracy} = \frac{\text{TP} + \text{TN}}{(\text{TP} + \text{TN} + \text{FP} + \text{FN})}$$

The percentage of accurately labeled samples relative to the total number of samples labeled is the accuracy. This metric is used to evaluate a model's performance as a whole:

$$\text{F1\_score} = 2 \times (\text{Sensitivity} \times \text{Precision})(\text{Sensitivity} + \text{Precision})$$

The F1 score combines precision and recall. Class F1 is a multiclass classification task's weighted average, where

TP: True positives refer to the exact count of picture specimens that were correctly detected as contaminated.
FP: False positives are the number of samples of images that were wrongly labeled as contaminated.
TN: The percentage of true negative images that were accurately labeled as healthy is denoted by the symbol TN.
FN: The percentage of erroneously classified clean images is denoted by the metric FN.

4 UAV (Unmanned Aerial Vehicle)

Today's drones are GPS-based autopilot vehicles, not radio-controlled devices. Size, landing, aerodynamics, number of rotors, altitude, range, and endurance are some ways to define drones. However, there are two primary categories of drones in this article, both of which are dependent on the presence or absence of wings or rotors. Some of the application of drone is shown in Table 4 [30, 53–55]. Drones can have either fixed or rotational wings. Helicopter, quadcopter, hexacopter, etc. are all names given to drones with different numbers of rotors. In this paper, we'll go into the specifics of how a hexacopter is built and how it can be used to spot leaf diseases. Octocopters, a type of helicopter, can carry the most spray since it has a larger payload than other helicopters.

Table 4 Modern drone application technology

Application	Description
Agriculture	This research discussed ML using a drone to identify plant diseases. The author used a quadcopter, AutoML, and Inception v3 to take these photographs
Agriculture	This page discusses leaf diseases like leaf minor, downy mildew, and yellow spot. GSM-enabled Raspberry Pi, K-means clustering, and a k-n classifier helped him diagnose
Agriculture	The paper describes a pesticide-spraying drone that can be operated from afar using a smartphone and a Wi-Fi module
Horticulture and agriculture	This idea proposes using a drone to spray pesticides only when needed to reduce pesticide consumption and enhance harvests. Drones and object recognition are used
Agriculture	The author used inception and convolutional neural networks to identify illnesses in apple leaves

4.1 Fixed-Wing UAV

These UAVs achieve the necessary lift with the use of aerofoil-shaped, fixed wings. In Fig. 3, we see a typical example of a fixed-wing UAV (Fig. 3a).

4.2 Helicopters

It generates lift and propulsion using a single set of horizontally revolving wings mounted to a middle pole, such as UAVs like the one depicted in Fig. 3b. A helicopter can take off and land vertically, move forward and backward, and hover in one location. For this reason, helicopters can be used in places that are too crowded or too far away for conventional aeroplanes to reach.

4.3 Multi-copters

UAVs can be lifted and maneuvered with the help of rotorcraft with numerous sets of horizontally revolving blades (usually four to eight), as seen in Fig. 3c. The processing of data derived from satellite images of tiny plants is challenging. Additionally, satellite photos are dependent on favorable lighting and weather conditions. UAVs (unmanned aerial vehicles) are a more efficient means of gathering this kind of visual information since they can take pictures at precisely the position, altitude, and frequency that the user specifies. Technology based on different drones is explained with its task in Table 5, and it also has the advantage of being an instantaneous data analyzer and a completely automated tool for weed and insect control.

Fig. 3 Types of UAVs: (**a**) Fixed-wing drones, (**b**) helicopters, (**c**) multi-copter drone

4.4 Discussion

In Table 6, researchers in this field use image processing, machine learning, and deep learning to identify and classify plant ailments. According to the results, the support vector machine [50, 56], the random forest [57, 58], the artificial neural network [39], and the convolutional neural network (CNN) [42, 43] are the methods most frequently used in research. Most studies have used data from PlantVillage, and histogram equalization has been employed to boost contrast, while median, Gaussian filter, and Gabor filters have been put to use for denoising and picture enhancement in the preprocessing stage [40]. In order to find the region of interest in an image, the k-means and fuzzy c-means algorithms perform color-based segmentation. Histograms of directed gradients, local binary patterns, and gray-level co-occurrence matrices can capture plant texture, form, and color (HOG), beginning with broad classifications.

Related Crop Disease Identification and Pesticide-Spraying Work

The group used many different types of neural networks, including artificial-neural networks, probabilistic neural networks, back-propagation neural networks, convolutional neural networks, and Inception v3.Disease categorization is a primary application of the SVM and NN. The fundamental benefit of NNs is that they are created from existing data and are also tolerant of noise. By mapping the input feature vector nonlinearly onto a high-dimensional space, the SVM provides excellent classification performance. Table 7 outlined the benefits and drawbacks of certain classifiers. However, SVMs should be avoided in cases of extremely noisy data and used the principal component analysis (PCA) dimensionality reduction method when the input vector contained several redundant variables [67]. In addition, the classification procedure has been entirely automated using the convolutional neural network (CNN), faster R-CNN, Vgg16, and ResNet50 models as well as using a novel approach called the extreme learning machine (ELM), which provides improved speed, performance, and generalization at a reduced computing cost [68].

In this context, we compare the effectiveness of ten deep learning models on the PlantVillage dataset: AlexNet, ResNet-101, GoogleNet, DenseNet201, Vgg16 Inceptionv3, Squeeze Net InceptionResNetv2, ShuffleNet, and MobileNets [69]. propose a 2020 autonomous UAV yellow rust disease monitoring system.

Table 5 Intelligent, efficient agricultural technology

References	Description	Crop	Disease	Leaf disease detection algorithm	Feature classifier	Drone technology		
						Type	Task	Camera
[59]	Machine learning and UAVs were discussed for plant disease diagnosis	Tomato	Bacterial spot	AutoML, Inception_V3	Not used any	Quadcopter	Image capturing	8 MP Sony camera
[60]	This paper covered rice disease detection and IoT/ML location mapping	Rice	Crown sheath rot	Convolutional neural networks (CNN)	SVM, GLCM classifiers, and HTTP protocol	Quadcopter	Position mapping and disease detection	Raspberry Pi camera
[61]	This research discusses image processing and convolutional neural networks for weed recognition in plants	Plant diseases	Cannabis, *Tridax*, goose grass	Convolutional neural networks and image processing	Not used any	DJI Mavic air 2	Weed identification	Hyperspectral X-ray imaging camera

Table 6 Identification of crop disease-related research

Title	Methodology	Advantages	Disadvantages
Disease detection in tomato plants through transfer learning and convolutional neural network-generated images [4]	We offer a DL-based solution to tomato sickness detection using C-GAN to generate artificial tomato leaf pictures for data improvement. Pretrained DenseNet121 models are fine-tuned with synthetic and real data to identify tomato leaf diseases	The proposed method was 99.51%, 98.65%, and 97.11% accurate. The C-GAN mitigates the effects of overfitting and improves the generalizability of the network	While leaves were utilized to diagnose tomato disease, stems and branches certainly played a role. The research focused on infected tomato plants
Color histogram and supervised classifier for oil palm leaf disease detection [22]	This study provides a new method for identifying healthy oil palm leaves. The 8-bin color channel histogram is divided into RGB, LAB, HSI, and HSV characteristics. K-means clustering segments leaves. ANN uses 41 PCA-selected features	100% specificity, 99.3% sensitivity, and 99.67% accuracy prove the suggested method's effectiveness. This method yields simpler, more robust features	The results of the classification indicate that an error occurred because a leaf was incorrectly identified as healthy Only oil palm leaf disease was studied
Diseases on rice leaves: using image processing to identify them	The suggested technique detects three leaf-damaging illnesses using IP and ML. This protects young plants for farmers. Before using Otsu thresholding for segmentation, photos are preprocessed to improve image quality and reduce distortions. LBP and HOG extract partitioned space features. SVM classifies these features with 94.6% accuracy	SVM + HOG using a polynomial kernel function may identify plant illness. The proposed work enhances earlier attempts' precision (94.6% vs. 92.5%)	The LBP fundamental operator can't capture certain trends
Using image processing for a farm-wide automatic plant disease detection and warning system [5]	We took a six-step strategy for IP. Raspberry Pi-connected webcam captured moving leaves. After preprocessing, segmenting, and k-means clustering, perimeter and illuminance were extracted. This analysis classified leaf diseases. A sickness alarm and buzzer alert the farmer	Disease affects Alternaria Alternate's graphical interface achieved 95.16313% precision. An illness alert system	The authors didn't mention any other diseases, and the results aren't clear

(continued)

Table 6 (continued)

Title	Methodology	Advantages	Disadvantages
IoT plant disease detection [62]	An IoT-based control system was built to monitor pests and illnesses on a three-tiered tree farm. Initial plant health was computed. A computerized system could also determine illness closeness. A disease detection mechanical framework incorporating humidity, temperature, and shadow sensors has been created. Arduino code analyzes sensor data to track plant growth. Data is transferred to the cloud over WI-FI for analysis after collection. These data are compared to evaluate the plant's overall health	Low-cost IoT model presented. Low-income farmers can buy it to stop the disease spread	Temperature, humidity, and leaf cover constituted the entire model. Inaccurate parameter assessments are another problem. From these unpredictable starting points, many characteristics developed. Uncategorized plant leaves make it impossible to identify infections
Digital image processing for leaf disease detection in plants [27]	This article discusses image analysis for early leaf disease detection. Automated disease detection minimizes farm monitoring. Various IP and ML approaches are utilized to identify diseases; a genetic optimization strategy was employed after k-means picture segmentation to optimize outcomes, and SVM was used for disease classification	The algorithm was evaluated on photos of infected leaves from five different plant species (corn, tomato, bell pepper, peach, and grape) with a 75% success rate	Commonplace approaches were used for the paper's research
Clustering algorithm for plant pest detection in digital images [63]	This approach uses K-means clustering to identify plant insects. Disease recognition requires acquiring, preprocessing, segmenting, and categorizing images. The median filter and boundary detection techniques were used to minimize the noise in RGB-to-HSV leaf images. K-means grouped photos	This study identifies damaged photos quickly and accurately. K-means clustering's accuracy and speed are evident benefits over other methods	This paper mostly focused on very standard and traditional methods

(continued)

Table 6 (continued)

Title	Methodology	Advantages	Disadvantages
Using image processing and multiclass support vector machine to detect and classify plant diseases [64]	Modeling plant disease with ML and IP. First, 148 leaf disease images are analyzed. There are two plant groups. The training set has 73 pictures; test set has 75. Segmentation identifies pathogenic leaf regions. 9 of 13 texture features were computed using the designated RGB subset. Grayscale images assess consistency, differentiation, vitality, and coherence. SVM separates healthy and diseased leaves	According to the findings, the suggested method has a diagnostic accuracy of 92.8571% for plant diseases	The system does not automatically select the infected segment from the three options displayed
Using evolutionary optimization of deep neural network features for plant leaf disease detection [39]	Black rot, Apple scab, and Cedar rust were just some of the apple leaf illnesses that a DNN was able to detect and categorize using GOA and Robust Accelerated Feature SURF. Disease categorization was accomplished by means of DNN after SURF feature extraction and GOA optimization	Model accuracy is increased by 18.03%, to a mean of 98.28%, when using the DNN-SURF method. A basic model's applicability is greater than that of a metric one	In this study, we focused solely on foliar diseases that can affect apples
Image processing for leaf disease detection [65]	IP and SVM were used to diagnose plant diseases. We grayscale an RGB image, improve it with AHE, extract 13 textural features with GLCM, and identify plant illnesses with SVM. More than 500 images with varying brightness levels (0–2500) were shot to train and test the system	The system detects leaf disease faster and cheaper than traditional methods	The algorithm's structure is complex, and its accuracy is not specified

(continued)

Table 6 (continued)

Title	Methodology	Advantages	Disadvantages
Automatic disease detection in plants utilize a CNN and a small amount of training data [3]	In this study, we developed strategies for disease diagnosis in plants with minimal resources. The networks used were DAML and CNN triplet topologies. Detection methods for novel diseases were trained on a large dataset consisting of anywhere from 5 to 20 images per disease	The model's accuracy drops to 81% from 99% if the origin and destination domains differ greatly from one another	In comparison to other methods, DAML error rates were 22.2 per 50 shots and 42.6 per 5 shots The fundamental model is beneath different strategies
Machine learning for identifying leaf diseases [66]	IP and ML determined whether employees were sick. Kaggle has 12,949 photos. The approach involves image segmentation, feature extraction (shape, color, texture), and SVM classification	The SVM's effectiveness in high-dimensional spaces gives it a leg up on competing classifiers with an 80% success rate was achieved by the SVM while	The presented approach, which is based on CNN, has a high degree of accuracy. However, it takes a lot of effort and time to train the model
Agricultural applications of IoT-enabled plant disease detection and classification [50]	The scientists described a way to use IoT to locate diseased banana trees. 80 hill banana plants are photographed at 256 × 256 pixels. Image preprocessing grayscales it. Histogram equalization is used when employing k-means clustering on a scaled gray image. The cloud analyses extracted GLCM features. RFC extracts traits to classify hill banana illnesses. Agriculture experts analyze data. Remote monitoring of soil humidity and temperature can minimize climate-related and infectious diseases	RFC-GLCM leaf disease classification obtained 99.99% accuracy on the hill banana dataset. Experts help farmers with plant disease and environmental changes	Lighting and camera angle in agriculture affect the system's precision

Table 7 Comparing different classifiers

Classifier	Advantages	Drawbacks
Artificial neural network (ANN)	More efficient and precise than KNN and MMC	Strict because there is only one possible category for the information
Random forest	Ability to accurately identify massive datasets	Time and space limitations in processing
Multiclass support vector machine	Aids in dividing up data into manageable chunks	Non-applicable in cases of noisy data
Least square SVM	Quick and easy to understand	A sparse appearance requires the use of pruning techniques
K-nearest neighbors (KNN)	There was zero investment in training	Each individual instance takes more time and money to test, and the system is more susceptible to noise and produces lower yields
Extreme learning machine (ELM)	Less time spent training and more consistent results	When a complicated model is overfitting, it loses
Naïve Bayes	Less information is needed for training. When an independent variable holds true, it outperforms competing methods	Confidence may suffer if conditions are independent of one another
Penalized discriminant analysis (PDA)	It helps when there are a lot of noisy features in the problem	Expensive computation
CNN/deep learning	Since feature extraction is no longer necessary, classification times are reduced	Training requires a lot of information, which is both time-consuming and costly to compute. They need more powerful hardware, like a GPU
Transfer learning	This facilitates the use of CNN on problems for which only limited data is available for training	Pretrained models don't always contain labeled classes of interest

Multispectral cameras collected data [70]. It recorded visible, NIR, UV, and green data. The system recommended using U-Net for semantic segmentation. More band usage improved picture segmentation. We used random forest for this image categorization challenge. Deep learning uses convolutional neural networks (CNNs). ML is used for data visualization and prediction. Figure 4 depicts the method. Raw data is gathered and processed in data acquisition and processing nodes. Clean and organize the raw data in this step. Deep learning models are well-written software.

Pesticides sprayed manually expose humans to cancer, hypersensitivity, asthma, and other illnesses [71]. Conventional methods carries various downsides, including higher chemical use, a lack of agricultural labor, inconsistent spraying leading to environmental damage, and insufficient coverage. The drone can spray fields using pre-mapped paths and a pesticide tank that holds up to 40 l. Drones have significant potential in areas where tractors and aeroplanes have trouble reaching the field. Figure 5 depicts some examples of the sprayer that can be attached to a drone. Kislaya Anand and Goutam built the AeroDrone to examine crops and apply pesticides in 2019 and reduce wasted insecticide and spraying time. The mission could

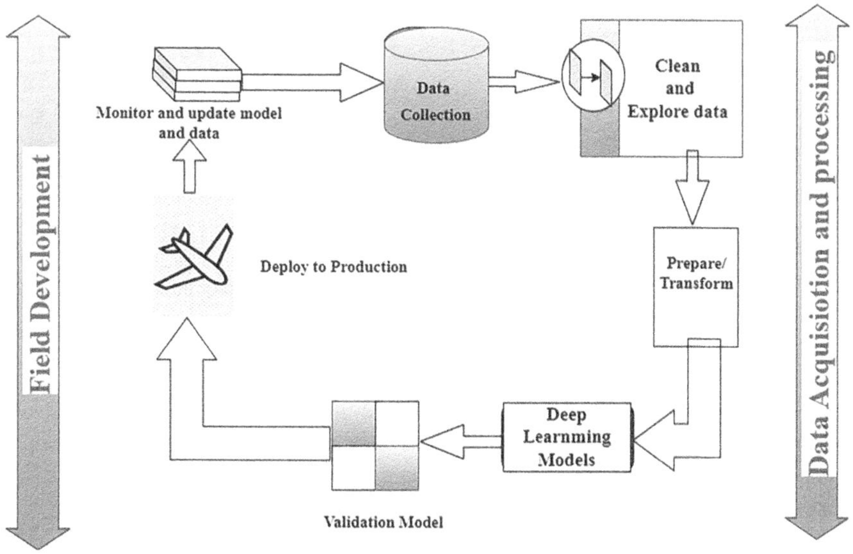

Fig. 4 A precision farming system based on deep learning

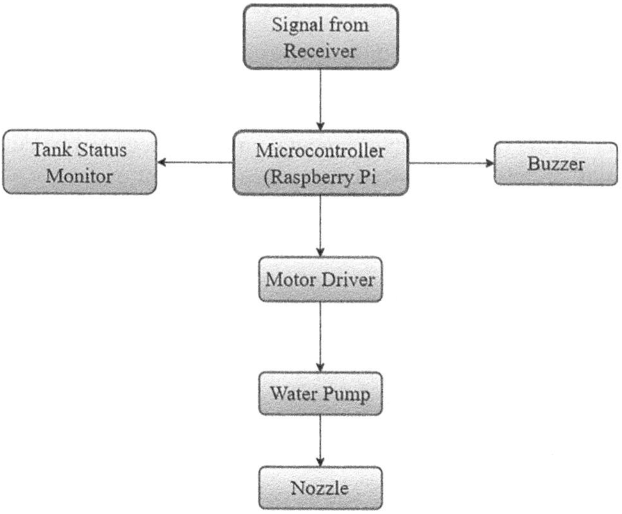

Fig. 5 Blocks with completely automated pesticide spraying

be assigned in the field utilizing a simulation platform to verify its sensitivity and accuracy. Figure 6 shows integrated system of quadcopters performed well and their flight times were identical. The concept worked well for a rectangular farm, but

Fig. 6 Spraying method constructed around UAVs was used in (a) paddy field, (b) tea crop, and (c) banana trees

other shapes may be better. Shaw et al. is designing lighter octocopters [72]. We used the tank's storage volume (6 l), fluid density, nozzles (fine spray), and pump to compute its payload. We used eight brushless DC motors (BLDCs), an ESC, a propeller, a 12-volt pump, an FPV camera, a video transmitter, and an LI-PO battery to hoist the cargo. This octocopter prototype with an AI pilot excelled in crop monitoring.

5 Challenges in the Crop Disease Detection Field

Several open questions and obstacles in the current literature must be addressed and conquered before reliable and practical crop disease detection systems that perform reliably under a wide range of environmental circumstances can be developed. The most significant of these emphasized difficulties are as follows:

5.1 Insufficient Data

The neural network (DL) models' use in plant identification of illnesses is hampered by the lack of variation and the quantity of inaccessible datasets [73]. Most plant disease identifications are done in sterile, controlled conditions, where only one disease exists. Because environmental factors are ignored, accuracy is higher than in practice [74]. Labeling images is a time-consuming process. Creating a reliable, complete dataset is difficult. Data augmentation, data sharing, citizen science, transfer learning, synthetic data, and few-shot learning can fill in the missing information.

5.2 Imbalanced Data

Popular agricultural disease identification datasets are scrubbed up, or its unevenness is ignored to avoid distracting training methods. In practice, discrimination or unbalance [75] between classes exists. Since most machine learning classification methods assume an equal number of samples for each class, predictive modeling may require resampling strategies. Some models' prediction performance declines, especially for minorities, who are more prone to be misidentified.

5.3 Vanishing Gradient Problem

Hochreiter identified the "vanishing gradient problem" with neural networks. The neural network adjusts its weights such that their ratios to error function partial derivatives remain constant [76]. If the gradient is near 0, the weights may not be changed. The neural network stops because gradient descent fails to converge.

5.4 Overfitting and Underfitting Problem

Because of the many interrelated components, overfitting and underfitting are key concerns while training a learning model. The model's test data efficiency diminishes. Underfitting means a learning system misses a data pattern. The results show the model is biased and lacks enough variance to describe the data. Insufficient data hampers reliable model formation, as does building a linear model from nonlinear data. Overfitting happens when a model is trained with too much data, leading to inferences based on noise and erroneous data. Too much information and noise in the input make the model inaccurate. It indicates a lack of bias and increased variance. Nonlinear and nonparametric learning algorithms have greater room to create an inaccurate model, making overfitting more likely. It would be great if models didn't have either, but it's hard. This is a common difficulty with the initial few iterations of training any model depending on efficient convolutional neural networks, which is their method.

5.5 Snapping Images

Requirements for capturing pictures (wind energy, illumination, and physical location and camera) ought to be lit in a similar manner. This may only be possible in a lab because it's tough to supervise capture circumstances. Identifying diseases from images is difficult since they may have surprising traits. Changing capture conditions hinder measuring citrus leaf canker [77] and identifying citrus diseases [78]. Invariant illumination strategies have been developed. Their achievements are limited [79].

5.6 Lighting Issue

Crops grow in naturally variable environments. Weather, lighting, and other elements affect photos. Illumination difficulties are inescapable and difficult to eliminate. We employ narrow-band imaging and polarising filters to identify citrus Huanglongbing disease in real-time. Specular lighting is the most difficult to solve [76]. Change the camera perspective or leaf location to reduce specular illumination in photos. Zhou noted that automatic captures make it more difficult to prevent lighting problems, which lead to specular reflections and shadows [80].

5.7 Camera

One of the most important aspects that directly affect the quality of an image is its resolution. Small lesions and spores can be spotted with improved resolution. They are also affected by the camera or other capture device's settings.

5.8 Image Preprocessing

More data is lost when the compression ratio increases during the preprocessing and storage of leaf photos. While this might not have much of an impact on the diagnosis of major lesions, it could drastically skew the interpretation of minor symptoms. Therefore, if the symptoms are mild, no or minimal compression should be applied.

5.9 Image Segmentation and Symptom Discrimination

In most cases, healthy and sick tissue are difficult to distinguish since healthy tissue symptoms fade gradually. This affects the threshold and extracted characteristics. Manual and visual representations fail to define edges, and machine-based representations are subjective. [81] and [82] addressed the subjective delineation of afflicted regions and the need for an external reference for validating illness identification systems and discovered that false negatives or positives are excessively high when using no reference for leaf powdery mildew [83]. Inconsistencies are fundamental to the process; hence, few solutions have been presented. Segmenting and locating ROIs present additional issues, including the following:

- A leaf could be tipped, covered in dew or dust, or overlapping another leaf or portion of the plant.
- Segmenting symptomatic regions of interest (ROIs) from images with complex backgrounds can be difficult and time-consuming.

5.10 Feature Selection and Extraction

Despite having similar appearances, several plant species have varied leaf forms. Symptoms may be masked by thick foliage, fruit, or flower petals or deep within the plant. The latter has less scholarly focus. Leaf disease detection from the above has been studied extensively. There is an advocated R-CNN for identifying tomato illnesses [29].

5.11 Disease Classification

When trying to identify plant diseases, a classifier may be rendered useless if multiple diseases share visually similar symptoms. These issues also have an impact on how severe the ailment is.

5.12 Differences in Disease Symptoms

Disease indicators vary in size, color, and shape, making precise diagnosis challenging. Many diseases might develop simultaneously, making it difficult to distinguish symptom combinations from individual symptoms [84]. We observed this issue when treating black drip sickness on banana leafs and powdery mildew on squash leaves [85].

5.13 Similarities in Manifestations Among Various Chaos Varieties

Diseases, phytotoxicity, the presence of parasites, and nutritional inadequacies are just a few examples of conditions that might cause symptoms that are visually similar to one another. Therefore, if only the visible spectrum is employed for identification, it might be extremely challenging to pinpoint the source of a symptom with certitude. This necessitates techniques that rely on subtle distinctions between symptoms. Several academics have pointed out that the clinical similarities between certain illnesses cause severe discrimination problems. According to Ahmad, a classifier they developed was unable to differentiate between symptoms caused by *Fusarium*, mosaic *Potyvirus*, *Alternaria*, and *Phomopsis* in soybean. Thus far, researchers have avoided this problem by focusing on diseases with quite different symptoms, but, even so, making the right call has been difficult [86].

6 Conclusions

Agriculture struggles with crop diseases. Crop diseases can be slowed if caught early. This book contains cutting-edge research on sickness identification systems. This study varies from others because it analyzes specific studies and approaches. It gives researchers a roadmap and resources. This paper discusses the automatic crop disease diagnosis technique and its primary elements, including fuzzy symptom boundaries, shifting imaging conditions, variable disease symptoms, similar disease symptoms, and simultaneous disease symptoms. All previous image processing and analysis technologies had these limitations. According to a study, image preparation influences segmentation accuracy. K-means clustering identified diseased leaf regions best. CNN models recognize visual patterns. Computer vision and AI are new to crop diagnostics; therefore, many of their alternatives and opportunities, which may aid, have not been investigated. Increasing processing power simplifies once-complex strategies. Based on this in-depth literature study, the researchers hope to build a method to diagnose crop illnesses from foliar images swiftly and affordably. This disease-detecting device will inform farmers through a smartphone app and identify plant diseases quickly. Farms utilize a hexacopter to photograph damaged foliage. Photos are further processed. Using image processing and deep learning, this project hopes to detect and diagnose plant diseases sooner. This method reduces operational costs because humans don't need to inspect plants for illnesses or diseases round-the-clock. This technology saves time and increases agricultural yield. Convolutional Neural Network has improved disease diagnosis to 71.042% and improve drone detection.

This research examines drone-based precision agriculture. This paper explores drone crop monitoring and pesticide spraying in precision agriculture. Design updates, data-gathering sensors, pesticide-spraying drones, deep learning, and AI are discussed. After 2017, precision farming drones will rise. UAVs are cheaper, lighter, and payload-capable. Drones patrol fields and herds. Drone size and price keep falling. Big payload unmanned planes spread insecticides and fertilizer. There are more pesticide-spraying planes, stabler multi-copter spraying, and smaller, lighter, higher-resolution drone cameras are available.

RGB cameras capture fewer details than multispectral cameras. Drones are controlled by Arduino Uno or Raspberry Pi AI. Autonomous drones use embedded electronics, data transmission, and processing. Farming drones employ AI. Agricultural drone technology has many problems. High costs, short battery life, vision damage, and little technological literacy are problems.

References

1. G.B. Lucas, C.L. Campbell, L.T. Lucas, Causes of plant diseases, in *Introduction to Plant Diseases*, (Springer, Berlin/Heidelberg, 1992), pp. 9–14. https://doi.org/10.1007/978-3-642-85063-9
2. J. Shirahatti, R. Patil, P. Akulwar, A survey paper on plant disease identification using machine learning approach, in *Proceedings of the 2018 3rd International Conference on Communication and Electronics Systems (ICCES)*, Coimbatore, India, 15–16 Oct 2018, pp. 1171–1174. https://doi.org/10.1109/CESYS.2018.8723881
3. A. Afifi, A. Alhumam, A. Abdelwahab, Convolutional neural network for automatic identification of plant diseases with limited data. Plants **10**, 28 (2021). [CrossRef] [PubMed]. https://doi.org/10.3390/plants10010028
4. V. Singh, A.K. Misra, Detection of plant leaf diseases using image segmentation and soft computing techniques. Inf. Process. Agric. **4**, 41–49 (2017). [CrossRef]. https://doi.org/10.1016/j.inpa.2016.10.005
5. P.K. Mugithe, R.V. Mudunuri, B. Rajasekar, S. Karthikeyan, Image processing technique for automatic detection of plant diseases and alerting system in agricultural farms, in *Proceedings of the 2020 International Conference on Communication and Signal Processing (ICCSP)*, Chennai, India, 28–30 July 2020, pp. 1603–1607. https://doi.org/10.1109/ICCSP48568.2020.9182065
6. H. Parikshith, S.N. Rajath, S.P. Kumar, Leaf disease detection using image processing and artificial intelligence—A survey, in *Proceedings of the International Conference on Computational Vision and Bio Inspired Computing*, Coimbatore, India, 25–26 Sept 2019, pp. 304–311. https://doi.org/10.1007/978-3-030-37218-7_35
7. P.K. Sethy, N.K. Barpanda, A.K. Rath, S.K. Behera, Deep feature-based rice leaf disease identification using support vector machine. Comput. Electron. Agric. **175**, 105527 (2020). [CrossRef]. https://doi.org/10.1016/j.compag.2020.105527
8. R. Balodi, S. Bisht, A. Ghatak, K. Rao, Plant disease diagnosis: Technological advancements and challenges. Indian Phytopathol. **70**, 275–281 (2017). [CrossRef]. https://doi.org/10.24838/ip.2017.v70.i3.72487
9. E. Bauriegel, W.B. Herppich, Hyperspectral and chlorophyll fluorescence imaging for early detection of plant diseases, with special reference to Fusarium spec. infections on wheat. Agriculture **4**, 32–57 (2014). [CrossRef]. https://doi.org/10.3390/agriculture4010032
10. P. Mishra, G. Polder, N. Vilfan, Close range spectral imaging for disease detection in plants using autonomous platforms: A review on recent studies. Curr. Robot. Rep. **1**, 43–48 (2020). [CrossRef]. https://doi.org/10.1007/s43154-020-00004-7
11. V. Suresh, D. Gopinath, M. Hemavarthini, K. Jayanthan, Mohana Krishnan, Plant disease detection using image processing. Int. J. Eng. Res. Technol. **9**(03) (2020). https://doi.org/10.17577/IJERTV9IS030114
12. S.D. Khirade, A.B. Patil, Savithribhai Phule, Plant disease detection using image processing. Pune University, IJERT2015. http://ijesc.org/
13. D. Varshney, D. Roy, J.V. Meenakshi, Impact of COVID-19 on agricultural markets: Assessing the roles of commodity characteristics, disease caseload and market reforms. Indian Econ. Rev. **55**, 83–103 (2020). https://doi.org/10.1007/s41775-020-00095-1
14. R.S. Rolle, G. Mrema, P. Soni, Agriculture G, *A Regional Strategy for Sustainable Agricultural Mechanization. Sustainable Mechanization across Agri-Food Chains in Asia and the Pacific Region* (RAP Publication, 2015). http://www.fao.org/
15. M. Ahmad, M. Abdullah, H. Moon, D. Han, Plant disease detection in imbalanced datasets using efficient convolutional neural networks with stepwise transfer learning. IEEE Access **9**, 140565–140580 (2021). [CrossRef]. https://doi.org/10.1109/ACCESS.2021.3119655
16. T.U. Ichiki, T. Shiba, K. Matsukura, T. Ueno, M. Hirae, T. Sasaya, Detection and diagnosis of rice-infecting viruses. Front. Microbiol. **4**, 289 (2013). https://doi.org/10.3389/fmicb.2013.00289

17. C. Lacomme, R. Holmes, F. Evans, Molecular and serological methods for the diagnosis of viruses in potato tubers, in *Plant Pathology*, (Springer, Berlin/Heidelberg, 2015), pp. 161–176. https://doi.org/10.1007/978-1-4939-2620-6_13
18. T. Akram, S.R. Naqvi, S.A. Haider, M. Kamran, Towards real-time crops surveillance for disease classification: Exploiting parallelism in computer vision. Comput. Electr. Eng. **59**, 15–26 (2017). https://doi.org/10.1016/j.compeleceng.2017.02.020
19. B. Garg, S. Aggarwal, J. Sokhal, Crop yield forecasting using fuzzy logic and regression model. Comput. Electr. Eng. **67**, 383–403 (2018). https://doi.org/10.1016/j.compeleceng.2017.11.015
20. K. Neupane, F. BaysalGurel, Automatic identification and monitoring of plant diseases using unmanned aerial vehicles: A review. Remote Sens. **13**, 3841 (2021). [CrossRef]. https://doi.org/10.3390/rs13193841
21. A. Abbas, S. Jain, M. Gour, S. Vankudothu, Tomato plant disease detection using transfer learning with C-GAN synthetic images. Comput. Electron. Agric. **187**, 106279 (2021). [CrossRef]. https://doi.org/10.1016/j.compag.2021.106279
22. H. Hamdani, A. Septiarini, A. Sunyoto, S. Suyanto, F. Utaminingrum, Detection of oil palm leaf disease based on color histogram and supervised classifier. Optik **245**, 167753 (2021). [CrossRef]. https://doi.org/10.1016/j.ijleo.2021.167753
23. M. Kaur, R. Bhatia, Leaf disease detection and classification: A comprehensive survey, in *Proceedings of the International Conference on IoT Inclusive Life (ICIIL 2019)*, Chandigarh, India, 19–20 Dec 2019, pp. 291–304. https://doi.org/10.1007/978-981-15-3020-3_27
24. V. Singh, A. Misra, Detection of unhealthy region of plant leaves using image processing and genetic algorithm, in *Proceedings of the 2015 International Conference on Advances in Computer Engineering and Applications*, Ghaziabad, India, 19–20 Mar 2015, pp. 1028–1032. https://doi.org/10.1109/ICACEA.2015.7164858
25. J. Karthika, M. Santhose, T. Sharan, Disease detection in cotton leaf spot using image processing. J. Phys. Conf. Ser. **1916**, 012224 (2021). [CrossRef]. https://doi.org/10.1088/1742-6596/1916/1/012224
26. A. Devaraj, K. Rathan, S. Jaahnavi, K. Indira, Identification of plant disease using image processing technique, in *Proceedings of the 2019 International Conference on Communication and Signal Processing (ICCSP)*, Chennai, India, 4–6 Apr 2019, pp. 0749–0753. https://doi.org/10.1109/ICCSP.2019.8698056
27. R.K. Mojjada, K.K. Kumar, A. Yadav, B.S.V. Prasad, Detection of plant leaf disease using digital image processing. Mater. Today Proc. (2020). [CrossRef]. https://doi.org/10.1016/j.matpr.2020.11.115
28. K.P. Ferentinos, Deep learning models for plant disease detection and diagnosis. Comput. Electron. Agric. **145**, 311–318 (2018). [CrossRef]. https://doi.org/10.1016/j.compag.2018.01.009
29. A. Fuentes, S. Yoon, S.C. Kim, D.S. Park, A robust deep-learning-based detector for real-time tomato plant diseases and pests' recognition. Sensors **17**, 2022 (2017). [CrossRef]. https://doi.org/10.3390/s17092022
30. J. Chen, J. Chen, D. Zhang, Y. Sun, Y.A. Nanehkaran, Using deep transfer learning for image-based plant disease identification. Comput. Electron. Agric. **173**, 105393 (2020). [CrossRef]. https://doi.org/10.1016/j.compag.2020.105393
31. M.A. Iqbal, K.H. Talukder, Detection of potato disease using image segmentation and machine learning, in *Proceedings of the 2020 International Conference on Wireless Communications Signal Processing and Networking (WiSPNET)*, Chennai, India, 4–6 Aug 2020, pp. 43–47. https://doi.org/10.1109/WiSPNET48689.2020.9198563
32. A.R. Srivastava, M. Venkatesan, Tea leaf disease prediction using texture-based image processing, in *Emerging Research in Data Engineering Systems and Computer Communications*, (Springer, Berlin/Heidelberg, 2020), pp. 17–25. https://doi.org/10.1007/978-981-15-0135-7_3
33. C.D. Whitmire, J.M. Vance, H.K. Rasheed, A. Missaoui, K.M. Rasheed, F.W. Maier, Using machine learning and feature selection for alfalfa yield prediction. AI **2**, 6 (2021). [CrossRef]. https://doi.org/10.3390/ai2010006

34. H. Zhu, H. Cen, C. Zhang, Y. He, Early detection and classification of tobacco leaves inoculated with tobacco mosaic virus based on hyperspectral imaging technique, in *Proceedings of the 2016 ASABE Annual International Meeting*, Orlando, FL, USA, 17–20 July 2016, p. 1. https://doi.org/10.1038/s41598-017-04501-2
35. H. Zhu, B. Chu, C. Zhang, F. Liu, L. Jiang, Y. He, Hyperspectral imaging for presymptomatic detection of tobacco disease with successive projections algorithm and machine-learning classifiers. Sci. Rep. **7**, 4125 (2017). [CrossRef] [PubMed]. https://www.nature.com/articles/s41598-017-04501-2#article-info
36. J. Ma, K. Du, F. Zheng, L. Zhang, Z. Gong, Z. Sun, A recognition method for cucumber diseases using leaf symptom images based on deep convolutional neural network. Comput. Electron. Agric. **154**, 18–24 (2018). [CrossRef]. https://agris.fao.org/agris-search/search.do?recordID=US201900091628
37. Y. Tian, G. Yang, Z. Wang, H. Wang, E. Li, Z. Liang, Apple detection during different growth stages in orchards using the improved YOLO-V3 model. Comput. Electron. Agric. **157**, 417–426 (2019). [CrossRef]. https://doi.org/10.1016/j.compag.2019.01.012
38. D. Singh, N. Jain, P. Jain, P. Kayal, S. Kumawat, N. Batra, PlantDoc: A dataset for visual plant disease detection, in *Proceedings of the 7th ACM IKDD CoDS and 25th COMAD*, Hyderabad, India, 5 Jan 2020, p. 249253. https://doi.org/10.1145/3371158.3371196
39. J.S.H. Al-bayati, B.B. Üstündag, Evolutionary feature optimization for plant leaf disease detection by deep neural networks. Int. J. Comput. Intell. Syst. **13**, 12–23 (2020). [CrossRef]. https://doi.org/10.2991/ijcis.d.200108.001
40. M. Arsenovic, M. Karanovic, S. Sladojevic, A. Anderla, D. Stefanovic, Solving current limitations of deep learning-based approaches for plant disease detection. Symmetry **11**, 939 (2019). [CrossRef]. https://doi.org/10.3390/sym11070939
41. J. Costa, C. Silva, B. Ribeiro, Hierarchical deep learning approach for plant disease detection, in *Proceedings of the Iberian Conference on Pattern Recognition and Image Analysis*, Madrid, Spain, 1–4 July 2019, pp. 383–393. https://doi.org/10.1007/978-3-030-31321-0_33
42. G. Geetharamani, A. Pandian, Identification of plant leaf diseases using a nine-layer deep convolutional neural network. Comput. Electr. Eng. **76**, 323–338 (2019). https://doi.org/10.1016/j.compeleceng.2019.04.011
43. R.G. De Luna, E.P. Dadios, A.A. Bandala, Automated image capturing system for deep learning-based tomato plant leaf disease detection and recognition, in *Proceedings of TENCON 2018–2018 IEEE Region 10 Conference*, Jeju Island, Korea, 28–31 Oct 2018, pp. 1414–1419. https://doi.org/10.1109/TENCON.2018.8650088
44. J.M. Bioucas-Dias, A. Plaza, G. Camps-Valls, P. Scheunders, N. Nasrabadi, J. Chanussot, Hyperspectral remote sensing data analysis and future challenges. IEEE Geosci. Remote Sens. Mag. **1**, 6–36 (2013). [CrossRef]. https://doi.org/10.3390/rs12061009
45. S. Mananze, I. Pôças, M. Cunha, Retrieval of maize leaf area index using hyperspectral and multispectral data. Remote Sens. **10**, 1942 (2018). [CrossRef]. https://doi.org/10.3390/rs10121942
46. C. Wei, J. Huang, X. Wang, G.A. Blackburn, Y. Zhang, S. Wang, L.R. Mansaray, Hyperspectral characterization of freezing injury and its biochemical impacts in oilseed rape leaves. Remote Sens. Environ. **195**, 56–66 (2017). [CrossRef]. https://doi.org/10.1016/j.rse.2017.03.042
47. W. Ng, B. Minasny, B.P. Malone, M. Sarathjith, B.S. Das, Optimizing wavelength selection by using informative vectors for parsimonious infrared spectra modelling. Comput. Electron. Agric. **158**, 201–210 (2019). [CrossRef]. https://doi.org/10.1016/j.compag.2019.02.003
48. W.L. Chen, Y.B. Lin, F.L. Ng, C.Y. Liu, Y.W. Lin, RiceTalk: Rice blast detection using internet of things and artificial intelligence technologies. IEEE Internet Things J. **7**, 1001–1010 (2019). [CrossRef]. https://doi.org/10.1109/JIOT.2019.2947624
49. M. Mishra, P. Choudhury, B. Pati, Modified ride-NN optimizer for the IoT based plant disease detection. J. Ambient. Intell. Humaniz. Comput. **12**, 691–703 (2021). [CrossRef]. https://link.springer.com/article/10.1007/s12652-020-02051-6

50. R.D. Devi, S.A. Nandhini, R. Hemalatha, S. Radha, IoT enabled efficient detection and classification of plant diseases for agricultural applications, in *Proceedings of the 2019 International Conference on Wireless Communications, Signal Processing and Networking (WiSPNET)*, Chennai, India, 21–23 Mar 2019, pp. 447–451. https://doi.org/10.1109/WiSPNET45539.2019.9032727
51. D. Cheng, Y. Shi et al., *Hybrid K-Means Clustering and Support Vector Machine Method for Via and Metal Line Detections in Delayered IC Images.* https://www.researchgate.net/journal/IEEE-Transactions-on-Circuits-and-Systems-II-Express-Briefs-1558-3791. https://doi.org/10.1109/TCSII.2018.2827044
52. P.K. Sethy, S.K. Behera, N. Kannan, et al., Smart paddy field monitoring system using deep learning and IoT. Concurr. Eng. Res. Appl. **29**(1), 16–24 (2021). https://doi.org/10.1177/1063293X21988944
53. M. Shankar, D.N. Mungal, Nandi wale Geetanjali, Mane Mahesh Jalinder, Green leaf disease detection using Raspberry Pi Maharashtra, IJERT-2018. https://doi.org/10.1109/ICIICT1.2019.8741508
54. S. Spoorthi, D.N. Shadaksharappa, S. Suraj, V.K. Manasa, FREYR Drone: A pesticide / fertilizer spraying drone-An agricultural approach. J. Common Phys. (2017). https://doi.org/10.1109/ICCCT2.2017.7972289
55. P. Jiang, Y. Chen, B. Liu, D. He, C. Liang, Real-time detection of apple leaf diseases using deep learning approach based on improved convolutional neural networks. IEEE Access **7** (2019). https://doi.org/10.1109/ACCESS.2019.2914929
56. M.E. Pothen, M.L. Pai, Detection of rice leaf diseases using image processing, in *Proceedings of the 2020 Fourth International Conference on Computing Methodologies and Communication (ICCMC)*, Erode, India, 11–13 Mar 2020, pp. 424–430. https://doi.org/10.1109/ICCMC48092.2020.ICCMC-00080
57. K.P. Panigrahi, H. Das, A.K. Sahoo, S.C. Moharana, Maize leaf disease detection and classification using machine learning algorithms, in *Progress in Computing, Analytics and Networking*, (Springer, Singapore, 2020), pp. 659–669. https://doi.org/10.1007/978-981-15-2414-1_66
58. H. Al-Hiary, S. Bani-Ahmad, M. Reyalat, M. Braik, Z. Alrahamneh, Fast and accurate detection and classification of plant diseases. Int. J. Comput. Appl. **17**, 31–38 (2011). [CrossRef]. https://doi.org/10.5120/2183-2754
59. *E-Agriculture in Action: Drones for Agriculture* (Food and Agriculture Organization of the United Nations and International Telecommunication Union Bangkok, 2018). http://handle.itu.int/11.1002/pub/8111728e-en
60. N. Kitpo, M. Inoue, Early rice disease detection and position mapping system using drone and IoT architecture, in *2018-12th South East Asian Technical University Consortium Symposium (SEATUC)*. https://doi.org/10.1109/SEATUC.2018.8788863
61. C.-M. Chao, Y.-J. Yang, W.-C. Liang, *Low-Cost Weed Identification System Using Drones – 2019 (CANDARW)*. https://doi.org/10.1109/CANDARW.2019.00052
62. A. Usman, T.F.N. Bukht, R. Ahmad, J. Ahmad, Plant disease detection using internet of thing (IoT). Plant Dis. **11**, 505–509 (2020). https://doi.org/10.14569/IJACSA.2020.0110162
63. K.A. Reddy, N.M.C. Reddy, S. Sujatha, Precision method for pest detection in plants using the clustering algorithm in image processing, in *Proceedings of the 2020 International Conference on Communication and Signal Processing (ICCSP)*, Chennai, India, 28–30 July 2020, pp. 894–897. https://doi.org/10.1109/ICCSP48568.2020.9182190
64. M.A. Khan, Detection and classification of plant diseases using image processing and multi-class support vector machine. Int. J. Comput. Trends Technol. **68**, 5–11 (2020). [CrossRef]. https://doi.org/10.14445/22312803/IJCTT-V68I4P102
65. R.I. Hasan, S.M. Yusuf, L. Alzubaidi, Review of the state of the art of deep learning for plant diseases: A broad analysis and discussion. Plants **9**, 1302 (2020). [CrossRef]. https://doi.org/10.3390/plants9101302
66. U.N. Fulari, R.K. Shastri, A.N. Fulari, Leaf disease detection using machine learning. J. Seybold Rep. **1533**, 9211 (2020). https://doi.org/10.3390/agriculture12010009

67. A. Kadir, L.E. Nugroho, A. Susanto, P.I. Santosa, Performance improvement of leaf identification system using principal component analysis. Int. J. Adv. Sci. Technol. **44**, 113–124 (2012). https://doi.org/10.1155/2023/3974346
68. M. Turkoglu, D. Hanbay, Recognition of plant leaves: An approach with hybrid features produced by dividing leaf images into two and four parts. Appl. Math. Comput. **352**, 1–14 (2019). [CrossRef]. https://doi.org/10.1016/j.amc.2019.01.054
69. L.C. Ngugi, M. Abelwahab, M. Abo-Zahhad, Recent advances in image processing techniques for automated leaf pest and disease recognition–A review. Inf. Process. Agric. **8**, 27–51 (2021). [CrossRef]. https://doi.org/10.1016/j.inpa.2020.04.004
70. S. Jinya, D. Yi, B. Su, Z. Mi, C. Liu, X. Hu, X. Xian, L. Guo, W.-H. Chen, Aerial visual perception in smart farming: Field study of wheat yellow rust monitoring. IEEE Trans. Industr. Inform., 1 (2020). https://doi.org/10.1109/tii.2020.2979237
71. C. Koc, Design and development of a low-cost UAV for pesticide applications. J. Agric. Fac. Gaziosmanpasa Univ. **34**, 94–103 (2017). Available from: https://doi.org/10.13002/jafag4274
72. K.K. Shaw, R. Vimalkumar, Design and development of a drone for spraying pesticides, fertilizers and disinfectants. Int. J. Eng. Res. **V9** (2020). https://doi.org/10.17577/IJERTV9IS050787
73. J.G. Barbedo, Factors influencing the use of deep learning for plant disease recognition. Biosyst. Eng. **172**, 84–91 (2018). [CrossRef]. https://doi.org/10.1016/j.biosystemseng.2018.05.013
74. S.P. Mohanty, D.P. Hughes, M. Salathé, Using deep learning for image-based plant disease detection. Front. Plant Sci. **7**, 1419 (2016). [CrossRef] [PubMed]. https://doi.org/10.3389/fpls.2016.01419
75. G. Sambasivam, G.D. Opiyo, A predictive machine learning application in agriculture: Cassava disease detection and classification with imbalanced dataset using convolutional neural networks. Egypt. Inform. J. **22**, 27–34 (2021). [CrossRef]. https://doi.org/10.1016/j.eij.2020.02.007
76. M. Wiwart, G. Fordonski, K. Zuk-Gołaszewska, E. Suchowilska, Early diagnostics of macronutrient deficiencies in three legumes species by color image analysis. Comput. Electron. Agric. **65**, 125–132 (2009). [CrossRef]. https://doi.org/10.1016/j.compag.2008.08.003
77. C. Bock, A. Cook, P. Parker, T. Gottwald, Automated image analysis of the severity of foliar citrus canker symptoms. Plant Dis. **93**, 660–665 (2009). [CrossRef] [PubMed]. https://doi.org/10.1094/pdis-93-6-0660
78. A. Pourreza, W.S. Lee, E. Raveh, R. Ehsani, E. Etxeberria, Citrus Huanglongbing detection using narrow-band imaging and polarized illumination. Trans. ASABE **57**, 259–272 (2014). https://doi.org/10.13031/trans.57.10147
79. W. Guo, U.K. Rage, S. Ninomiya, Illumination invariant segmentation of vegetation for time series wheat images based on decision tree model. Comput. Electron. Agric. **96**, 58–66 (2013). [CrossRef]. https://doi.org/10.1016/j.compag.2013.04.010
80. R. Zhou, S.I. Kaneko, F. Tanaka, M. Kayamori, M. Shimizu, Image-based field monitoring of Cercospora leaf spot in sugar beet by robust template matching and pattern recognition. Comput. Electron. Agric. **116**, 65–79 (2015). [CrossRef]. https://doi.org/10.1016/j.compag.2015.05.020
81. J.W. Olmstead, G.A. Lang, G.G. Grove, Assessment of severity of powdery mildew infection of sweet cherry leaves by digital image analysis. HortScience **36**, 107–111 (2001). [CrossRef]. https://doi.org/10.21273/HORTSCI.36.1.107
82. E. Moya, L. Barrales, G. Apablaza, Assessment of the disease severity of squash powdery mildew through visual analysis, digital image analysis and validation of these methodologies. Crop Prot. **24**, 785–789 (2005). [CrossRef]. https://doi.org/10.1016/j.cropro.2005.01.003
83. R. Oberti, M. Marchi, P. Tirelli, A. Calcante, M. Iriti, A.N. Borghese, Automatic detection of powdery mildew on grapevine leaves by image analysis: Optimal view-angle range to increase the sensitivity. Comput. Electron. Agric. **104**, 1–8 (2014). [CrossRef] Agriculture 2022, 12, 9 29 of 29. https://doi.org/10.1016/j.compag.2014.03.001

84. A. Camargo, J. Smith, An image-processing based algorithm to automatically identify plant disease visual symptoms. Biosyst. Eng. **102**, 9–21 (2009). [CrossRef]. https://doi.org/10.1016/j.biosystemseng.2008.09.030
85. C. Xie, Y. Shao, X. Li, Y. He, Detection of early blight and late blight diseases on tomato leaves using hyperspectral imaging. Sci. Rep. **5**, 16564 (2015). [CrossRef]. https://doi.org/10.1038/srep16564
86. I.S. Ahmad, J.F. Reid, M.R. Paulsen, J.B. Sinclair, Color classifier for symptomatic soybean seeds using image processing. Plant Dis. **83**, 320–327 (1999). [CrossRef] [PubMed]. https://doi.org/10.1094/pdis.1999.83.4.320

Empowering Agriculture: Blockchain's Revolution in Smart Farming

N. A. Natraj, Sundaravadivazhagan Balasubaramanian, K. B. Gurumoorthy, A. Purushothaman, and P. Kannan

1 Introduction

Modern agriculture is at a crossroad, confronted with tremendous problems caused by global population increase, resource restrictions, climate change, and the pressing need for sustainable practices. In this setting, smart farming has arisen to address these multidimensional difficulties and ensure food security, environmental stewardship, and economic viability for future generations. A novel paradigm at the nexus of agriculture and technology known as "smart farming" has arisen as a revolutionary force in the contemporary agricultural landscape. The demand for effective, sustainable, and data-driven farming practices has never been greater due to the world population's steady increase and the escalating environmental problems. In order to transform conventional agricultural practices, smart farming makes use of cutting-edge technology including Internet of Things (IoT) devices, data analytics, artificial intelligence (AI), and automation. Smart farming's main goals are to

N. A. Natraj (✉)
Symbiosis Institute of Digital and Telecom and Management, Symbiosis International (Deemed University), Maharashtra, India
e-mail: natraj@sidtm.edu.in

S. Balasubaramanian
Department of Information and Technology, University of Technology and Applied Sciences-Al Mussana, Al Mussana, Oman

K. B. Gurumoorthy
Department of ECE, KPR Institute of Engineering and Technology, Coimbatore, India

A. Purushothaman
Hindusthan Institute of Technology, Coimbatore, India

P. Kannan
Department of ECE, Francis Xavier Engineering College, Tirunelveli, India

S. Balasubaramanian et al. (eds.), *Intelligent Robots and Drones for Precision Agriculture*, Signals and Communication Technology,
https://doi.org/10.1007/978-3-031-51195-0_11

maximize resource use, boost productivity, and lessen environmental effect. Farmers can make informed decisions regarding irrigation, fertilization, pest management, and other crucial aspects of agriculture by combining real-time data from sensors, satellites, and drones. This data-driven approach promotes sustainable practices, safeguarding vital resources for future generations while also boosting output and lowering waste. The need for smart agricultural practices is addressed in the following sections:

1.1 Population Growth and Food Demand

The relationship between population expansion and food consumption is a critical concern in the twenty-first century, highlighting the sustainability of global agricultural systems. As the world's population grows to an anticipated 9.7 billion by 2050, as forecasted by the UN, the demand on food resources becomes more apparent. This growing population necessitates a significant increase in dietary demand. People require food to survive, and the demand for food goes beyond quantity to include nutritional quality, diversity, and accessibility [1]. Adequate nutrition is more than just a matter of satiation; it is the foundation of health, development, and societal well-being. The complexities of supplying this need become more complex as urbanization accelerates and nutritional tastes shift [2]. The modern customer seeks not only food but also ethical considerations such as ethically sourced food and lower environmental effect. This complex matrix of elements puts significant strain on present farming practices. The task is not simply to produce more food but to do so while protecting key resources. Arable land is limited, and water resources are becoming increasingly scarce. Climate change exacerbates agricultural output instability, resulting in yield swings and crop failures. The need to close the gap between food demand and availability is more than a theoretical worry; it has worldwide implications, affecting economies, social stability, and human dignity. As a result of this obstacle, creativity emerges as a ray of hope. Precision agriculture, smart farming, and vertical farming are examples of technology-driven solutions that use data, automation, and sustainable practices to maximize yields and resource utilization. Farmers may make more informed decisions, increase production, and reduce waste by combining artificial intelligence, remote sensing, and data analytics. Furthermore, the importance of sustainable agricultural practices becomes unmistakable. Crop rotation, agroecological approaches, and lower chemical inputs become more popular, creating a harmonious relationship between farming and the environment. Efforts to address food demand are gaining traction, from local initiatives to worldwide cooperation, with a focus on resilience, adaptation, and equitable distribution. The dynamic interaction between population increase and food consumption is a serious challenge that requires an overhaul of paradigms in global agriculture.

1.2 Scarcity of Resources and Efficiency

The growing world population, which is expected to reach 9.7 billion by 2050, exacerbates concerns about the shortage of key resources and emphasizes the importance of agricultural efficiency. With limited arable land and water, meeting the world's growing food demand becomes increasingly difficult. Statistical data underscores the seriousness of resource scarcity. According to the World Resources Institute, agriculture currently covers around 38% of the world's land area and uses approximately 70% of freshwater withdrawals. This utilization, however, is far from equitable, with poor practices resulting in significant waste. According to the Food and Agriculture Organization (FAO), one-third of all food produced globally, worth about $1 trillion, goes to waste each year. The world's expanding population, which is anticipated to reach 9.7 billion by 2050, exacerbates concerns about a lack of critical resources and highlights the necessity of agricultural efficiency. Meeting the world's expanding food demand is becoming increasingly difficult due to the limited arable land and water. Statistical data emphasizes the gravity of resource shortage. Agriculture currently spans around 38% of the world's land surface and consumes approximately 70% of freshwater withdrawals, according to the World Resources Institute. However, this utilization is far from equitable, with poor practices resulting in enormous waste. According to the Food and Agriculture Organization (FAO), one-third of all food produced globally each year, worth around $1 trillion, is wasted. Precision agriculture, a pillar of efficiency-driven innovation, tackles these issues square on. Precision agriculture optimizes resource allocation by utilizing real-time data from sensors, satellites, and drones. Precision solutions, such as variable rate technology and site-specific management, can cut input use by up to 20%, reducing waste and improving sustainability, according to the American Society of Agronomy. Furthermore, emerging technologies such as vertical farming and aquaponics hold enormous promise. According to MarketsandMarkets, the vertical farming industry will expand from $2.5 billion in 2020 to $7.3 billion by 2025, indicating a growing appreciation for their ability to maximize yield per unit of land and water. Aquaponics, a hybrid of fish farming and hydroponics, provides an appealing solution by recycling nutrient-rich water while reducing consumption. Climate change, which exacerbates resource scarcity, emphasizes the importance of agricultural efficiency even more. Changes in precipitation patterns and the increased frequency of extreme weather events, according to the Intergovernmental Panel on Climate Change (IPCC), endanger agricultural output, emphasizing the significance of optimizing resource utilization to counteract these impacts. Scarcity of resources, particularly arable land and water, poses a significant threat to global food security. The potential for efficiency-driven solutions propelled by technical innovation, on the other hand, is huge.

1.3 Necessity to Adapt According to Climate Change Issues

The urgent need to adapt agricultural practices to climate change challenges has arisen as a critical concern in guaranteeing global food security and environmental sustainability. Traditional farming systems face significant challenges as the Earth suffers increasingly irregular weather patterns, rising temperatures, and altered precipitation regimes. Statistical data emphasizes the importance of this adaptability. According to the Intergovernmental Panel on Climate Change (IPCC), climate change is already hurting food production, with a projected 25% reduction in staple crops like rice and maize by 2050. Furthermore, the Food and Agriculture Organization (FAO) believes that climate-related variables have already deteriorated around 20% of worldwide agricultural land. Climate change adaptation necessitates a fundamental overhaul of agricultural practices. Among the methods gaining traction are the adoption of drought-resistant agricultural types, the improvement of water management systems, and the use of agroforestry techniques. A study published in *Nature Climate Change* journal found that changing to drought-tolerant crops might boost global crop production by up to 10% under climate change scenarios. Furthermore, precision agriculture is emerging as a critical component of climate adaption initiatives. Precision agriculture improves resource allocation, reduces waste, and optimizes yields by leveraging data from satellites, drones, and sensors. According to the World Economic Forum, precision agriculture could reduce water usage by up to 50% while increasing crop yields by up to 30%. Climate adaptation requires collaboration among farmers, governments, and researchers. Investment in research and development, extension services, and climate-resilient infrastructure can help spread new practices. The combination of indigenous knowledge and traditional farming techniques with modern technologies provides a comprehensive approach. Climate change imperatives need a paradigm shift in agricultural practices. Not only is adaptation a guarantee for food security, but it is also a critical step in lessening the effects of climate change on vulnerable ecosystems and communities [3]. The agricultural industry can manage the difficulties of climate change and create a resilient future for future generations by combining technology innovation, sustainable practices, and collaborative efforts.

1.4 Shortages in Labor and Changing Demographics

The worldwide agricultural sector is facing a serious challenge: labor shortages and shifting demographics that are transforming the farming workforce dynamics. As rural populations decline and aging demographics predominate, the need for new solutions to sustain agricultural output and assure food security becomes more pressing. Statistical data underscores the intensity of the situation. According to the United Nations, by 2050, the share of the global population living in rural areas would fall to roughly 45%, down from 56% in 1950. Furthermore, the Food and

Agriculture Organization (FAO) notes that the average age of farmers is progressively growing in many nations, as the youth population migrates away from rural areas in quest of urban possibilities. For instance, the average age of farmers in Japan is over 67, highlighting the imminent demographic shift. To preserve the long-term viability of agricultural production, innovative solutions are required to address the interaction of workforce shortages and shifting demographics. Particularly promising for technological advancement are automation and robots. A study in the journal *Agricultural Economics* found that the use of agricultural robots might significantly alleviate labor shortages and increase production effectiveness. Robots ranging from precise weeders to automated harvesters can fill manpower shortages and boost output. Additionally, the generational divide between younger and older farmers may be closed through digital tools and mobile applications. The World Bank claims that farmers may get critical information through mobile devices, such as market prices, weather forecasts, and sustainable practices. This provides experienced farmers and newbies the information they need to make sound decisions. Governments, educational institutions, and the corporate sector must work together to overcome these difficulties. Investments in agricultural education, vocational training, and capacity-building for young farmers have the potential to revitalize the industry and attract a new generation of agricultural enthusiasts. Labor shortages and changing demographics offer significant difficulties to global agriculture. As a disruptive strategy, the integration of technology, automation, and targeted instruction arises. The agricultural industry can negotiate the complexity of a moving workforce, secure food production, and preserve the vitality of rural communities in an ever-changing globe by harnessing these instruments and encouraging collaboration.

1.5 Necessity in Adopting Eco-friendly Practices in Agriculture

As the global world grapples with the rising issues of environmental degradation and climate change, the obligation to embrace eco-friendly agricultural practices has taken on enormous significance. Statistical data underscores the importance of moving to sustainable agricultural methods, emphasizing the need to balance food production with environmental preservation. Agriculture, according to the Food and Agriculture Organization (FAO), contributes significantly to greenhouse gas emissions, accounting for around 14% of world emissions. Furthermore, the Intergovernmental Panel on Climate Change (IPCC) report emphasizes that the current trajectory of agricultural emissions is incompatible with the objective of keeping global warming well below 2 °C. Eco-friendly practices provide an appealing answer. Organic farming, conservation tillage, and integrated pest control are examples of agroecological practices that can dramatically reduce agriculture's environmental imprint. According to a study published in the journal *Nature Communications*, organic agricultural practices can lower greenhouse gas emissions by up to 30% when compared to traditional approaches. Water conservation is also an important issue. According to the World Wildlife Fund (WWF), agriculture accounts for 70% of worldwide freshwater withdrawals [4]. Drip irrigation and rainwater collection are two eco-friendly irrigation strategies that not only save water but also improve water use efficiency. According to a study conducted by the International Water Management Institute, adopting water-efficient practices might enhance agricultural yields by up to 50%. Adopting environmentally responsible practices has economic ramifications. According to the Organic Trade Association, the global market for organic products has grown significantly, reaching $124.6 billion in 2019. This trend highlights the growing consumer desire for food that is produced in a sustainable manner. Collaboration between governments, farmers, and the commercial sector is critical in making the shift to environmentally friendly practices. Incentive programs, subsidies for ecologically responsible agricultural methods, and capacity-building activities can all help to spread the use of environmentally responsible approaches. The need of implementing eco-friendly agricultural practices is emphasized by the urgency of mitigating climate change, conserving natural resources, and guaranteeing long-term food security. Incorporating sustainable practices not only decreases environmental effect but also opens up economic prospects. As the world's issues rise, adopting environmentally friendly practices appears as a critical step toward building a healthy relationship between agriculture and the environment.

1.6 Increasing Market Demand and Quality Assurance Measures

The agricultural industry is transforming as rising market demand and the need for quality assurance procedures collide to influence the future of food production. Statistical data emphasizes the critical need of addressing consumer preferences and guaranteeing product integrity in order to fulfill the growing needs of a discriminating global market. According to the World Bank, population growth, changing eating patterns, and urbanization are expected to propel the global food market to $10 trillion by 2025. Quality and safety have taken front stage as customers have become increasingly attentive of their food choices. According to the World Health Organization (WHO), over 600 million people become ill each year as a result of foodborne infections, emphasizing the importance of strong quality assurance methods. Quality assurance encompasses all aspects of authenticity, traceability, and ethical and environmentally friendly production [5]. As a powerful tool for enhancing quality assurance throughout the supply chain, blockchain technology has recently come into its own. Blockchain can increase traceability and transparency by enabling customers to confirm their food's place of origin and travel history, according to a Deloitte study. This technology promotes confidence and trust, both of which are essential for satisfying consumer demand. Standards and certifications play a significant role in quality assurance. Approximately 1.5 million businesses are accredited to the Global Food Safety Initiative's (GFSI's) internationally established food safety standards. These standards ensure that food products meet established quality standards, boosting consumer confidence and market accessibility. Furthermore, the organic and non-GMO sectors demonstrate strong examples of rising market demand driven by quality concerns. According to the Organic Trade Association, the worldwide organic food market topped $100 billion in 2018, demonstrating consumer preferences for chemical-free, sustainably produced products. Similarly, the non-GMO project estimates that non-GMO-verified product sales in the United States will exceed $30 billion by 2020. Technology integration is critical to negotiate the complexity of quality assurance and market demand. Advanced analytics, sensor technology, and AI-driven solutions make real-time monitoring and data-driven decision-making possible. These instruments improve manufacturing efficiency, reduce waste, and meet quality standards. The convergence of rising market demand and quality assurance requirements reshapes the agricultural environment. A comprehensive solution based on technology, certifications, and transparent supply chains is required to meet customer preferences for safe, authentic, and sustainably produced food. As the global market expands, employing quality assurance procedures is critical to ensuring the agricultural industry's lucrative and sustainable future.

1.7 Economic Stability and Profitability Aspects

The pursuit of economic stability and profitability is a major priority in agriculture, where the intersection of market dynamics, resource management, and technological innovation shapes the financial environment. Statistical data emphasizes the importance of these factors, emphasizing their deep impact on the agricultural sector's sustainability and viability. Agriculture is critical to the global economy. According to the World Bank, agriculture accounts for around 4% of global GDP and provides a primary source of income for more than 40% of the world's population. Furthermore, the sector's economic importance extends beyond production to include commerce, employment, and rural development. The global economy greatly depends on the agricultural sector. More than 40% of the world's population relies primarily on agriculture, which, according to the World Bank, contributes about 4% of the global GDP. Additionally, the industry's economic significance extends beyond production to encompass commerce, employment, and rural development. Resource management initiatives are intrinsically related to efforts to sustain economic stability and profitability. For instance, a major problem is the lack of water. The United Nations estimates that 70% of the world's freshwater withdrawals come from agriculture. For instance, drip irrigation and precise application technologies can optimize water use while bringing down costs and protecting a priceless resource. Technological innovation has emerged as a critical component in achieving economic stability and profitability. Precision agriculture, made possible by data analytics, remote sensing, and Internet of Things (IoT) devices, improves resource allocation, reduces waste, and increases yields. According to a study published in the journal *Science of the Total Environment*, precision farming can lower input costs by up to 15% while increasing yields by up to 13%. A strong agricultural sector is built on the foundations of economic stability and profitability. The sector's importance is underscored by its contribution to global GDP, jobs, and livelihoods. The agricultural community can negotiate market dynamics, optimize resource utilization, and secure a successful future by adopting technology, sustainable practices, and collaborative relationships. As the agricultural sector navigates a complicated environment characterized by population increase, resource depletion, climate unpredictability, and shifting customer expectations, the abovementioned problems highlight the importance of smart farming. By embracing technology-driven advances, smart farming offers a transformative path to a more resilient, sustainable, and wealthy agricultural future.

Blockchain is a pivotal asset in smart farming, where technology is critical. Its inherent openness, security, and decentralized data management capabilities can considerably improve smart agricultural practices. By smoothly incorporating blockchain, the agricultural environment gains increased traceability, greater data exchange, and more confidence throughout the supply chain. This technology enables farmers to adopt sustainable practices, provides customers with accurate product information, and fosters collaborative ecosystems. As a cornerstone of innovation, blockchain holds the key to ushering in a new era of efficient, resilient, and transparent smart farming practices. This book chapter on "blockchain technology in smart farming" is organized as follows: The second successive chapter deals with blockchain applications in smart farming. The third chapter explores the benefits and advantages of blockchain technology in smart farming practices. Further, the challenges and opportunities of blockchain technology in smart farming are analyzed, and the final chapter concludes with a future look at the implementation of blockchain technology in smart farming practices.

2 Blockchain Technology: An Introduction

The emergence of blockchain, a groundbreaking technology in conjunction with the advent of the digital era, has fundamentally transformed our perception and engagement with concepts such as trust, transparency, and the secure management of data. Fundamentally, blockchain is an inherently decentralized and distributed digital ledger that maintains a transparent and resistant-to-tampering record of transactions. The significance of blockchain technology extends beyond its affiliation with cryptocurrencies such as Bitcoin, embracing a diverse range of applications across several industries.

The core principle behind blockchain technology is the establishment of a sequential chain of blocks, wherein each block encompasses a set of transactions. The aforementioned transactions undergo a process of secure encryption, and upon inclusion in a block, they get interconnected with the preceding transaction by cryptographic hashes, so establishing a continuous and uninterrupted sequence of data. The design of this system guarantees that any modification to the information within a block necessitates updating the entire subsequent chain, hence enhancing the system's resistance against unauthorized adjustments.

The decentralized nature of blockchain is widely recognized as one of its most significant characteristics. Traditional centralized systems rely on intermediaries like banks or governments to validate transactions. In contrast, the blockchain technology functions within a decentralized network, commonly known as a peer-to-peer network, wherein members, typically referred to as nodes, collaboratively authenticate transactions by means of consensus procedures. The elimination of intermediaries not only serves to address the need for their involvement but also contributes to the reinforcement of security measures and the mitigation of potential vulnerabilities associated with single points of failure.

The influence of blockchain technology extends across multiple industries. In the field of finance, advancements have facilitated the development of streamlined and secure cross-border payment systems, along with the implementation of programmable smart contracts that autonomously carry out predetermined activities upon the fulfillment of specified circumstances. The field of supply chain management derives advantages from the use of blockchain technology, which possesses the capability to track and authenticate the source and trajectory of commodities. This feature serves to augment transparency within the supply chain and effectively address the issue of counterfeit goods. Healthcare systems utilize blockchain technology to ensure the secure administration of patient data and facilitate interoperability between various healthcare providers.

The advent of blockchain technology has revolutionized the methods through which digital information is managed, secured, and exchanged. The concept gained early prominence through its affiliation with cryptocurrencies such as Bitcoin; nevertheless, its scope extends well beyond virtual currencies, spanning diverse domains such as supply chain management, healthcare, banking, voting systems, and others. Fundamentally, blockchain is an inherently decentralized and distributed digital ledger that provides unparalleled levels of transparency, security, and immutability. Figure 1 shows the overview of blockchain functionality.

The following is a full exposition elucidating the operational mechanics of blockchain technology:

1. Decentralization and Peer-to-Peer Network: The blockchain technology functions through a decentralized network comprising multiple members, commonly referred to as nodes. Every node possesses a complete replica of the entire blockchain, so guaranteeing redundancy and eliminating the necessity for a centralized governing entity. The utilization of a peer-to-peer network architecture serves to augment both security and robustness, since it eliminates the presence of a singular point of control or vulnerability.
2. Transactions and Verification: The blockchain process commences when a participant initiates a transaction. The transaction may encompass the transfer of

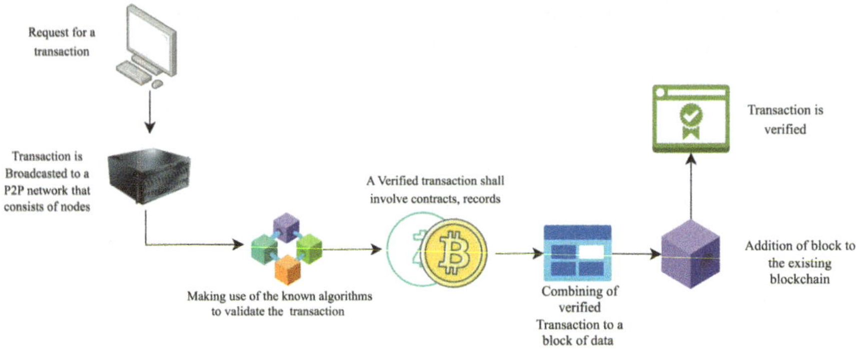

Fig. 1 Overview of blockchain functionality

Bitcoin, the establishment of asset ownership records, or the execution of smart contracts. The act of transmitting transactions is disseminated to the entirety of the network.

3. Blocks and Data Structures: The process of organizing verified transactions involves grouping them into blocks. Each block consists of two primary components: a header and a body. A block's header in a blockchain system has essential metadata, including a timestamp, a reference to the hash of the preceding block, and other pertinent details. The physical structure of the body contains the factual information about the transaction.
4. The process of transforming the data within each block into a fixed-length string of characters is achieved by the utilization of cryptographic hash algorithms. The hash generated for each block's data is distinct and functions as a digital fingerprint. The inclusion of the hash of the preceding block in the header of the current block establishes a sequential chain of blocks.
5. Immutability and Security: The incorporation of hash functions in the process of linking blocks in a blockchain guarantees that once a block is appended to the chain, modifying its contents or the data of any preceding block necessitates updating all subsequent blocks. The inherent characteristics of the blockchain render it exceptionally safe and resistant to tampering. The computational complexity and detectability associated with altering even a solitary piece of data make it practically impossible to do so.
6. Consensus Mechanisms: Consensus procedures are utilized to protect the integrity of the blockchain in a decentralized fashion. One widely recognized approach is known as proof of work (PoW), in which miners engage in solving intricate mathematical puzzles to authenticate transactions and incorporate blocks. The initial miner that successfully resolves the problem disseminates the solution, while other nodes subsequently authenticate its accuracy. After undergoing the verification process, the newly generated block is appended to the existing blockchain. Additional consensus mechanisms that are commonly employed in blockchain systems are proof of stake (PoS), delegated proof of stake (DPoS), and practical Byzantine fault tolerance (PBFT).

7. Merkle Trees for Data Integrity: The organization of transaction data within a block commonly involves the utilization of a Merkle tree structure to ensure data integrity. Within this hierarchical structure, transactions are organized into groups, and their respective hashes are merged together to generate a singular hash that encapsulates the entirety of the transaction set. The utilization of the Merkle root hash facilitates the expeditious verification of the integrity of transactional data.
8. Smart Contracts and Decentralized Applications (DApps): Smart contracts refer to contractual agreements that are encoded with specific terms and conditions within computer code, enabling them to execute automatically. Contractual agreements can be automated and enforced without the involvement of intermediaries. The introduction of smart contracts by Ethereum has facilitated the creation of decentralized applications (DApps) that make use of the blockchain's inherent attributes of transparency and security.
9. Chain Extension and Longest Chain Rule: The process of adding new blocks to the blockchain results in the extension of the chain. In instances where many miners concurrently discover valid solutions, it is possible for competing chains to emerge. According to the longest chain rule, the chain that possesses the highest number of accumulated computational work, as shown by the most proof-of-work (PoW) solutions, is deemed to be the legitimate chain. This mechanism facilitates the achievement of consensus and convergence among nodes.
10. Privacy and Permissioned Blockchains: The utilization of private or permissioned blockchains imposes limitations on access to only authorized participants, in contrast to public blockchains which are accessible to anybody. Privacy-enhancing mechanisms such as zero-knowledge proofs and private transactions are employed inside the blockchain ecosystem to safeguard confidential data.

To summarize, the transformative impact of blockchain technology is founded on its capacity to provide trust, transparency, and security in a decentralized digital world. Blockchain technology ensures tamper-resistant data storage, safe transactions, and the possibility for a wide variety of transformational applications by utilizing cryptographic hashes, consensus procedures, and decentralized networks. Its continued development can transform entire industries, rethink traditional paradigms of trust, and spark innovation worldwide.

2.1 *Benefits of Blockchain Technology*

The blockchain technology offers a wide variety of benefits that have the potential to revolutionize businesses and alter the way we carry out transactions, manage data, and build trust in the digital world. These benefits have the ability to revolutionize industries and transform how we conduct transactions, manage data, and establish trust. The following is a list of some of the primary benefits offered by blockchain technology:

Decentralization and security blockchain technology rely on a distributed network of nodes to perform its operations, doing away with the requirement for a single, centralized authority. This decentralization increases safety by lowering the likelihood that a single point of failure will occur or that an attack will be launched against a central institution. Because consensus techniques are used to verify transactions, it is exceedingly difficult for hostile actors to modify data.

Accountability and Transparency: Because of the decentralized nature of blockchain technology, all participants are able to examine the same data at the same time, so providing a single source of truth. This openness increases accountability because any modifications or transactions are recorded and available to all participants of the network. As a result, the likelihood of fraud or disputes is diminished as a result of this.

Immutability and Data Integrity: Once data has been added to the blockchain, it becomes virtually hard to alter the data without the consensus of the majority of network participants. This ensures that the data retains its original integrity. Because of the cryptographic ties that exist between blocks, the system is extremely difficult to being tampered with because any modifications to a single block would need the modification of all following blocks.

Traditional financial systems and intermediaries frequently include processes that take a lot of time and result in hefty costs. Modern alternatives provide significant time and money savings. As a result of removing the need for middlemen through the facilitation of direct peer-to-peer transactions, blockchain technology makes it possible to do business more quickly and at a cheaper cost, particularly when it comes to international payments and remittances.

Transactions Can Happen Rapidly: Blockchain-based transactions can happen rapidly, especially when compared to traditional banking systems, which might involve delays owing to processing times, working hours, and different time zones. Transactions Might also Take Place Across Borders: Blockchain-based transactions can also take place across borders. This speed is especially useful for business dealings that take place across international borders.

Reduced Intermediates: The use of intermediates, which can include financial institutions, payment processors, and escrow services, is essential for the completion of transactions in a variety of business sectors. The decentralized and peer-to-peer nature of blockchain technology eliminates the need for third parties, which streamlines operations and may result in cost savings.

Traceability and Supply Chain Management: Blockchain technology makes it possible to track products and things from beginning to end and enables parties in a supply chain to monitor each stage of its progression. This transparency aids in the prevention of fraud and counterfeiting and assures that the products being sold are real.

Contracts that Execute Themselves and Automation: Smart contracts are contracts that have their predetermined rules explicitly encoded into code. These contracts are also known as "self-executing contracts." When certain criteria are satisfied, they will carry out predetermined behaviors on their own. Because of this automation, there is less of a need for human intervention, which both boosts productivity and cuts down on errors.

Privacy and Ownership of Data: Blockchain technology has the potential to provide a safe environment for the administration and exchange of sensitive data. The level of control that individuals have over their own personal information can be increased, allowing them to give access to certain parties while still retaining ownership.

Applications in Multiple Industries: Because of its adaptability, blockchain technology is useful to a wide variety of industries in addition to banking. The transparency and security properties of blockchain could be beneficial to a variety of industries, including healthcare, real estate, supply chain, voting systems, and intellectual property rights, to name just a few.

Accessibility on a Global Scale: Blockchain software runs on the Internet, making it possible to connect with users located all over the world. This accessibility is especially significant in areas that have restricted access to the traditional financial infrastructure.

Innovation and Collaboration: Blockchain has helped foster innovation by giving a platform for developers to construct decentralized applications (DApps) and new business models. This has led to more collaboration between companies. It does this by enabling players to develop on top of preexisting blockchain systems, which in turn stimulates collaboration among members.

In spite of the numerous advantages it offers, it is essential to keep in mind that blockchain technology also has a number of disadvantages. These disadvantages include problems with scalability, worries about energy consumption (particularly in PoW-based blockchains), regulatory uncertainties, and the requirement for user education. Despite this, the technology is continuing to advance, which means it still has a significant capacity to transform entire industries and give people more agency.

3 Blockchain in Smart Farming

The agricultural sector is currently confronted with many issues and complexities, many of which are amenable to being efficiently solved through blockchain technology. Blockchain's one-of-a-kind characteristics can be of great use to smart farming, an approach that uses cutting-edge technology to improve the efficiency of agricultural processes [6].

Traceability is the ability to follow the transit of food products from the farm to the consumer's plate using blockchain technology. This can help to assure the safety of food as well as its ability to be traced. A farmer, for instance, might use blockchain technology to monitor the transit of their crops from the field to the supermarket using the system. Consumers would be able to trace the origins of their food and understand the production process as a result of this.

Blockchain technology has the potential to increase transparency throughout the agricultural supply chain. This can assist farmers, retailers, and customers develop a stronger sense of trust in one another. For instance, a food distributor may utilize blockchain technology to monitor the flow of food products all the way from the farm to the supermarket using RFID tags. Because of this, the distributor would be able to spot any possible difficulties in the supply chain and take the necessary corrective actions. Blockchain technology may be able to assist in making the supply chain for agricultural products more efficient. The cost of food and its availability could be reduced due to this. A farmer, for instance, may use blockchain technology to monitor the amount of water and fertilizer that is applied to their crops. Using these data could lead to improvements in irrigation and fertilization procedures [7]. Figure 2 shows the various applications of blockchain in smart farming.

Blockchain is a secure technology that can help to protect data from being fraudulently altered or altered in any other way. In the agricultural industry, where there is a great deal of sensitive data, such as crop yields and production costs, this is a crucial consideration to take. Compliance blockchain technology can assist farmers in complying with rules, such as those pertaining to the safety of food and the protection of the environment. A farmer, for instance, might use blockchain technology to monitor the use of fertilizers and pesticides to his or her crops. After then, the data could be utilized to provide evidence of conformity with the regulations. The agricultural industry may become more environmentally friendly with the use of blockchain technology. For instance, the carbon footprint of various food products might be tracked using blockchain technology. After gathering this information, decisions about the production and consumption of food can be made that are more environmentally friendly [8].

Blockchain technology can be implemented in smart farming for multiple applications. It will ease out multiple processes involved in smart farming. In upcoming sections, they are explained in detail.

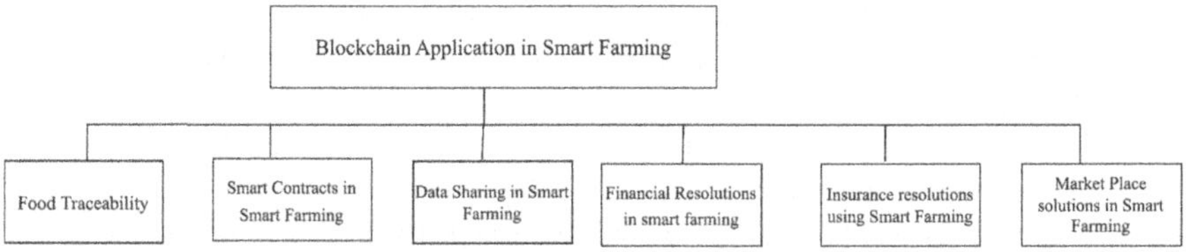

Fig. 2 Application of blockchain in smart farming

3.1 *Food Traceability*

Food traceability refers to the act of following the path of food products from their point of origin all the way through the various stages of production, processing, and distribution, whether they are being sold or consumed at the end of the chain. It is an essential component in the whole process of maintaining the safety, quality, and openness of the food supply chain. Traceability is of paramount importance in today's increasingly interconnected and globalized food market since it allows businesses to allay consumer fears over the authenticity and safety of the food they purchase. Traceability systems make it possible to quickly identify and mitigate potential hazards, such as outbreaks of foodborne illnesses or contamination, by giving complete information on the source, handling, and movement of food products. These risks include foodborne illnesses and contamination. For the purpose of establishing and preserving food traceability, cutting-edge technology such as blockchain, barcodes, radio frequency identification, and quick response (QR) codes is utilized. These technologies make it possible to collect and store data regarding the origin of materials, production procedures, transit routes, storage conditions, and a variety of other factors [9]. In particular, blockchain provides an immutable and transparent ledger that records each transaction in the supply chain. This ensures that data cannot be altered or tampered with in any way. Consumers are provided with unparalleled transparency into the origins, handling, and quality of the food that they consume as a result of the utilization of blockchain technology, which enables each step of the route that a food product takes to be securely recorded and confirmed:

(a) Farm-to-Fork Tracking: The ability to trace the movement of food products from the time they are grown or produced on the farm to the time they are served on a consumer's plate is at the core of this system. Every step of the process, from harvesting to processing to transportation to distribution, is documented as a transaction on the blockchain. These steps include harvesting, processing, transportation, and distribution.
(b) Immutable Records: The decentralized nature of blockchain technology means that once data is uploaded to the distributed ledger, it cannot be changed or tampered with in any way. This immutability provides a robust solution for maintaining the integrity of information relating to the origin and safety of food because it cannot be changed.
(c) Data Collection and Integration: The blockchain has the potential to combine a number of different technologies, such as sensors, RFID tags, and Internet of Things (IoT) devices, in order to collect real-time data regarding conditions such as temperature, humidity, and how it is handled. Following this step, the corresponding digital records of the food goods are added to the blockchain and correlated with the data.
(d) Intelligent Contracts and Automated Decision-Making: Intelligent contracts can be used to automate actions and decisions based on predetermined parameters. For instance, if the temperature inside a refrigerated vehicle that is

transporting perishable items increases above a safe threshold, the blockchain may trigger an alert to be sent to the relevant parties and initiate corrective procedures.

(e) Transparency for Customers: Customers can access the blockchain-based system by scanning product labels with their mobile devices using QR codes or other mobile applications. This allows users to read specific information about the journey the food has taken, such as the farm where it was grown, the day it was harvested, the transit route, and any certifications pertinent to the voyage [5].

(f) Food Safety and Product Recalls: In the unfortunate event of a foodborne illness outbreak or product recall, blockchain enables quick identification of batches that are implicated. Companies are able to establish the primary cause of the problem and reduce the scope of the recall if they can track the impacted product all the way back to its original source.

(g) Verification of Claims: Blockchain technology enables the verification of claims relating to organic, non-GMO, fair trade, and sustainable business practices. Customers can have faith that the information that is presented on the blockchain is true and is supported by a record that is difficult to manipulate.

(h) Efficiency of the Supply Chain: **Blockchain's ability to provide real-time visibility into the movement of items enables it to discover inefficiencies and bottlenecks in the supply chain. By having access to this information, stakeholders are given the ability to make decisions that are driven by data in order to improve efficiency and cut waste.

(i) Ethical Sourcing: Blockchain technology has the potential to support ethical sourcing by giving evidence of ecologically responsible production methods and fair labor practices [10]. Displaying their products' whole path is one way businesses demonstrate their dedication to sustainable practices.

(j) Regulatory Compliance: Blockchain technology makes it easier to adhere to standards governing the safety of food and the criteria for labeling it. Records that are easily accessible and can be audited make it easier to report on regulatory requirements and assist avoid penalties.

Consumers can gain trust in the safety, authenticity, and ethical production practices of the food they choose by using blockchain-based food traceability mechanisms, while producers and supply chain partners can operate more efficiently, transparently, and responsibly.

3.2 Smart Contracts in Smart Farming

The innovative innovation that represents blockchain technology is called a smart contract. Smart contracts enable the automation of complicated agreements and transactions in a way that is safe and resistant to tampering. Smart contracts have the potential to revolutionize relationships between farmers, suppliers, and retailers in the context of the agricultural supply chain. This would result in the simplification of processes, a reduction in costs, and an increase in the efficiency of operations. Here, we have explained about the usage of smart contracts for smart agriculture process:

1. Defining Smart Contracts: Smart contracts are agreements that can carry out their own terms and conditions and have those terms and conditions directly encoded in the code. The contract will automatically carry out the delegated tasks after the predetermined conditions have been satisfied. There will be no need for any intermediaries or human participation throughout this process.
2. Automating Agricultural Transactions: The agricultural supply chain is comprised of several stakeholders, including farmers, suppliers, distributors, and retailers. These stakeholders engage in a variety of transactions with one another. The purchase of crops, the transfer of ownership of products, the arranging of delivery times, and the processing of payments are all processes that are involved in these transactions.
3. Increased Productivity and Decreased Expenses: In the past, completing these types of transactions required a significant amount of paperwork, verification by hand, and the use of middlemen to make payments. These inefficiencies can be eliminated through the use of smart contracts, which automate the entire process. Because of this, the execution takes place more quickly, there is less of a stress placed on administrative staff, and there are cost savings for all parties concerned.
4. Automating the Payment Process: An excellent illustration of the implementation of smart contracts is the process of automating payments to farmers. The smart contract will direct an automated payment to the farmer's bank account after the conditions of the agreement have been satisfied, such as the delivery of the agreed-upon quantity of crops. This decreases the likelihood of payment disputes and eliminates the delays that are typically associated with conventional means of payment.
5. Trust and Transparency: The blockchain, which is a distributed and immutable record, is the platform on which smart contracts are executed. Because of this transparency, it is guaranteed that all parties are aware of the terms of the contract and how it is being carried out. As a consequence of this, there is a high level of trust among the participants, as they are able to independently verify the actions and the outcomes.
6. Getting Rid of the Need for Intermediaries: Smart contracts do away with the requirement that transactions be supervised and verified by intermediaries such as banks and escrow services. This not only lowers the costs of the transactions but also speeds up the process because there are no delays caused by the processing times of the intermediaries.
7. A Decrease in the Occurrence of Errors and Disputes: The use of smart contracts to automate financial transactions results in a significant reduction in the chance of human errors occurring during the manual data entry or processing of financial transactions. In addition, the fact that the execution of the contract is based on predefined rules helps to reduce the number of disagreements that can arise from misconceptions.

8. Adaptability and Versatility: The unique requirements of a transaction can be programmed into a smart contract so that it can be modified to satisfy those needs. Because of this flexibility, stakeholders are able to tailor the contract terms, payment conditions, and triggers to meet their specific requirements.
9. Capacity to Scale and Maintain Consistency: Because blockchain technology is naturally distributed, smart contracts have the capacity to expand to handle a high volume of transactions without sacrificing their performance. In addition, the execution of smart contracts is standardized across all parties, which ensures uniformity and eliminates conflicts between transactions.
10. Optimization of the Supply Chain: Smart contracts can be implemented in a variety of supply chain processes in addition to payment processing. They are able to automate the placing of orders, the management of stock, and the quality inspections and even trigger automatic restocking when stock levels fall below a given threshold.

3.3 Supply Chain Management in Smart Farming Using Blockchain

When it comes to supply chain management, blockchain technology is emerging as a disruptive force that has the ability to revolutionize the ways in which goods and materials are monitored, traced, and controlled over the entirety of the supply chain lifecycle. Firms are able to assure the safe and transparent flow of products by integrating blockchain technology into the operations that make up their supply chains. In addition, this allows the firms to improve safety, cut waste, and increase overall efficiency:

1. The Complexity of Modern Supply Chains: Modern supply chains are complicated, encompassing various stakeholders, phases, and geographical locations. This presents a challenge for supply chain management. It might be difficult to maintain quality control over the items as they progress through each of these stages. Food safety risks and excessive waste are frequently the results of problems such as fraud, counterfeiting, inefficient operations, and a lack of openness in the food industry.
2. Using Blockchain to Track Items: The decentralized and tamper-proof nature of blockchain technology makes it an ideal solution for tracing items and materials across the supply chain. Transactions can be safely recorded on the blockchain at each stage with data pertaining to the movement, handling, and conditions of the products.
3. Increased Food Safety: The use of blockchain technology in supply chain management has the ability to increase food safety, which is one of the most important benefits of this technology. For example, blockchain technology can enable real-time visibility into the temperature conditions that exist during the transit and storage of perishable food products, such as fruits and vegetables. In the case that the appropriate temperature range is exceeded, the parties responsible

can be immediately notified, which will prevent the food from going bad and reduce the likelihood of any associated health hazards.

4. Less Food Waste Blockchain: It has the potential to drastically cut down the amount of food that is wasted by making real-time monitoring and transparency possible. Products that are at an increased risk of expiration or spoilage can be discovered early, which provides stakeholders with the opportunity to take corrective action, such as rerouting or prioritizing the distribution of the affected products.
5. Transparency in the Supply Chain: Blockchain's inherent transparency makes it possible for all parties involved in the supply chain to share access to information that is both accurate and current at all times. This transparency helps stakeholders discover inefficiencies, bottlenecks, and potential sources of risk, which in turn develops trust among the stakeholders.
6. Quick Problem Detection and Resolution: Using a food distributor as an example, blockchain technology may provide real-time insights into the transit of food products from farm to store. It is possible to identify any delays, deviations, or anomalies as soon as they occur, allowing for appropriate action to be done to prevent or lessen the impact of any potential interruptions.
7. Putting a Stop to Fraud and Counterfeiting: Fake goods can present substantial dangers to customers, as well as to the reputation of a business and its potential earnings. A secure record of each product's trip may be kept using blockchain technology. This verifies that the products are authentic and have not been altered or replaced at any point along their path.
8. Improving the Productivity of the Supply Chain: The automation capabilities of blockchain can help streamline the operations involved in the supply chain. The use of smart contracts can eliminate the need for manual intervention and speed up business operations by automating payments, orders, and shipments depending on predetermined circumstances.
9. Collaborative Work and Data Sharing: Blockchain technology encourages supply chain partners to work together by making it possible for them to share data in a way that is both safe and subject to permissions. As a result of this collaborative environment, quicker decision-making is possible since all parties have access to information that is reliable and up to date.
10. Compliance with Regulatory Standards: In businesses with severe regulatory standards, such as the food and pharmaceutical sectors, blockchain technology can facilitate compliance by providing a transparent record of product movement, handling, and conditions. This is especially useful for industries like the food and pharmaceutical sectors.

The use of blockchain technology into management of supply chains presents a chance to improve the safety, transparency, and effectiveness of the movement of goods and commodities via complicated supply chains. Blockchain technology contributes to a more resilient and responsible supply chain ecosystem by ensuring safe handling, decreasing waste, eliminating fraud, and improving food safety. This ecosystem benefits businesses, consumers, and the environment, all of whom stand to gain from its implementation.

3.4 *Effective Data Sharing Using Blockchain in Smart Farming*

The agricultural sector is on the verge of undergoing a digital transformation, and blockchain technology is emerging as a potent instrument that can stimulate collaboration, transparency, and sharing of data among the various parties that make up the sector. Blockchain's decentralized and secure nature presents a viable alternative to improve information sharing between farmers, academics, government agencies, and other critical actors as the demand for environmentally friendly and productive agricultural practices grows. The following section explains the benefits of using blockchain for data sharing in smart agriculture:

1. The Current Landscape of Data Sharing in Agriculture: Agriculture is a complex ecosystem that comprises a large number of stakeholders, each of whom generates valuable data. Farmers gather data on a variety of topics, including the qualities of the soil, the yields of their crops, and the weather patterns. Researchers come up with novel solutions based on the findings of scientific research, while government bodies work to design regulations that encourage the expansion of the sector. It is necessary for these parties to effectively share data with one another in order to maximize the utilization of resources, propel innovation, and guarantee sustainable practices.
2. Using Blockchain to Improve Decision-Making: Using blockchain's decentralized ledger provides a safe and tamper-resistant platform for sharing data, which can help improve decision-making. Stakeholders are able to submit information and access it in a transparent manner, which helps to ensure the information's correctness and integrity. This pooled information can provide agricultural decision-makers across the board with the opportunity to make educated decisions that have the potential to affect production, sustainability, and profitability favorably.
3. Collaboration Between Farmers and Researchers: Farmers and researchers both play an important part in the progression of agricultural practices. Farmers are able to share data with academics about their experiences, issues, and observations in a safe and secure manner because of blockchain technology. After that, researchers can use this data to improve processes, create new crop types, and offer individualized recommendations that improve production while reducing resource waste.
4. Formulation of Government Policy: In order to formulate policies that promote sustainable agriculture and solve issues pertaining to food security, government agencies rely on data that is both reliable and up to date. Farmers are able to make their data immediately available to the government through the use of blockchain technology, which results in the creation of a complete database that is beneficial to the formulation of efficient regulations.
5. Real-Time Information for Agriculturalists: Sharing of data between farmers, made possible by blockchain technology, gives farmers access to real-time

insights that might help them make decisions. For instance, sharing weather data on a blockchain might assist farmers in making more educated decisions regarding planting, irrigation, and pest control, ultimately leading to increased crop yields.

6. Supply Chain Transparency: Blockchain's data-sharing capabilities extend to supply chain transparency, making it a useful tool for that purpose. Stakeholders are able to verify the genuineness and safety of agricultural products by exchanging information regarding the products' point of origin, how they are handled, and the quality of the items. Consumers are able to trace the origins of their food items, all the way from the farm to their tables, which promotes trust and encourages responsible consumption.
7. Validation and Reproducibility of Research: Researchers can take advantage of the immutability of blockchain technology to confirm the authenticity of their results and benefit from its reproducibility. Sharing research data on a blockchain makes it possible for peers to more easily validate the data, ensuring the reproducibility of tests and outcomes.
8. Data Ownership and Ownership Blockchain: It enables data owners, such as farmers, to maintain ownership over their information while at the same time giving them the ability to keep it private. They have the ability to determine who has access to their data and under what circumstances that access is granted. This answers concerns over data ownership and the protection of personal information.
9. The Potential for the Monetization of Data and the Provision of Incentives: Blockchain technology opens the door to the possibility of monetizing data through the use of tokens. Tokens might be given as a form of payment to farmers and other data providers as an incentive for them to share useful information, which would result in the creation of new economic models within the agricultural ecosystem.
10. International Cooperation and Innovation: Blockchain technology's decentralized and transnational nature makes it easier for countries to work together. Farmers and academics from many regions of the world are able to work together on projects, share their thoughts, and contribute collectively to addressing global agricultural concerns.

To summarize, the data-sharing capabilities of blockchain technology have the potential to revolutionize the agricultural industry into an ecosystem that is more productive, innovative, and environmentally responsible. Blockchain enables decision-makers to create positive change, optimize resource allocation, and promote responsible agricultural practices that benefit both current and future generations by fostering collaboration among farmers, academics, government agencies, and other stakeholders. This is accomplished through the use of distributed ledger technology (blockchain).

3.5 *Financial Resolutions Using Blockchain Technology in Smart Farming*

The financial options available to farmers, particularly those working on a smaller scale, could undergo a dramatic shift as a result of the introduction of blockchain technology. Farmers that have difficulty gaining access to traditional sources of credit may benefit from using blockchain technology, which enables financial institutions, microfinance organizations, and even peer-to-peer lending platforms to provide farmers with funding solutions that are both efficient and secure. This new technology has the potential to increase agricultural productivity, bolster the nation's defenses against hunger, and revitalize rural economies:

1. The Difficulty in Obtaining Financial Support for Agriculture: The acquisition of the necessary capital to engage in their agricultural endeavors is frequently a challenge for farmers operating on a smaller scale. The borrower typically has a short credit history, no collateral, and a banking infrastructure that is difficult to access, all of which act as common hurdles that prevent them from obtaining loans from conventional financial institutions. Because of the unequal availability of finance, farmers are unable to make investments in vital resources like crops, equipment, and technology that might otherwise help them increase their output.
2. Empowering Small-Scale Farmers with Blockchain Technology: Blockchain technology offers a platform that is both decentralized and transparent. This platform has the potential to revolutionize the method in which financial assistance is provided to farmers. Financial institutions are able to develop a trustworthy ecosystem that overcomes the issues associated with traditional lending because of the unchangeable and auditable nature of blockchain. This is made possible by exploiting the technology.
3. A Clear Representation of Your Credit History: Blockchain technology logs every transaction on a distributed ledger that is both safe and open. A transparent credit history for farmers can be created by keeping a record of their financial behavior and making loan repayments on time. Even for people who don't have a traditional credit history, this can be a credible reference for determining whether or not someone is creditworthy.
4. Digital Identity and Reputation: Blockchain technology has the potential to provide farmers with a protected digital identity, which will allow them to construct and cultivate their reputation within the ecosystem of financing. This identity is comprised of both personal and transactional data, which helps to cultivate a sense of trust between lenders and borrowers.
5. Smart Contracts for Automated Payments: Smart contracts on the blockchain have the ability to automate loan disbursements and repayments based on predefined criteria. This feature was introduced in version 1.0 of the Ethereum platform. This removes the need for intermediaries, lowers the cost of administrative overhead, and guarantees that repayments will be made on time.

6. Peer-to-Peer Lending: Blockchain technology has the potential to enable peer-to-peer lending platforms, which will make it possible for anyone to lend money directly to farmers without the need for intermediaries. This opens up access to funding for more people while simultaneously lowering the fees that are often connected with the more traditional lending methods.
7. Remittances and Cross-Border Transactions: Blockchain technology facilitates cross-border transactions and remittances, bridging geographical and financial divides. This is especially useful for farmers working in places with restricted access to banking services.
8. Tokenization and Agricultural Assets: The tokenization features of blockchain can be used to represent agricultural assets like cattle, crops, or land. Due to the fact that these tokens may be put up as security for loans, it is now possible for farmers to obtain finance that was previously inaccessible to them.
9. Reducing the Risk for Lenders: Because of blockchain's transparency, lenders are able to keep an eye on how their money is being spent and the development of their investments. Because of this visibility, the danger of default is decreased, which in turn boosts the trust of lenders to give credit to farmers.
10. Microfinance Transformation: Microfinance institutions may embrace blockchain to streamline operations, decrease costs, and expand their reach into underserved rural areas. Microfinance institutions may embrace blockchain to streamline operations, decrease costs, and expand their reach into underserved rural areas. This transition improves financial inclusion and contributes to the expansion of the economy in rural areas.

In summary, there is a significant possibility that blockchain may revolutionize the financing of agricultural activities. Blockchain gives small-scale farmers the ability to invest in their enterprises, so increasing their production and contributing to greater food security. This is accomplished through democratizing access to credit, lowering administrative costs, and producing credit records that are public. This technology is expected to continue maturing and gaining greater usage, and when it does so, it has the promise of radically altering the financial landscape for farmers, boosting economic empowerment, and pushing the development of sustainable agriculture.

3.6 Revolutionizing Agricultural Insurance Through Blockchain in Smart Farming

In the field of agriculture, where unexpected factors such as weather, pests, and market swings pose substantial risks, blockchain technology has the potential to revolutionize the landscape of agricultural insurance. This might be a game-changer for the industry. Farmers can be protected from the financial losses that can arise from crop failures, natural disasters, and other unanticipated events by utilizing blockchain technology, which provides solutions for insurance that are transparent, efficient, and safe. This not only protects their means of subsistence but also provides an incentive for the adoption of cutting-edge technologies and business practices:

1. Providing Crop Insurance to Farmers: Failure of crops to be harvested as a result of poor weather conditions or pests can have a devastating effect on a farmer's income. Farmers now have a safety net in the form of blockchain-based insurance, which enables them to protect their crops against the hazards that could affect them. When certain predetermined conditions, such as minimum crop output requirements, are satisfied, smart contracts have the capacity to be programmed to immediately initiate insurance payouts.
2. Transparent Risk Assessment: Insurers are able to more precisely estimate risks connected with different areas and crops because of the inherent transparency that blockchain technology provides. A blockchain can be used by an insurance firm, for instance, to record meteorological data, previous crop performance, and other important information. Insurers are able to calculate the chance of crop failure using this data, and they may then adjust the premiums for their policies appropriately.
3. Encouraging the Adoption of Risk-Reducing Best Practices: Blockchain-based insurance not only rewards farmers for losses but also encourages them to adopt risk-reducing best practices, which can help minimize the likelihood of future losses. When farmers know they have insurance coverage to rely on in the event that unforeseen difficulties arise, they are more inclined to make financial investments in agricultural technologies such as drought-resistant seeds, precision agriculture, and management of soil health.
4. Efficient Handling of Claims: Historically, the handling of claims in agricultural insurance has been known to be a time-consuming process that is also prone to disagreements. This procedure can be streamlined with the use of blockchain technology because it offers a safe and tamper-proof record of all pertinent information. This includes everything from the data on crop health and damage to the initial policy issuing. Because of this transparency, the examination of claims is completed more quickly, and fair reimbursement is guaranteed.
5. Peer-to-Peer Insurance Networks: Due to the decentralized nature of blockchain technology, it is possible to create peer-to-peer insurance networks. Farmers have the ability to combine their resources and act as a group to insure each other against potential losses. The fund pool is managed using smart contracts, which also automate the claim procedure and cut down the administrative burden.
6. Intelligent Data Collection: Internet of Things (IoT) devices, such as weather sensors and drones, are able to collect data in real time regarding the health of crops and the conditions in which they are grown. This information may be safely maintained on the blockchain, which enables insurers to obtain precise insights for risk assessment.

7. The Simplification of Regulatory Compliance: Blockchain technology has far-reaching implications for agricultural regulatory compliance, even outside the realm of insurance. Record-keeping needs to be as accurate and open as possible because of the increasingly stringent standards governing food safety, environmental protection, and the use of chemicals.
8. Creating an Immutable Ledger of Chemical Applications: Blockchain technology gives farmers the ability to record the amount of pesticides and fertilizers they use, thereby producing an immutable ledger of chemical applications. This data can be used to establish compliance with rules and offer authorities and customers concerned about product safety with clear information.
9. Tamper-Proof and Readily Auditable Records: Blockchain technology makes it possible to readily audit and verify records that are related to regulatory compliance. This removes any concerns regarding the manipulation of data or the fabrication of records, which in turn strengthens the credibility of compliance efforts.
10. Increasing Trust and Holding Stakeholders Accountable: The transparency blockchain technology provides increases trust among consumers, regulatory authorities, and other stakeholders. Accountability is improved, and the credibility of the entire agricultural supply chain is bolstered when farmers can submit data that can be independently verified in support of their claims that they comply.

To summarize, there is a tremendous amount of potential for blockchain technology to revolutionize agricultural insurance and regulatory compliance. Blockchain technology shields farmers from financial losses, promotes environmentally responsible practices, and fuels innovation all at once by providing insurance solutions that are both effective and transparent. The development of a more responsible, efficient, and secure agricultural ecosystem that is to the mutual benefit of farmers, consumers, and the environment is brought about by blockchain's ability to simultaneously ensure compliance with rules through the use of secure and immutable data.

3.7 Enhancing Marketplace Experience Through Blockchain Technology

Introducing a decentralized platform that directly connects farmers and customers through blockchain technology has the potential to restructure the market for agricultural products completely. This forward-thinking method eliminates the need for the conventional intermediaries that are normally involved, so facilitating a trade that is both open and effective in nature. Farmers can sell their products at higher prices thanks to this decentralized marketplace. At the same time, consumers benefit from having access to higher-quality produce at more affordable costs.

The decentralized marketplace does away with the need for intermediaries. It makes it possible for buyers and sellers to communicate directly with one another using blockchain technology's characteristics, such as its transparency, immutability, and security. Farmers can present their products, including information on farming practices, certifications, and origins, while consumers can make informed purchasing decisions based on transparent and validated information.

In this setting, the use of smart contracts is of critical importance. They will automate the transaction process, including payment and delivery, based on the circumstances that have been set. The use of a smart contract ensures that a transaction is carried out without any manual interventions or documentation being required once a buyer has placed an order and the requirements that have been outlined have been satisfied.

Farmers are given more agency through decentralization since it enables them to determine their own prices and interact with a larger pool of customers. This democratization of pricing enables them to make more money and decreases their reliance on middlemen, who can eat up a major percentage of a company's profit margin. Farmers also can form direct relationships with customers, which can develop trust and loyalty in the customer base.

The benefits are just as great for the end users, the consumers. They have access to fresh produce and higher-quality produce due to the items moving directly from the farm to the consumer's table, reducing the time spent in storage and shipping. In addition, consumers can make purchasing decisions that align with their preferences for organic, sustainable, or locally sourced items thanks to the fact that blockchain technology gives information that can be verified regarding the origins of products and the manufacturing methods used.

In addition, because of blockchain's inherent openness, worries about counterfeiting and incorrect labeling are rendered moot. As a result of consumers being able to verify the legitimacy of products and the promises made by farmers, the marketplace has become more open and honest.

To summarize, a decentralized marketplace that is supported by blockchain technology has the potential to transform the agricultural trading scene completely. This platform empowers farmers and consumers by facilitating direct contact between the two groups. This leads to fairer pricing, increased access to products of higher quality, and a heightened sense of trust and accountability among all parties involved. Decentralized marketplaces promise to change the agriculture business into an ecosystem that is more transparent and efficient and centered on the consumer as blockchain technology advances.

4 Challenges and Limitations of Using Blockchain in Smart Farming

The application of blockchain technology to smart farming presents several difficulties and constraints, all of which need to be thoughtfully examined to guarantee a successful implementation and adoption of the technology. Although blockchain technology has the potential to revolutionize the agricultural industry completely, its implementation will require careful planning and careful consideration of various factors before it can be successfully implemented.

4.1 Technical Challenges Involved in Blockchain for Smart Farming

Blockchain, even though it holds a great deal of potential, is fraught with a number of technical difficulties that may prove to be obstacles during implementation and use. Because it is a complex and ever-evolving technology, the adoption of this technology necessitates careful analysis of these issues and the development of strategies to mitigate them in order to assure successful integration into diverse industries, including agriculture:

1. Scalability: Blockchain networks may experience problems with scalability, particularly in public blockchains, which may have slower transaction speeds than private blockchains. It is possible that the performance of the network would deteriorate as the number of participants and transactions increases. This could result in slower confirmation times and increased costs. When dealing with large-scale agricultural supply chains that involve a big number of partners and a high volume of transactions, the importance of this difficulty cannot be overstated.
2. Energy Consumption: The consensus processes used by many blockchain networks, particularly those based on proof-of-work, result in considerable quantities of energy being used by the networks. Because of this, there is the potential for adverse effects on the environment as well as increased operational expenses, which might reduce the profitability of blockchain solutions in resource-conscious industries such as agriculture.
3. Interoperability: The environment of blockchain technology includes a variety of platforms and protocols, some of which may not be intrinsically interoperable with one another. It might be difficult to integrate already existing systems and databases with blockchain technology, which calls for standardized protocols and exhaustive testing to ensure that data can be transferred without any interruptions.
4. Data Privacy and Security: Although blockchain technology provides increased security via encryption and immutability, it can be difficult to maintain data privacy on a public ledger. Especially when dealing with sensitive information like farmer data or patented crop technologies, it is a key concern to find a balance between transparency and secrecy. Finding this balance can be a substantial challenge.
5. Compliance with Regulatory Frameworks: Utilizing blockchain technology in agriculture necessitates an awareness of the various legal and regulatory frameworks. It is necessary to handle issues such as data ownership and liability, as

well as the flow of data across international borders, in order to guarantee compliance with a variety of regional rules.

6. The User Experience: Blockchain user interfaces and applications aren't always intuitive to end users because of the technology's novelty. It is essential to create an experience that is user-friendly in order to stimulate acceptance and usage, particularly among farmers and other stakeholders who may not be familiar with technology.
7. Education and Skill Gap: A successful adoption of blockchain technology demands the expertise of trained individuals who are familiar with the complexities of the technology. The closing of the knowledge and skill gaps that exist among potential users, developers, and administrators can be a difficult and time-consuming task.
8. Network Security: Even though blockchain technology in and of itself is thought to be safe, the broader ecosystem surrounding it, which includes wallets, smart contracts, and decentralized applications, might be susceptible to malicious assaults. In order to prevent breaches and access by unauthorized parties, it is vital to ensure the security of these components.
9. Governance and Consensus: It might be difficult to set up consensus procedures and governance models that are tailored to meet the requirements of all of the stakeholders. In decentralized systems, reaching a consensus on protocol upgrades, transaction validation, and decision-making procedures can be a difficult and time-consuming task.
10. Long-Term Sustainability: Continual development, regular updates, and active participation from the community are required in order to ensure the long-term viability and continued relevance of blockchain networks. In order to prevent blockchain initiatives from becoming irrelevant, it is vital to find a solution to the problem of maintaining a vibrant ecosystem.

In conclusion, although the blockchain technology has enormous potential for revolutionizing industries such as agriculture, it is vital to recognize and overcome the technical problems that can emerge during the process of putting it into practice. These issues can be reduced by investing in research, innovation, and collaboration, which will pave the way for an agricultural ecosystem that is more efficient, transparent, and resilient.

4.2 Regulatory Challenges Involved in Implementing Blockchain in Smart Farming

The regulatory environment for blockchain technology in the agricultural industry provides a dynamic and constantly changing challenge. As blockchain becomes increasingly recognized as a potentially game-changing technology in the agricultural sector, regulatory agencies all over the world are struggling with the necessity of establishing clear frameworks that address the applications of blockchain technology and the repercussions it may have.

The distributed and global nature of blockchain technology presents regulatory problems that span a wide range of areas. These areas include consumer protection, intellectual property rights, data privacy, and liability concerns. The fact that the technology has the ability to disrupt conventional business models, supply chains, and the relationships between various stakeholders is the source of the complexity.

Data ownership and privacy are one of the most important considerations. Because of the immutability and transparency of blockchain technology, certain data protection standards are incompatible with it. This raises problems regarding who has control over the data that is stored on the blockchain and how it ought to be maintained in order to comply with privacy laws.

In addition, there are concerns regarding liability and the use of smart contracts that come into play. Within the context of the currently applicable legal frameworks, determining culpability can be difficult in the event that an automated smart contract carries out an action that results in unintended effects.

As the number of applications that make use of blockchain technology grows in the agricultural industry, regulatory agencies have been tasked with the responsibility of either revising current regulations or drafting whole new ones in order to strike a balance between encouraging innovation and protecting the rights, safety, and privacy of all parties involved. To navigate these regulatory hurdles and develop a framework that maximizes the benefits of blockchain technology while addressing potential dangers and uncertainties, technological innovators, legislators, legal experts, and industry stakeholders need to collaborate on their efforts.

4.3 Adoption Challenges of Blockchain Technology

The application of blockchain technology in the agricultural industry faces a number of hurdles as a result of the innovation's status as a relatively recent development that is still in the process of maturing. While it is clear that blockchain technology could have beneficial applications, the widespread adoption of this technology in agricultural practices presents challenges that must be overcome for the shift to be successful:

1. A Lack of Awareness and Understanding: Many stakeholders within the agricultural sector, such as farmers, cooperatives, and participants in the supply chain, may have a limited awareness of the possibilities of blockchain and how it might address industry difficulties. This is a problem because blockchain has the potential to address many of these challenges. Education and awareness initiatives are absolutely necessary in order to close this knowledge gap and encourage the informed adoption of new technologies.
2. Complexity and Requirement for Technical Competence: In order to effectively develop, implement, and administer blockchain technology, one must have a

particular level of technical competence due to the system's deep principles and intricacies. Farmers and other participants might not have the necessary technical abilities to navigate blockchain interfaces and comprehend the possibilities offered by the technology.

3. Obstacles Presented by Integration: Blockchain technology can be difficult to incorporate into preexisting agricultural systems, procedures, and databases. It is necessary to engage in careful planning and extensive customization in order to achieve seamless compatibility and data exchange between blockchain and legacy systems. This can be a time-consuming process that demands a significant investment of resources.
4. Expenses and the Distribution of Resources: The implementation of blockchain solutions requires a financial investment, both in terms of the development of technology and the allocation of training resources. These expenses can be a barrier to entry for small-scale farmers and resource-constrained organizations, restricting their ability to embrace blockchain-based practices and making it more difficult for them to use the technology.
5. Interoperability: The landscape of blockchain technology sometimes lacks standardized protocols and platforms, which can make it more difficult for different types of systems to communicate with one another. The obstacle that needs to be surmounted is making it so that blockchain networks can effortlessly communicate with one another as well as with technologies that are already in use.
6. Resistance to Change: The adoption of blockchain technology frequently necessitates the modification of preexisting procedures and workflows. Despite the potential advantages of blockchain solutions, their adoption could be hampered by resistance to change on the part of stakeholders who are accustomed to more conventional approaches.
7. Anxiety Regarding Regulation: The ever-changing regulatory landscape for the use of blockchain technology in agriculture has the potential to inspire anxiety among stakeholders. Dealing with complex regulations that are always being updated can be difficult and could potentially slow down adoption attempts.
8. Provide a Proof of Value: It is essential to provide evidence of the real benefits that may be gained by using blockchain technology. There is a possibility that stakeholders will be hesitant to commit to adopting the technology if there are not clear examples of successful deployments and demonstrable returns on investment.
9. Effects of the Network: Blockchain's value typically rises in tandem with the expansion of the number of participants using the network. Beginning the process of adoption can be difficult if there is not currently a critical mass of stakeholders utilizing the technology.
10. Industry Partnership: In order to achieve widespread use of blockchain technology, a partnership between a variety of stakeholders is required. These stakeholders include farmers, industry associations, technology suppliers, and legislators. The process of coordinating efforts and aligning interests might be difficult, but it is vital in order to create an atmosphere that is favorable to adoption.

In conclusion, despite the fact that blockchain technology has the potential to radically alter the agricultural industry, there are still a number of obstacles that must be cleared before it can be successfully implemented. Stakeholders can pave the way for the adoption of blockchain technology that promotes efficiency, transparency, and sustainability across the agricultural value chain by addressing these problems through education, simplified interfaces, collaboration, and unambiguous demonstrations of value. These challenges can be addressed by education, simplification of interfaces, and collaboration.

In spite of these obstacles, the use of blockchain technology in smart farming could potentially result in major benefits. There is a significant potential for blockchain technology to bring about a revolution in the agricultural industry by making it more equal, sustainable, and efficient.

5 Future Scope of Blockchain in Smart Farming

The application of blockchain technology in the future of smart farming offers enormous promise and presents a journey that can completely reshape the agricultural environment. This will bring about a revolution. Blockchain is poised to play a major role in transforming the way agriculture is practiced, managed, and experienced as a result of the difficulties that are being addressed and innovations that are being accepted:

1. More Efficient and Open Supply Chains: The capacity of blockchain technology to trace and verify each step of the supply chain for agricultural products will result in operations that are both more efficient and open to public scrutiny. Consumers will have access to information that can be relied upon on the origin, quality, and journey of the food products they purchase, which will develop both trust and responsible consumption.
2. The Revolution of Precision Agriculture: Data sharing enabled by blockchain technology will provide farmers with real-time information that will empower them to engage in precision agriculture. The monitoring of the soil's health and the optimization of irrigation are two of the intelligent farming practices that will be improved, which will lead to better yields, increased resource efficiency, and a reduced impact on the environment.
3. Decentralized Markets and Fair Compensation: Decentralized markets that are enabled by blockchain technology will do away with the need for intermediaries, making it possible for farmers to interact directly with consumers. As a result of this democratization of trade, farmers will receive more just recompense, and consumers will have better access to affordable goods.
4. Environmentally Good Practices and Traceability: Blockchain technology will verify and incentivize environmentally good practices as sustainability becomes an essential component of the agricultural industry. It will be clear to customers that the production of their food adheres to moral standards, which will

encourage a food supply chain that is more accountable and environmentally conscientious.

5. Financial Inclusion and Empowerment: Blockchain's financial services will bring previously underserved farmers into the financial fold, enabling access to loans, insurance, and other necessary resources. Because of this empowerment, there will be a rise in productivity as well as an improvement in people's standard of living.
6. Collaborative Efforts and Innovation: The open and safe dissemination of data made possible by blockchain technology will encourage cooperative efforts among farmers, academics, policymakers, and consumers. This approach to collaboration will speed up the process of developing and implementing creative solutions to solve the difficulties faced by the agricultural sector.
7. Regulatory Evolution: The regulatory environment will adapt to accept blockchain in agriculture. The regulatory environment will adapt to accept the one-of-a-kind qualities of blockchain in agriculture. The potential of blockchain technology will eventually be fully realized without jeopardizing its ability to comply with applicable laws as rules continue to evolve to protect data privacy, consumers, and fair trade.
8. Continuous Evolution and Adaptation: The path of blockchain technology in smart farming will be marked by continuous evolution and adaptation. This is the eighth and final point on the path of blockchain technology in smart farming. The development of the technology will lead to the discovery of new applications and solutions, which will further improve the efficacy, sustainability, and inclusiveness of agricultural practices.

6 Conclusion

In conclusion, the implementation of blockchain technology into smart farming has the potential to usher in a new era within the agricultural industry that is characterized by increased efficiency, transparency, and sustainability. Blockchain technology presents a number of advantages, one of which is the provision of a decentralized and secure platform for the management of data, transactions, and collaboration. These advantages have the potential to radically alter the way farming is practiced, managed, and experienced. There is a lot of potential upside to utilizing blockchain technology in agricultural settings. Transparency in the supply chain can be improved as a result, giving customers more information with which to make decisions about the items they buy. The ability of blockchain to trace and verify each stage of the supply chain helps to assure the genuineness and quality of agricultural products, which in turn contributes to an increase in the overall level of food safety. In addition, because blockchain technology makes it possible for farmers and consumers to communicate directly with one another, it possesses the potential to give rise to more equitable remuneration arrangements and do away with the need for middlemen, so promoting greater justice within the sector. This technology also has the potential to empower small-scale farmers by increasing their access to funding, streamlining their transactions, and allowing them to participate in global markets. The implementation of blockchain technology in smart farming, however, is not without its share of obstacles and restrictions. Complexities on the technical front, problems with scalability, unpredictability in regulatory matters, and the requirement for education and awareness all provide obstacles that must be surmounted for successful implementation. In order to successfully integrate blockchain technology into already existing agricultural systems, careful planning and adaptation are required, and it is vital to address issues over data protection and ensure compliance with ever-evolving legislation. When looking into the future, the use of blockchain technology to smart farming looks to have a bright future. The agricultural industry is positioned to become more efficient, transparent, and robust as the technology continues to develop and more people begin to adopt it. The practices of precision agriculture will undergo a revolution, which will ensure the most efficient use of resources and the preservation of the environment. Both farmers and consumers will have more power thanks to decentralized marketplaces, and the exchange of data based on blockchain technology will encourage innovation and collaboration among stakeholders. To summarize, there is significant potential for blockchain technology to bring about a revolution in smart farming. Even though obstacles now exist, overcoming them and embracing the revolutionary potential of blockchain technology could result in a future agricultural system that is more effective, egalitarian, and environmentally friendly. The goal of a blockchain-powered smart agricultural ecosystem that is to the advantage of farmers, consumers, and the planet is getting closer to becoming a reality as stakeholders from across the industry work together to find solutions to the difficulties that they face.

References

1. N.M. Noor, N.A.M. Razali, S.N.S.A. Sham, K.K. Ishak, M. Wook, N.A. Hasbullah, Decentralised access control framework using blockchain: Smart farming case. Int. J. Adv. Comput. Sci. Appl. **14**(5) (2023). https://doi.org/10.14569/IJACSA.2023.0140560
2. D. Kumar, R.K. Dwivedi, Blockchain and IoT based smart agriculture and food supply chain system, in *2023 International Conference on Intelligent and Innovative Technologies in Computing, Electrical and Electronics (IITCEE)*, (IEEE, 2023), pp. 755–761
3. E.M. Quafiq, A. Chehri, R. Saadane, Advanced analytics for smart farming in a big data architecture secured by blockchain and pBFT, in *International KES Conference on Human Centred Intelligent Systems*, (Springer Nature Singapore, Singapore, 2023), pp. 13–23
4. G.S. Sajja, K.P. Rane, K. Phasinam, T. Kassanuk, E. Okoronkwo, P. Prabhu, Towards applicability of blockchain in agriculture sector. Mater. Today: Proc. **80**, 3705–3708 (2023)
5. M.N. Ahmad, Emerging crop traceability systems in smart farming: A review, in *Proceedings of the International Conference on Computer, Information Technology and Intelligent Computing (CITIC 2022)*, vol. 10, (Springer Nature, 2023), p. 182
6. T. Kushartadi, A.E. Mulyono, A.H. Al Hamdi, M.A. Rizki, M.A. Sadat Faidar, W.D. Harsanto, et al., Theme mapping and bibliometric analysis of two decades of smart farming. Information **14**(7), 396 (2023)

7. S. Alam, Security concerns in smart agriculture and blockchain-based solution, in *2022 OPJU International Technology Conference on Emerging Technologies for Sustainable Development (OTCON)*, (IEEE, 2023), pp. 1–6
8. M.R. Anwar, F. Dewanta, F. Fardan, Implementasi Blockchain Pada Sistem smart farming Berbasis internet of things. eProc. Eng. **9**(6), 2994–3005 (2023)
9. S. Sarowa, V. Kumar, B. Bhanot, M. Kumar, Enhancement of security posture in smart farming: Challenges and proposed solution, in *2023 International Conference on Device Intelligence, Computing and Communication Technologies, (DICCT)*, (IEEE, 2023), pp. 1–5
10. Y. Arkeman, N.J. Hidayah, A. Suharso, F. Adhzima, T. Kusuma, Implementation of artificial intelligence and Blockchain in agricultural supply chain management. Int. Soc. Southeast Asian Agric. Sci., **29(1), 135 (2023)**

5G Technology in Smart Farming and Its Applications

S. R. Raja, B. Subashini, and R. Selwin Prabu

1 Introduction

Agriculture, the cornerstone of human sustenance, has witnessed transformative revolutions throughout history, each driven by technological advancements. From the British Agricultural Revolution to the Green Revolution, these shifts have enabled farmers to feed burgeoning populations and increase crop yields. However, in the face of growing global food demand, resource constraints, climate change, and labor shortages, a new era of agriculture has emerged—smart farming or precision agriculture. This paradigm shift leverages modern technologies, including the Internet of Things (IoT), sensors, and automation, to optimize agricultural practices.

The proliferation of smart farming practices has given rise to a new agricultural landscape where data is the currency of productivity. In this digital era, data-driven decision-making has become paramount, and agriculture's reliance on technology has never been more pronounced. At the heart of this agricultural renaissance lies the fifth-generation (5G) wireless network, a catalyst for revolutionizing the industry. 5G technology, celebrated for its high-speed connectivity, low latency, and extensive coverage, has seamlessly integrated with the IoT, creating an ecosystem where every aspect of farming can be monitored, analyzed, and optimized in real time. This research paper embarks on a journey through the intersection of 5G technology and smart farming, exploring the multifaceted impacts and implications of this convergence.

S. R. Raja (✉)
Saveetha College of Liberal Arts and Science, Chennai, India
e-mail: rajasr.sclas@saveetha.com

B. Subashini
Thiagarajar College, Madurai, India

R. S. Prabu
CSE, K.P.R. Institute of Engineering and Technology, Coimbatore, India

S. Balasubaramanian et al. (eds.), *Intelligent Robots and Drones for Precision Agriculture*, Signals and Communication Technology,
https://doi.org/10.1007/978-3-031-51195-0_12

The symbiotic relationship between 5G and smart farming is poised to redefine agriculture as we know it. 5G technology, characterized by its blazing speed and near-instantaneous data transfer capabilities, ushers in an era of precision and efficiency previously unattainable in agriculture. This paper dives deep into the specific contributions of 5G technology, emphasizing its role in bridging the digital divide in rural farming communities. Real-world case studies and examples illustrate the practical applications of 5G in agriculture, from real-time monitoring of crops and livestock to autonomous farm machinery. As the agriculture sector grapples with multifaceted challenges, ranging from resource scarcity to the need for sustainable practices, 5G emerges as a beacon of hope—a technological enabler that promises to enhance productivity, reduce waste, and promote sustainable farming practices.

In the following sections, we explore the pivotal role of IoT in smart farming, dissecting its sensor-based data collection mechanisms and data-driven decision-making capabilities. Various applications of IoT in agriculture are highlighted to underscore its transformative potential. Moreover, we delve into the impact of 5G on rural areas, where improved connectivity is not merely a technological upgrade but a socioeconomic lifeline. Through case studies of 5G deployment in rural farming communities, we unveil the economic and social implications of enhanced connectivity. As we navigate the dynamic intersection of 5G technology and smart farming, this research paper aims to shed light on the transformative potential of this alliance, offering insights and recommendations for policymakers, stakeholders, and farmers alike.

2 Literature Survey

Modern farming practices have evolved to a critical point, which is the integration of fifth-generation (5G) wireless technology into the agricultural industry. Its ramifications have been thoroughly studied and discussed in academic circles, illuminating the complex effects and possible advantages of this convergence.

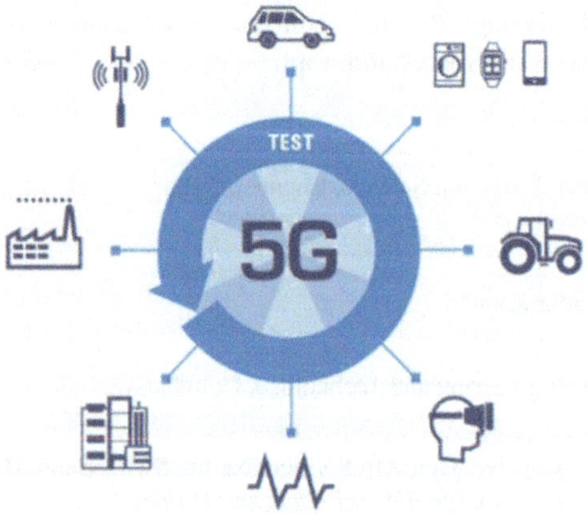

2.1 5G Technology and Smart Farming

Numerous studies have emphasized the transformative potential of 5G technology in the context of smart farming or precision agriculture. The research by Li et al. (2021) elucidates how 5G's high-speed connectivity and low latency facilitate real-time data collection and decision-making in farming operations. Similarly, Zhang et al. (2020) delve into the role of 5G in enabling autonomous agricultural machinery and its impact on labor efficiency.

2.2 IoT and Agricultural Data

There has been much research done on the critical role that the Internet of Things (IoT) plays in smart farming. IoT sensors provide data-driven decision-making in agriculture, optimizing crop management and resource utilization, as demonstrated by research by Khan et al. (2020). Furthermore, the Magomadov (2019) study emphasizes the importance of IoT in detecting plant diseases early and minimizing agricultural losses.

2.3 Digital Divide and Rural Connectivity

Bridging the digital divide in rural farming communities through 5G has garnered scholarly attention. A study by Tong et al. (2019) analyzes the economic and social implications of 5G deployment in rural areas, highlighting its role in enhancing the quality of life for farmers. Furthermore, Zhao et al. (2016) explore the challenges of rural connectivity in the context of 4G and the promises offered by 5G technology.

2.4 Sustainability and Resource Management

Sustainable farming practices are central to addressing global challenges, including resource scarcity and climate change. Research by Bekele and Drake (2003) discusses the adoption of precision agriculture techniques, enabled by IoT and 5G, to optimize resource utilization and reduce environmental impact. Additionally, Steenwerth et al. (2014) examine how automation and robotics in agriculture, driven by 5G technology, enhance resource efficiency.

2.5 *Economic Implications and Future Prospects*

The economic ramifications of 5G in agriculture have been a focal point of inquiry. The report by Smart Agriculture Market (2020) estimates substantial market growth, emphasizing the increasing adoption of smart farming practices. Furthermore, telecommunications experts (GSMA, 2018) project diverse use cases for 5G technology, with agriculture emerging as a sector poised for significant benefits.

3 The Role of 5G in Smart Farming

Smart farming, also known as precision agriculture, has emerged as a beacon of hope for the agricultural sector, promising increased productivity, sustainability, and resource efficiency. At the heart of this agricultural revolution lies the fifth-generation (5G) wireless technology, a technological marvel celebrated for its high-speed connectivity, low latency, and extensive coverage. The convergence of 5G and smart farming is reshaping the landscape of agriculture in unprecedented ways, ushering in a new era of precision and efficiency.

3.1 *Unleashing Real-Time Monitoring*

Enabling real-time asset monitoring in agriculture is one of the primary contributions of 5G to smart farming. Farmers may get instantaneous data on critical elements like soil moisture, temperature, and crop health by deploying 5G-connected sensors and devices around the farm. With the use of this real-time data, farmers are able to optimize pest management, fertilization, and irrigation while acting quickly and decisively. Drones with high-resolution cameras can survey large areas and transmit data instantly, making crop analysis and early pest identification easier. With older wireless technology, this kind of real-time monitoring was unthinkable.

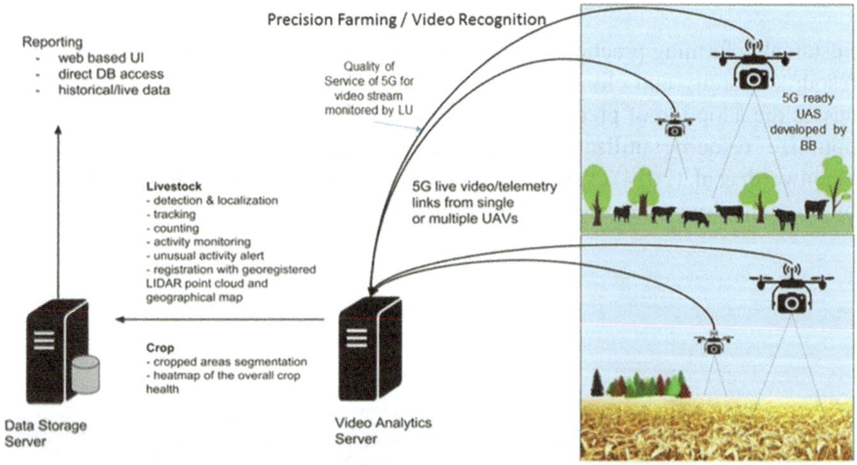

3.2 *Autonomous Farm Machinery*

5G technology is the linchpin in the deployment of autonomous farm machinery. Connected by 5G networks, tractors, combines, and other agricultural equipment can operate with unprecedented precision and efficiency. Autosteering features, for instance, enable these machines to navigate fields autonomously, reducing overlap and minimizing resource wastage. As Kevin Butt, an agriculture professor, notes, autosteering not only enhances efficiency but also contributes to farmers' mental well-being by reducing the cognitive load associated with manual operation. Additionally, these autonomous machines can be remotely controlled through 5G networks, allowing farmers to manage operations from their tablets or smartphones.

3.3 *Optimizing Resource Utilization*

Sustainable farming practices have become imperative in the face of environmental challenges. 5G technology plays a pivotal role in optimizing resource utilization in agriculture. Through real-time data analytics and IoT sensors, farmers can precisely tailor their resource allocation. For instance, instead of uniformly applying fertilizers across vast fields, farmers can use soil data to identify specific areas requiring nutrients, minimizing waste and environmental impact. Water resources are also managed judiciously, with 5G-enabled sensors gauging soil moisture levels to determine optimal irrigation schedules. This resource optimization is a cornerstone of sustainable and eco-friendly farming practices.

3.4 *Fostering Connectivity in Rural Areas*

The impact of 5G in agriculture extends beyond the farm gate. It addresses a critical challenge in rural areas—connectivity. Rural communities often face limited access to high-speed Internet, hampering not only farming operations but also quality of life. 5G technology promises to bridge the digital divide, offering robust connectivity even in remote farming regions. This connectivity doesn't just facilitate data transmission; it opens the doors to telemedicine, online education, and e-commerce, enhancing the overall well-being of rural communities.

4 Challenges in Agriculture

While the integration of fifth-generation (5G) technology in agriculture promises substantial benefits, it also presents several challenges that must be addressed to harness its full potential. These challenges encompass technical, economic, and social aspects, shaping the landscape of smart farming in profound ways.

4.1 High Infrastructure Costs

One of the foremost challenges in adopting 5G technology in agriculture is the high infrastructure costs. Establishing a robust 5G network infrastructure in rural areas, where many farms are located, demands significant investment. The deployment of 5G cell towers and network equipment is capital-intensive, and in many cases, it may not be economically viable for smaller farms. This issue raises concerns about equitable access to 5G benefits, as large commercial farms may have a competitive advantage in implementing this technology.

4.2 Data Security and Privacy

As smart farming relies heavily on data collection and sharing, data security and privacy become paramount concerns. The vast amounts of sensitive data generated by IoT sensors, drones, and autonomous machinery are susceptible to cyberattacks and breaches. Farmers need assurance that their data, including crop yields, soil conditions, and equipment performance, remains secure and private. Robust cybersecurity measures and data encryption are essential to protect against potential threats.

4.3 Network Dependability in Remote Areas

Agriculture often takes place in remote and rural regions with limited connectivity infrastructure. Ensuring reliable 5G network coverage in these areas is a significant challenge. Dead zones or areas with weak signals can disrupt critical operations that rely on real-time data transmission, such as autonomous machinery control and remote monitoring. Extending dependable 5G coverage to these remote agricultural landscapes is essential to maximize the benefits of smart farming.

4.4 Compatibility and Integration

Another challenge lies in ensuring the compatibility and seamless integration of 5G technology with existing agricultural systems. Many farms already employ various technologies, such as GPS-guided tractors and IoT sensors. Ensuring that these systems can operate harmoniously with 5G-connected devices and machinery requires careful planning and technical expertise. Compatibility issues could lead to inefficiencies and hinder the adoption of 5G in agriculture.

4.5 Skills and Training

The successful implementation of 5G technology in agriculture necessitates a skilled workforce capable of managing and troubleshooting advanced equipment and systems. Farmers and agricultural workers may require training to effectively utilize 5G-connected devices, interpret real-time data, and troubleshoot technical issues. Bridging the digital skills gap in rural areas is essential to ensure that all stakeholders can reap the benefits of smart farming.

4.6 Regulatory and Policy Frameworks

The regulatory and policy frameworks governing 5G in agriculture must evolve to address new challenges. Issues related to spectrum allocation, data ownership, and liability in the event of system failures or accidents require careful consideration. Governments and regulatory bodies must collaborate with the agricultural industry to create a conducive environment for 5G adoption while safeguarding the interests of all stakeholders.

5 Benefits of Smart Farming with 5G

The convergence of fifth-generation (5G) technology and smart farming presents a plethora of benefits that are poised to revolutionize agriculture. These advantages extend across various facets of farming, from resource management to environmental sustainability, ushering in an era of unprecedented efficiency and productivity.

5.1 High Data Transfer Capacity and Low Latency

Perhaps the most prominent benefit of 5G in smart farming is its high data transfer capacity and low latency. The ultra-fast data transmission allows for real-time communication and data exchange between agricultural devices, sensors, and machinery. This real-time connectivity enables farmers to make prompt and informed decisions, optimizing resource allocation and minimizing waste. Whether it's monitoring soil conditions or controlling autonomous tractors, the low latency of 5G ensures that actions are executed swiftly and accurately.

5.2 *Extensive Connectivity*

5G networks boast an exceptional connection density, allowing a vast number of devices to be interconnected simultaneously. This extensive connectivity enables the deployment of an intricate network of sensors, drones, and autonomous machinery throughout the farm. As a result, farmers can collect comprehensive data on every aspect of their operations, leading to a holistic understanding of crop health, soil conditions, and equipment performance.

5.3 *Spectral Efficiency Improvement*

5G technology is characterized by spectral efficiency, which means that it can transmit more data over the same frequency spectrum compared to its predecessors. This efficiency is particularly advantageous in agriculture, where data-intensive applications like high-resolution imaging and remote monitoring are essential. Spectral efficiency ensures that data transmission remains reliable even in congested agricultural environments.

5.4 *Smooth Communication Performance*

In smart farming, seamless communication is critical for the effective coordination of various operations. 5G's consistent and reliable communication performance ensures that data flows smoothly across the agricultural ecosystem. This reliability is essential for tasks such as coordinating autonomous machinery, tracking livestock, and managing irrigation systems.

5.5 *Resource Optimization*

Smart farming with 5G facilitates resource optimization on multiple fronts. With real-time data from IoT sensors, farmers can tailor their irrigation schedules, ensuring that water is applied precisely where and when it is needed. Similarly, data analytics help in determining the optimal distribution of fertilizers and pesticides, reducing overuse and environmental impact. The result is increased crop yields, reduced input costs, and enhanced sustainability.

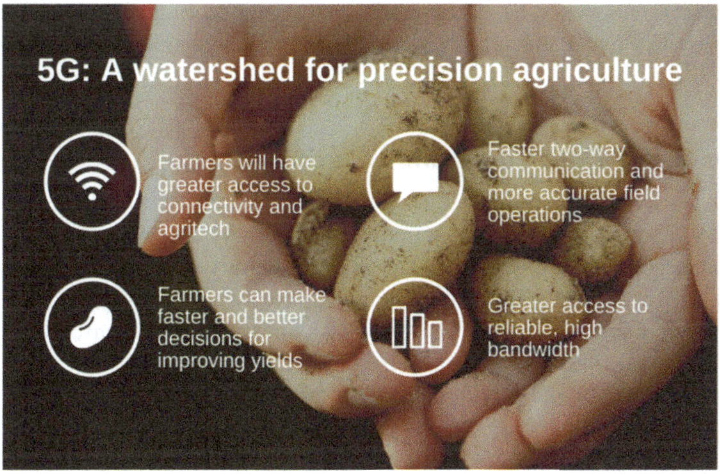

5.6 *Environmental Impact Reduction*

Globally, sustainable agriculture is essential, and smart farming enabled by 5G technology makes a major contribution to lessening the environmental effect of farming operations. 5G-enabled farms function more environmentally friendly by applying precise procedures, minimizing waste, and optimizing resource utilization. Some advantages of smart farming for the environment are less water and chemical use, as well as less energy usage.

5.7 *Promoting Sustainable Practices*

The integration of 5G technology fosters the adoption of sustainable farming practices. These practices include precision agriculture, which leverages data-driven insights to make informed decisions, and conservation tillage, which minimizes soil disturbance. Additionally, 5G facilitates the implementation of organic farming methods and the use of environmentally friendly pest control measures.

6 IoT in Smart Farming

The Internet of Things (IoT) has emerged as a transformative force in the realm of smart farming, offering unprecedented opportunities for data-driven decision-making, automation, and precision agriculture. IoT technology, integrated with fifth-generation (5G) networks, has the potential to revolutionize farming practices and optimize resource management in ways previously unimaginable.

6.1 Sensor-Based Data Collection

At the heart of IoT in smart farming are a multitude of sensors deployed across the agricultural landscape. These sensors are designed to monitor various parameters critical to farming, including soil moisture, temperature, humidity, light intensity, and even the health of crops and livestock. With 5G connectivity, these sensors continuously transmit real-time data to centralized systems, providing farmers with comprehensive insights into the conditions of their farms.

6.2 Precision Agriculture

IoT sensors enable precision agriculture, a key component of smart farming. By collecting data on soil quality, nutrient levels, and weather conditions, farmers can precisely tailor their farming practices. For example, irrigation systems can be automated based on soil moisture data, ensuring that crops receive the exact amount of water required. This not only conserves water but also enhances crop yields.

6.3 Crop Monitoring and Management

IoT-enabled devices extend beyond sensors to include drones and smart cameras. Drones equipped with imaging technology can fly over fields, capturing high-resolution images. These images can be analyzed using artificial intelligence (AI) algorithms to detect early signs of diseases, pests, or nutrient deficiencies in crops. With 5G connectivity, these images can be transmitted and analyzed in real time, enabling prompt intervention and minimizing crop losses.

6.4 Livestock Monitoring

IoT is essential to livestock management since it allows for real-time behavior and health monitoring of animals. Livestock can have wearable sensors and GPS trackers connected to them to monitor their whereabouts, identify sickness, and establish the best feeding regimens. Farmers may receive instant notifications in the case of abnormalities, such a cow wandering from the herd, enabling prompt action.

6.5 Supply Chain Optimization

IoT technology extends beyond the farm gate, influencing the entire agricultural supply chain. Sensors can be placed in storage facilities to monitor temperature and humidity, ensuring the quality and safety of stored crops. Additionally, IoT-enabled tracking systems can provide end-to-end visibility into the transportation of agricultural products, reducing spoilage and food waste.

6.6 Data Analytics and Decision Support

The sheer volume of data generated by IoT devices in smart farming necessitates robust data analytics. Advanced analytics and machine learning algorithms process this data to generate actionable insights. Farmers can receive recommendations on planting times, optimal harvesting periods, and the precise application of fertilizers and pesticides. This data-driven decision support empowers farmers to maximize productivity while minimizing inputs.

6.7 Energy Efficiency

On farms, IoT technology also helps with energy efficiency. Farmers can monitor and regulate their energy use with the aid of smart grids and IoT-enabled energy management systems. For instance, automated lighting systems in greenhouses can lower power costs by adjusting intensity based on real-time light data.

7 5G's Impact on Rural Areas

The rollout of fifth-generation (5G) wireless technology is often associated with urban areas, promising high-speed connectivity and transformative experiences in cities. However, the impact of 5G extends far beyond urban centers and holds significant potential for rural areas. In rural communities, 5G technology can bridge the digital divide, enhance economic opportunities, and transform essential services such as healthcare and education.

7.1 Improved Connectivity

One of the primary challenges in rural areas has been limited access to high-speed Internet connectivity. 5G technology offers a solution by providing fast and reliable Internet access to rural communities. This connectivity is essential for enabling businesses to thrive, supporting distance learning, and ensuring that rural residents have access to essential online services.

7.2 Rural Economic Development

5G has the potential to spur economic growth in rural areas. When rural businesses have access to high-speed Internet, they can compete globally. Rural craftsmen may reach a wider audience with their products, and small agricultural businesses can access Internet marketplaces. Furthermore, precision agriculture provided by 5G can increase agricultural production, supporting the expansion of the rural sector.

7.3 Agricultural Advancements

The influence of 5G on agriculture is particularly noticeable in rural areas. Thanks to 5G technology, farmers can monitor and manage their crops with never-before-seen accuracy thanks to smart farming. Drones, autonomous vehicles, and Internet of Things sensors all provide real-time data that farmers may use to make informed decisions regarding crop health, fertilization, and irrigation. Increased yields, less resource waste, and sustainable farming practices result from this.

7.4 Telemedicine and Healthcare

Rural healthcare faces challenges related to access and resources. 5G facilitates telemedicine by providing high-quality video conferencing and remote monitoring capabilities. Patients in remote areas can access consultations with specialists without traveling long distances, improving healthcare outcomes and reducing costs. Additionally, IoT-enabled medical devices can transmit patient data in real time, enabling faster responses in emergencies.

7.5 Distance Learning

Rural schools stand to gain a great deal from 5G connection. Students in rural areas may interact with classmates globally, engage in virtual classes, and access online educational materials thanks to high-speed Internet. This improves learning chances and closes the achievement gap between kids in rural and urban areas.

7.6 Smart Infrastructure

5G enables the development of smart infrastructure in rural areas. Smart grids can enhance energy management and reduce costs, while IoT sensors can monitor water quality and environmental conditions. This leads to more efficient resource utilization and greater sustainability.

7.7 Emergency Services

Rural areas often face challenges in emergency response due to limited communication infrastructure. 5G networks provide first responders with high-speed, reliable communication, enabling quicker response times and better coordination during emergencies.

7.8 Tourism and Cultural Preservation

Rural areas often possess unique cultural and natural attractions. 5G connectivity can boost tourism by providing visitors with immersive experiences through augmented and virtual reality (AR/VR). Additionally, high-quality live streaming and virtual tours can promote cultural preservation efforts and generate tourism revenue.

7.9 Entrepreneurship and Innovation

5G connectivity empowers rural entrepreneurs and innovators. It opens doors to remote work opportunities, encourages tech startups, and supports e-commerce ventures. Rural communities can foster innovation and entrepreneurship, contributing to economic diversification.

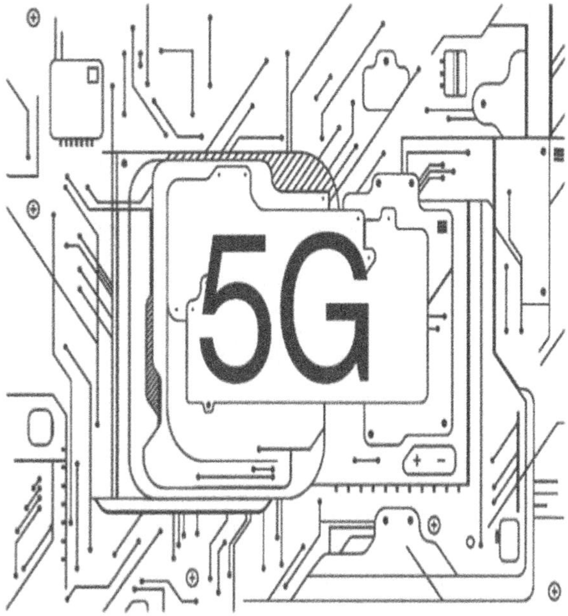

8 Future Prospects and Challenges of 5G in Agriculture

As fifth-generation (5G) technology continues to unfold its transformative capabilities in agriculture, several exciting prospects and associated challenges shape the future landscape of smart farming.

8.1 Prospects

AI-Driven Agriculture The integration of 5G with artificial intelligence (AI) will drive autonomous decision-making on the farm. AI algorithms will analyze vast datasets generated by IoT sensors and drones, providing real-time insights into crop health, weather patterns, and optimal planting times.

Edge Computing Edge computing, coupled with 5G, will enable data processing closer to the source, reducing latency and enhancing the responsiveness of autonomous agricultural machinery. This will pave the way for real-time, localized decision-making.

Augmented Reality (AR) and Virtual Reality (VR) AR and VR applications in agriculture will become more prevalent with 5G. Farmers can receive training, access repair manuals, and visualize crop data in three dimensions, improving efficiency and reducing errors.

Remote Monitoring 5G networks will allow farmers to remotely monitor and manage their farms from anywhere. This capability will not only save time but also reduce the need for physical labor, making farming more accessible to an aging population.

Global Connectivity Farmers in remote areas can access global markets and agricultural expertise, breaking down geographical barriers and expanding opportunities for collaboration and trade.

8.2 Challenges

Infrastructure Expansion Deploying 5G infrastructure in rural and remote farming regions remains a challenge. The cost of infrastructure development, including the installation of small cells and base stations, may be prohibitive in some areas.

Data Security The vast amount of data generated by IoT sensors and drones raises concerns about data security and privacy. Farmers must adopt robust cybersecurity measures to protect sensitive agricultural data from breaches and cyberattacks.

Digital Divide While 5G holds immense promise for rural areas, the digital divide persists. Ensuring equitable access to 5G technology and the necessary devices is critical to prevent leaving certain farming communities behind.

Interoperability As the agricultural technology ecosystem expands, ensuring interoperability among various devices, sensors, and software platforms becomes crucial. Standards and protocols must be established to facilitate seamless data exchange.

Environmental Impact The energy consumption associated with 5G infrastructure and data centers raises environmental concerns. Balancing the benefits of 5G with sustainability considerations will be an ongoing challenge.

Regulatory Hurdles Regulations governing the deployment of 5G networks can vary by region. Navigating regulatory frameworks and securing necessary permits for infrastructure development can be time-consuming and complex.

Skill Gap Farmers and agricultural professionals must acquire the skills to harness the full potential of 5G and associated technologies. Education and training programs are essential to ensure that users can effectively leverage these tools.

9 Application of 5G in Precision Farming

The application of fifth-generation (5G) technology in precision farming is revolutionizing the agricultural sector, offering a wide range of capabilities that enhance productivity, sustainability, and resource management. Here, we delve into the diverse applications of 5G in precision farming, showcasing how this technology is reshaping modern agriculture.

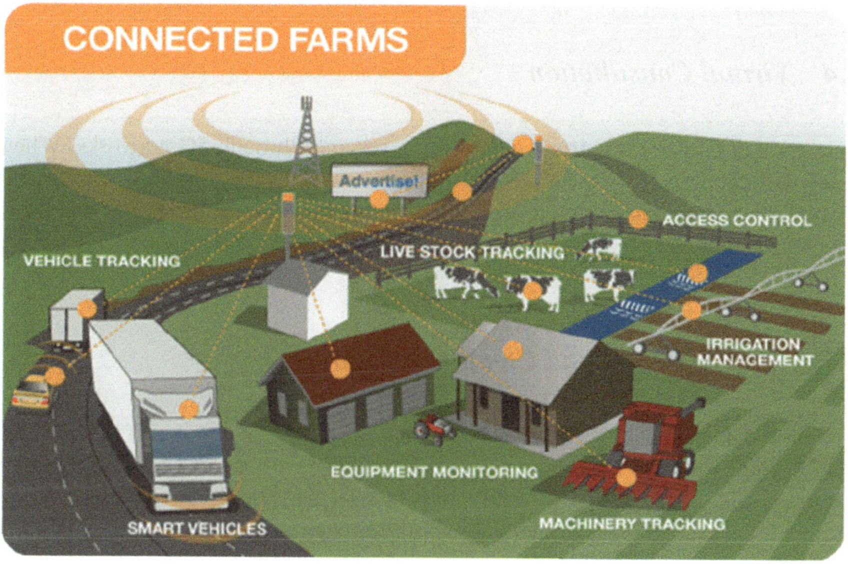

9.1 *Real-Time Monitoring*

Real-time monitoring of several agricultural indicators, such as crop health, temperature, humidity, and soil moisture, is made possible by 5G. IoT sensors positioned all throughout the farm send data instantaneously, enabling farmers to act quickly and decisively. For example, they can optimize water consumption by adjusting irrigation levels depending on measurements of soil moisture.

9.2 *Autonomous Machinery*

Autonomous farming equipment becomes a reality with 5G connection. Drones, harvesters, and tractors with sophisticated cameras and sensors can work extremely precisely. Without human assistance, they can traverse fields, sow seeds, apply fertilizer, and even handle pests, increasing productivity and cutting labor expenses.

9.3 Predictive Analytics

The combination of 5G and artificial intelligence enables predictive analytics in agriculture. Machine learning algorithms process vast datasets, including historical weather patterns and crop performance, to predict future conditions and crop yields. Farmers can proactively plan for adverse weather events or optimize planting schedules.

9.4 Virtual Consultation

Farmers can now access virtual consultation services through 5G networks. When faced with crop issues or uncertainties, they can connect with agronomists or experts in real time via video conferencing. This instant access to expertise enhances problem-solving and reduces crop losses.

9.5 Data Analytics and Cloud Repositories

The high data transfer speeds of 5G facilitate the seamless transfer of agricultural data to cloud repositories. These cloud-based platforms store and analyze massive datasets, allowing farmers to gain insights into their farming operations over extended periods. Advanced analytics tools provide actionable recommendations for improving crop management.

9.6 Precision Irrigation

Enhanced irrigation techniques are made possible by 5G-powered precision farming. Precision control of irrigation systems is made possible by IoT sensors, which continually measure the moisture content of the soil. Water waste is decreased, and crop health is improved since water is given precisely where and when it is required.

9.7 Crop Scouting with Drones

Drones equipped with high-resolution cameras and multispectral sensors fly over fields to scout for pests, diseases, or nutrient deficiencies. The data collected is transmitted in real time via 5G networks, allowing farmers to identify problems early and take targeted corrective measures.

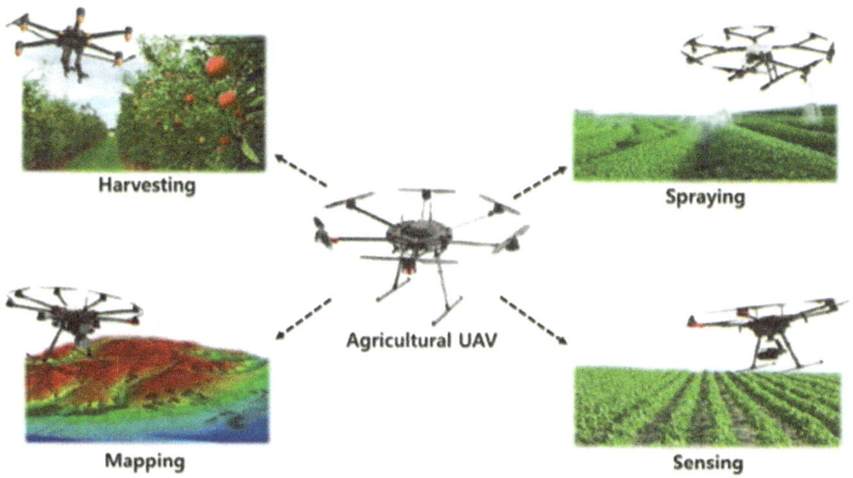

9.8 Smart Greenhouses

5G connectivity transforms traditional greenhouses into smart ones. Sensors monitor temperature, humidity, light levels, and CO2 concentration, ensuring optimal growing conditions for crops. Automated systems can adjust environmental parameters based on real-time data, maximizing crop yields.

9.9 Livestock Management

Precision farming extends to livestock management with 5G. IoT devices can track the health and location of individual animals, ensuring their well-being. This technology enables efficient herd management and early detection of diseases.

9.10 Market Access

5G networks provide rural farmers with direct access to online markets. They can sell their produce, receive orders, and manage transactions digitally, eliminating intermediaries and increasing profitability.

10 5G's Impact on Rural Areas

Beyond traditional connection limitations, the implementation of fifth-generation (5G) technology in rural areas has far-reaching ramifications across several industries, including agriculture. The improvement of connection is one of 5G's most important effects in rural areas. Historically, obtaining dependable high-speed Internet has been difficult in rural locations. Nevertheless, these communities now have smooth, fast Internet connection thanks to the development of 5G networks. Rural farmers now have fast access to online information, agricultural apps, and real-time data thanks to this increased connection. It essentially creates parity between them and their urban colleagues, empowering them to make well-informed choices and maximize their farming techniques.

Furthermore, 5G's impact extends beyond agriculture into other crucial areas. Rural education benefits greatly from 5G connectivity. Students in remote farming communities gain access to online resources and virtual classrooms, bridging the educational divide between rural and urban areas. Simultaneously, 5G enables telemedicine services, improving healthcare access in rural regions. Residents can consult with healthcare professionals remotely, reducing barriers to medical expertise and treatment options, thus promoting the overall well-being of rural communities.

Additionally, 5G encourages economic diversification in rural areas. It fosters entrepreneurship, supports e-commerce ventures, and promotes small business growth. This economic expansion contributes to rural sustainability and reduces dependency on traditional agriculture. In addition to economic diversification, 5G facilitates the development of smart infrastructure in rural areas. This includes smart grids, water management systems, and environmental monitoring solutions, leading to more efficient resource utilization and sustainable practices.

The deployment of 5G in rural areas signifies a transformative shift, empowering rural communities to embrace modern agricultural practices, diversify their economies, enhance education and healthcare access, and improve overall quality of life. As 5G continues to expand its footprint in rural regions, the synergy between technology and agriculture holds promise for a sustainable and prosperous future.

11 Future Prospects

As we gaze into the future of smart farming, the potential benefits offered by 5G technology become increasingly evident. One of the most promising prospects lies in the realm of enhanced precision farming. With 5G's high-speed, low-latency connectivity, farmers will gain access to real-time data that can revolutionize their practices. This means precise control excessive use of pesticides, fertilizers, and irrigation, which boosts crop production while making effective use of scarce resources.

Furthermore, automation and robotics are set to transform agriculture. 5G's reliability and connectivity will make it possible for autonomous farming equipment to become ubiquitous on the modern farm. Smart tractors, drones, and robotic harvesters will work tirelessly, reducing labor demands and operational costs. These machines will perform tasks with unmatched precision, ensuring that every aspect of farming is optimized for efficiency.

The Internet of Things (IoT) will be integrated and will be crucial in addition to automation. 5G networks will provide seamless connectivity between sensors and devices, enabling continuous monitoring and data collection on crop health, weather patterns, and soil conditions. This wealth of data will be channeled into advanced analytics systems, providing farmers with invaluable insights. These insights will enable them to fine-tune their farming practices, making them more sustainable and environmentally friendly.

The future of smart farming with 5G also entails a shift toward data-driven decision-making. With an abundance of data generated by 5G-connected devices, farmers will increasingly rely on sophisticated analytics and AI algorithms. These tools will help them make informed decisions, from anticipating crop diseases and predicting weather patterns to adjusting production to meet market demands effectively.

Moreover, the ability for remote monitoring and management will become a defining feature of future farming. Farmers will have the power to oversee and control their operations from virtually anywhere. Whether it's adjusting machinery settings, fine-tuning environmental conditions in greenhouses, or conducting virtual farm tours for educational or marketing purposes, the possibilities are boundless.

12 Challenges

However, as we embrace these promising prospects, we must not underestimate the challenges that come with the integration of 5G technology in agriculture. One of the foremost hurdles is the necessity for substantial infrastructure investment, particularly in rural and remote farming areas. Comprehensive 5G coverage is essential to ensure that all farmers can access and benefit from the technology.

Another critical challenge pertains to data security. With the increasing reliance on data-driven farming practices, safeguarding sensitive agricultural data from cyber threats becomes paramount. Ensuring data privacy and implementing robust security measures will be an ongoing battle.

Moreover, there may be a significant skill gap within the agricultural workforce. To fully harness the potential of 5G and its associated technologies, farmers and agricultural workers may require training and upskilling. Bridging the digital divide and ensuring that all stakeholders possess the knowledge and expertise to effectively utilize these advanced tools is of utmost importance.

Cost considerations also come into play. While 5G technology offers tremendous potential, the initial costs associated with adopting new technologies can be prohibitive, particularly for small-scale farmers. Finding ways to make these innovations financially accessible and ensuring that the benefits are distributed equitably will be a crucial aspect of future adoption.

Lastly, navigating the regulatory landscape is vital. The regulatory framework surrounding 5G in agriculture must be well-defined, addressing issues such as spectrum allocation, data ownership, and compliance with environmental and safety standards. Clarity in regulations will provide a stable foundation for the growth of smart farming with 5G technology.

13 Conclusion

To sum up, the incorporation of fifth-generation (5G) technology into the domain of smart farming signifies a noteworthy advancement in farming methodologies. This revolutionary combination of agriculture and connectivity has the power to completely change how we manage resources, grow food, and maintain rural communities. Through a thorough exploration of the role of 5G in smart farming, we have uncovered a multitude of benefits and opportunities. The improved connectivity, low latency, and high data transfer capacity of 5G empower farmers to embrace precision agriculture with real-time data analytics. This results in higher crop yields, reduced resource waste, and enhanced sustainability.

IoT in smart farming has emerged as a critical component, enabling the seamless integration of sensors, drones, and robotics. These technologies work in harmony, providing precise insights into soil conditions, crop health, and weather patterns. AI-driven decision-making further augments farming practices, allowing for proactive disease management and efficient resource allocation. The advantages of 5G extend beyond the agricultural sector, reaching into rural education, healthcare, and economic diversification. Students gain access to online resources, healthcare services become more accessible, and entrepreneurial opportunities thrive.

However, we must not overlook the challenges that accompany this technological revolution. Infrastructure investment, data security, skill gaps, costs, and regulatory considerations demand our attention. Bridging the digital divide, safeguarding data, and making 5G technology financially accessible to all farmers are paramount. As we gaze into the future, we envision a landscape where smart farming with 5G becomes the norm. It offers a path toward sustainable agriculture, increased food production, and improved livelihoods for rural communities. The synergy between technology and agriculture is poised to shape a brighter and more resilient future, where data-driven decisions empower us to feed a growing global population while preserving our precious resources. To achieve this vision, collaboration and innovation will be our guiding principles, ensuring that the promise of 5G in smart farming is fulfilled.

For citations of references, we prefer the use of square brackets and consecutive numbers. Citations using labels or the author/year convention are also acceptable. The following bibliography provides a sample reference list with entries for journal articles [1], an LNCS chapter [2], a book [3], proceedings without editors [4], as well as a URL [5].

References

1. N.J.H. Marcano, M. Nørremark, R.H. Jacobsen, Wireless communications for internet of farming: An early 5G measurement study. IEEE Access **10**, 105263–105277 (2022). https://doi.org/10.1109/ACCESS.2022.3211096
2. N.N. Misra, Y. Dixit, A. Al-Mallahi, M.S. Bhullar, R. Upadhyay, A. Martynenko, IoT, Big Data, and artificial intelligence in agriculture and food industry. IEEE Internet Things J. **7**(8), 6305–6324 (2020). https://doi.org/10.1109/JIOT.2020.2998584
3. J. Song, Q. Zhong, W. Wang, C. Su, Z. Tan, Y. Liu, FPDP: Flexible privacy-preserving data publishing scheme for smart agriculture. IEEE Sensors J. **20**(18), 17430–17438 (2020). https://doi.org/10.1109/JSEN.2020.3017695
4. S.D. Panjaitan, Y.S.K. Dewi, M.I. Hendri, R.A. Wicaksono, A drone technology implementation approach to conventional paddy fields application. IEEE Access **10**, 120650–120658 (2022). https://doi.org/10.1109/ACCESS.2022.3221188
5. T. Li, D. Li, Prospects for the application of 5G Technology in Agriculture and Rural Areas, in *Proceedings of the 2020 International Conference on Measurement, Control and Computer Engineering (ICMCCE)*, (IEEE, 2020), pp. 472–476. https://doi.org/10.1109/ICMCCE51767.2020.00472
6. S.B. Damsgaard, N.J.H. Marcano, M. Nørremark, R.H. Jacobsen, Wireless communications for internet of farming: An early 5G measurement study. IEEE Trans. Industr. Inform. **10**, 105263 (2021)
7. M.S. Farooq, S. Riaz, A. Abid, K. Abid, M.A. Naeem, A survey on the role of IoT in agriculture for the implementation of smart farming. IEEE Access **7**, 156237–156271 (2019). https://doi.org/10.1109/ACCESS.2019.2949703. IEEE
8. T. Yu, S. Dananjayan, C. Hou, Q. Guo, S. Luo, Y. He, A survey on the 5G network and its impact on agriculture: Challenges and opportunities. Comput. Electron. Agric. **180**, 105895 (2021)
9. M.-A. Kourtis, M. Batistatos, G. Xylouris, A. Oikonomakis, D. Santorinaios, C. Zarakovitis, I. Chochliouros, Energy efficiency in agriculture through tokenization of 5G and edge applications. Energies **16**, 5182 (2023)
10. S. Marios, J. Georgiou, *Precision Agriculture: Challenges in Sensors and Electronics for Real-Time Soil and Plant Monitoring* (IEEE, 2017)

11. K. Mukherjee, S. Mukhopadhyay, S. Roy, A. Biswas, Application of IoT-enabled 5G network in the agricultural sector, in *Smart Agriculture Automation Using Advanced Technologies. Transactions on Computer Systems and Networks*, ed. by A. Choudhury, A. Biswas, T.P. Singh, S.K. Ghosh, (Springer, Singapore, 2021) First Online: 01 January 2022
12. D. Boursianis, M.S. Papadopoulou, P. Damantoulakis, A. Karampatea, P. Doanis, D. Geourgoulas, A. Skoufa, *Advancing Rational Exploitation of Water Irrigation Using 5G-IoT Capabilities: The AREThOU5A Project* (IEEE, 2019)
13. M. Keshtgari, A. Deljoo, A wireless sensor network solution for precision agriculture based on Zigbee technology. Wirel. Sens. Netw. **4**(1), 16586 (2012)
14. S.B. Damsgaard, N.J. Hernández Marcano, M. Nørremark, R.H. Jacobsen, Wireless communications for internet of farming: An early 5G measurement study. IEEE Access **10**, 105263 (2022)
15. L. Tomaszewski, R. Kołakowski, M. Zagórda, Application of mobile networks (5G and beyond) in precision agriculture, in *Artificial Intelligence Applications and Innovations. AIAI 2022 IFIP WG 12.5 International Workshops. AIAI 2022. IFIP Advances in Information and Communication Technology*, ed. by I. Maglogiannis, L. Iliadis, J. Macintyre, P. Cortez, vol. 652, (Springer, Cham, 2022)
16. C. Najjuuko, G.K. Ayebare, R. Lukanga, E. Mugume, D. Okello, *A Survey of 5G for Rural Broadband Connectivity* (IEEE, 2021)
17. M. van Hilten, 5G in Agri-food: A review on current status, opportunities, and challenges. Comput. Electron. Agric. **201**, 107291 (2022)
18. H. Meng, Y. Cheng, *Research on Key Technologies of Intelligent Agriculture Under 5G Environment* (IOP Publishing Ltd, 2019)
19. K. EKhujamatov, T.K. Toshtemirov, A.P. Lazarev, Q.T. Raximjonov, *IoT and 5G Technology in Agriculture* (IEEE, 2021)
20. W. Tan, X. Xu, C. Wang, Z. Li, D. Li, From smart farming towards unmanned farms: A new mode of agricultural production. Agriculture **11**, 145 (2021)
21. F. Berto, C. Ardagna, M. Torrente, D. Manenti, E. Ferrari, A. Calcante, R. Oberti, *A 5G-IoT Enabled Big Data Infrastructure for Data-Driven Agronomy* (IEEE, 2022)

Smart Organic Agriculture in Traditional South Indian-Based Farming System

Rakesh Gnanasekaran, Sandhya Soman, Gnanasankaran Natarajan, and Sabah Ali AL'Abd AL-Busaidi

1 Introduction

Agricultural farming contributes food and fabrics in a worldwide scenario. Farming is an essential need to human race existence. India is a land of rich heritage and a global agricultural power house of farming process compared to the rest of the world. India is the second largest producer of rice cultivation in worldwide scenario in recent days. Moreover 58% of the population of India is doing agriculture process. Indian farming began in 9000 BCE as per the archaeological reports. In India, farming is done through traditional seven-step process such as ploughing, sowing, nutrient supply, irrigation, crop protection, harvesting and storage. Most of this process is done through man-made process with the help of livestock assistance. As years passed by modern time is evolved the need of agricultural modernization is inevitable. It helps the farming practice to increase agricultural efficiency, and also it minimizes the loss of natural resources. The modern farming increases efficiency and productivity while decreasing environmental impact. The proposed research work implements modernization of agricultural activity in an eco-friendly manner.

R. Gnanasekaran (✉) · G. Natarajan
Department of Computer Science, Thiagarajar College, Madurai, Tamil Nadu, India

S. Soman
Department of Computer Science, GITAM (Deemed to be) University, Bangalore, Karnataka, India

S. Ali AL'Abd AL-Busaidi
Department of Information Technology, University of Technology and Applied Science, Al Mussanah, Oman
e-mail: sabah.albusaidi@act.edu.om

S. Balasubaramanian et al. (eds.), *Intelligent Robots and Drones for Precision Agriculture*, Signals and Communication Technology,
https://doi.org/10.1007/978-3-031-51195-0_13

1.1 Agricultural Farming

Food, clothing and shelter are the basic needs of a human being, and food is essential for human survival. If there is no food, there is no human being, and agriculture plays a primary role in bringing that food to humans. Although humans consume seafood and meat, the primary food source is food obtained through agricultural farming.

1.1.1 Farming Need

Since the need for agriculture has been felt more, agriculture has been highlighted by humans since then. Plant-based foods such as fruits, vegetables, grains and legumes play an important role in providing a human with the nutrients needed to function. Although seafood and meat can be used as food, they are high in fat, so consuming them in excess can be harmful to the body. Since the need for agriculture has been felt more, agriculture has been highlighted by humans since then. Plant-based foods such as fruits, vegetables, grains, and legumes play an important role in providing a human with the nutrients and nutrients needed to function. Therefore, the most important task of agriculture is to produce healthy plant food products. In addition, paddy rice plays a major role in the food habit of Tamils. If there is no farming industry to produce that rice, there will be a situation of going without food.

1.1.2 Old Agricultural Farming System

All the civilizations considered to have lived in historical man were located along the banks of rivers, indicating that early humans lived by agriculture. They practiced agriculture using river water and produced what they needed and lived a self-sufficient economy. People who lived in other highlands other than the banks of the river observed the changes in the climate, calculated the periods of rain and carried out agriculture accordingly. At times, unexpected changes in climates made it impossible for them to continue farming. So, they constructed massive ponds and dams to collect rainwater and use it for farming. Ancient agricultural practices often involved the use of animals and collective production of what they needed [1].

1.1.3 Evolution of Farming System

Although the ancient farming methods were sufficient to meet the food demand of that time, the farming methods have gradually changed to suit the increasing population and the needs of the people, and today the world has progressed to the point where agriculture is carried out using modern technologies. Initially, agriculture was carried out using only cattle and a large amount of human labour. Now human labour is less, and machine labour is more used. For example, during paddy cultivation in the early days, oxen ploughed the field using ploughs, and humans were used for weeding and harvesting.

1.1.4 Technology in Farming

At present, technological machines are being used like ploughs, ploughing the field and harvesting using machines. This makes farming easier and faster. It can be cultivated on a large scale and can be exported to other countries. Thus, agriculture gradually developed and today has become technological. Many technological methods are being used in agriculture today. Not only technologically developed equipment but also methods of farming without the use of basic requirements of agriculture such as soil, fertilizer, etc. have been introduced as the highest innovation. For example, in countries such as Arabia, Japan and the United States, the method of floating seeds on water and using the nutrients found in water to grow crops has been successfully implemented.

1.2 Organic Farming

Today most people are turning towards nature like organic farming and organic food. The changing nature and climate may have warned people globally. As a result, organic farming need is essential in today's perspective. It is very important for everyone to know about organic farming, not just farmers. It is also necessary to change the chemical soil to lead healthy life style. In organic farming, all organisms benefit us in one way or another. The important point thing to remember that farming taken place in a natural way without affecting the five basic elements like earth, water, sky, air and fire (i.e. pancha bootham in Tamil). The thing to remember when farming has taken place in a natural way without affecting the five basic elements is the gift that today's generation of farmers giving out next generation with a safe farming system and healthy food.

1.3 Fundamentals of Organic Farming

1.3.1 Preparation of Arable Land

Preparing the land for growing all kinds of crops is the first step in agriculture. So, the land should be ploughed well to make the soil easy to plough and soft like cotton. Organic farming can be started anytime. Even 50 years of artificial fertilizers can restore the fertility of the land in 6 months through organic farming.

1.3.2 Crop Rotation System

Farmers avoid cultivating the same types of crops throughout the year in their farmlands and choose crops in rotation to get additional yield. Apart from that, the land loses its fertility due to continuous cultivation of the same crop. So, by crop rotation, the lost fertility of the land can be restored. Crop rotation can be done according to the nature of the cropping land and the amount of water.

1.3.3 Mixed Cropping and Intercropping Cultivation

In organic farming, crop yield is increased by intercropping cultivation. By doing this, the number of weeds can be controlled to a large extent, and the attack of insects can be greatly reduced. With natural insect repellents, we are giving our next generation a safe farming system and healthy food.

1.3.4 Cover-Up

The cover-up is used for mulching to increase yield. For this purpose, leaves, straw and sugar cane are mulched between the crops. This will secured the moisture of the root parts, which will allow silkworms to grow, and control weed growth and maintenance of soil quality.

1.3.5 Use of Natural Fertilizers and Crop Growth Promoters

Natural compost used like vermicomposting, cow dung compost, cow manure and green leaf compost. For the crops to grow well, more natural crop growth promoters in South India do farm for some traditional crops such as Gunapachalam, Coconut Balmor, Amritkaraisal and Panchagavya that should be used widely.

1.3.6 Pacing Between Crops

Our forefathers divided the spacing of each crop according to the saying "Nandota" for rice, "Eroda" for sugarcane, "Cartioda" for banana and "Oroda" for coconut, to yield quality and organic farming to the society.

1.4 Farming in South India

Agricultural farming was the primary part of the people living in southern region of India. It was considered essential to life and therefore took precedence over all occupations. Peasants were at the top of the social ladder. As they were producers of food grains, they lived with self-respect. Agriculture was primitive in the early stages of the Sangam period, but advanced and efficient in irrigation, ploughing, manuring, storage and distribution. Ancient South Indian people knew different types of soil and different types of irrigation suitable for their respective regions. In the era of king rule, they had lots of land, but he did not own the entire land because he gifted land to poets, brahmins, schools, hospitals and temples. Majority of the peasants cultivated their own lands. They were known by different names according to the soil.

Apart from the traditional landowners and cultivators, there were also the Sella landlords. There are various instances of kings donating land to poets, brahmins, educational institutions, hospitals, etc. The land given to brahmins was known as Brahmadeya. In lands that were granted to brahmins and peasants, agricultural work was often left to tenants or farm labourers. The regulations regarding such cultivation are not known. Sometimes labourers called adiyars were employed on other people's lands for wages. Large landowners who owned large tracts of land were food producers and had a greater sense of pride than an ordinary farmer who owned a small piece of land [2].

1.4.1 Farming Products in South India

In ancient southern part of India, rice, sugarcane, small grains, pepper, pulses, coconut, cotton, banana, tamarind, sandalwood, etc. were widely cultivated. Paddy was the main crop. Varieties of paddy crops such as Vennel, Sennel, Pudunel, Ivananel, Thorai, etc. were cultivated in. Every house had trees like jackfruit, coconut, palm and betel nut. Yellow plants were grown in front of the houses and flower gardens behind the houses.

1.4.2 Farming System in Early Era

Cultivation was done in a very systematic way during the early period. It was known that if ploughing, sowing, fertilizing, weeding, irrigation and crop protection are done properly, they need to do all these activities carefully to get a good yield. Paddy fields were ploughed with the help of bullocks. Farmers stomped the leaves under their feet and drowned them. After the seedlings grow, they are transplanted. They were harvested when the crop matured. Weeding was done periodically in the Middle Ages. The harvested paddy was brought to the field and threshed into the ground to separate the kernels. Paddy beads were collected, weighed and stored in proper containers. Small grains were grown in dry lands. Crop rotation was followed, for example, cotton and small grains were grown in the same season, followed by paddy.

Various implements were made which were required for ploughing, harvesting, etc. in agriculture. The basic tool, air melee, was also called plow in Nanji. The harrow was made of wood, iron or steel frame and used for ploughing the soil with a sharp unit locked into it. Locked to a cow or buffalo, it was used to loosen the soil and stir it up and down. A wooden pole or tree was used to level the cultivated land. Cultivation (harrowing) was used to remove weeds and reduce crop stress. It consisted of metal or wooden tines mounted on a wooden frame, and this frame was used to clear the field of weeds with the help of cattle. In the early period, people depended heavily on rain for their water needs for agriculture. But the growing population and the correspondingly increased demand for food created the conditions for improving irrigation systems. Major water storage systems like ponds, lakes and dams are created for this purpose. They constructed sluices and dams to regulate water for irrigation. Sometimes earthen embankments were built to control floods in the river and to divert water for irrigation.

1.5 Smart Farming System

Digital technology encapsulates each and every living to non-living thing globally. Therefore, the technology is already invaded in farming system. In that some of the technology aspects such as IoT, smart applications and smart, intelligent drug delivery to farming is unavoidable in recent days. Here in this article, a small note on smart farming technology [3] along with Fig. 1 represents smart organic farming system as follows.

1.5.1 IoT in Farming

The Internet of Things (IoT) is formed as a heartbeat of technology in smart farming system. Internet of Things is a process of object that communicates with another object over the Internet and exchanges necessary information needed for framing activity. With water scarcity in today's era, the use of water needed for agriculture can be reduced by a margin of 50–75% through IoT, and this IoT has found a way to make smart farming easier without wasting water. Also, this technology is very useful for us to forecast whether the crop needs nutrients and give the required fertilizers at the required time and in the required amount. Figure 2 shows IoT drone used for smart agricultural farming [4].

Fig. 1 Smart organic farming

Fig. 2 Drones used for agriculture

1.5.2 Smart Application in Farming

Smart applications provide a precise solution for ultra-model farming systems in terms of metric measurement for crop yielding processing. Some of the smart application uses for smart farming system are as follows: crop metrics is an important application in precision agriculture. It provides precisely the solutions required for ultra-modern farming systems. A crop measurement system uses sensors in IoT technology to analyse soil moisture and determine what types of crops can be grown; precision agriculture is used to make many important decisions such as improving yield. Ground-based drones and aerial drones are used in agriculture for applications such as crop yield assessment, irrigation, crop monitoring, crop spraying, planting and soil and field analysis. Using technology aspect drones, the fertility of the crops can be monitored from the location [5]. Also, these drones play an important role in agriculture for easy analysis of crop characteristics and better planning. Therefore, the role of smart applications in the current emerging technology is immense help for smart farming system. Figure 3 shows smart application used for farming system.

1.6 Smart Organic Farming Model

1.6.1 Existing Model

Organic smart farming solution uses IoT solution that is built for monitoring the crop and field with aid of sensors for monitoring light, humidity, temperature, soil moisture and crop health. With the help of smart application, farmers can monitor the farming field conditions from anywhere and store all relevant information in cloud platform [6]. In Fig. 4, existing smart organic farming model is shown as follows.

1.6.2 Proposed Model

This research work facilitates organic smart farming by implementing machine learning logic to IoT solution and its relevant application to do organic smart farming in an optimized manner to promote agricultural farming system in an effective manner. Figure 5 shows proposed smart organic farming model as follows.

The organic smart farming system encompasses with optimistic approach contribute to get maximize yield and also it promotes smart way of organic farming process are as follows:

- IoT solutions
- IoT devices and sensors monitoring process
- Smart applications

Fig. 3 Smart applications

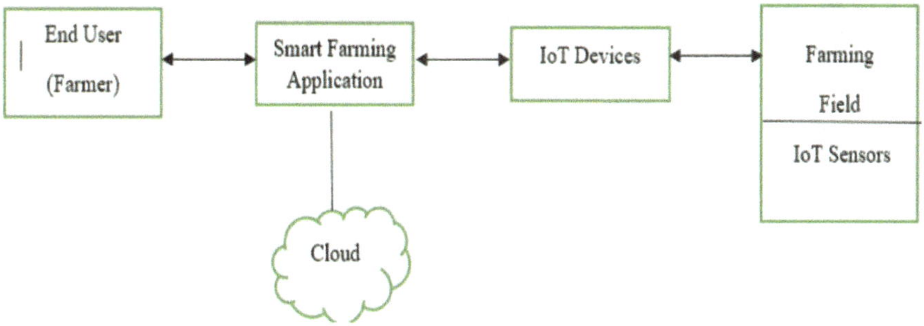

Fig. 4 Smart organic farming model

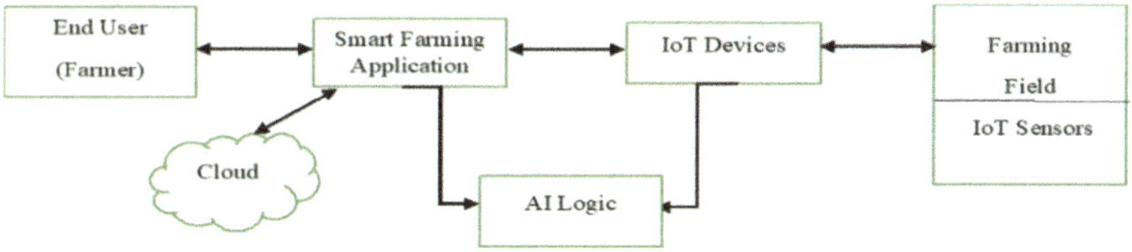

Fig. 5 Proposed smart organic farming model

1.7 IoT Solutions for Proposed Smart Organic Farming Model

The proposed system encompasses with IoT devices like drones and sensors for smart organic farming process. Drone is an unmanned aerial vehicle normally used to monitor the field and crop yield measurements. It acts as an observer in smart farming system [7]. Drones are typically used to survey the farm field status and irrigation process needed for farming. Therefore, farmers get a higher productivity and efficient use of irrigation land, water and organic fertilizer. The following are the benefits of using drones in smart farming:

- Increased yields
- Time saving
- Crop monitoring
- Optimization water usage needed for farming
- Sensors for climate condition monitoring

1.8 Proposed IoT Devices and Sensor for Organic Smart Farming

Organic-based smart farming works on crop rotation, companion planting and farming with natural fertilizer. These IoT devices with machine learning computation process help the farming system with the following functions:

- Climate condition predictions – IoT sensors in the field
- Water sensors – IoT sensor in the field
- Optical sensors - IoT sensor in the field

1.8.1 Optimized Monitoring Approach Using IoT Device and Sensors for Climate Predictions

In traditional-based organic farming methods, crops and vegetables have their own cultivating period such as some may grow in humid condition and some may grow in warm conditions; some grow in a high attitude for growth and development. Based on this organic farming cultivation done in South India by the following process:

- Intercrops and crop rotation
- Growing season
- Soil
- Crop duration
- Crop varieties

Here in this research work, the IoT device and sensor devices used for smart farming constitute with above process in an efficient manner. Figure 6 is one of the IoT sensor devices used for organic smart farming with computational process inclusion.

In Fig. 6, the IoT sensor device used for climate prediction installed on the field and frequently observes the field condition and updates it to centralized cloud storage about the climate conditions for cropping.

Figure 7 is the proposed monitoring method for climate condition prediction process for organic crop cultivation. The IoT sensor device captures the soil, wind, rainfall and temperature state and sends the relevant data using the help of sim module to cloud environment for permanent storage process.

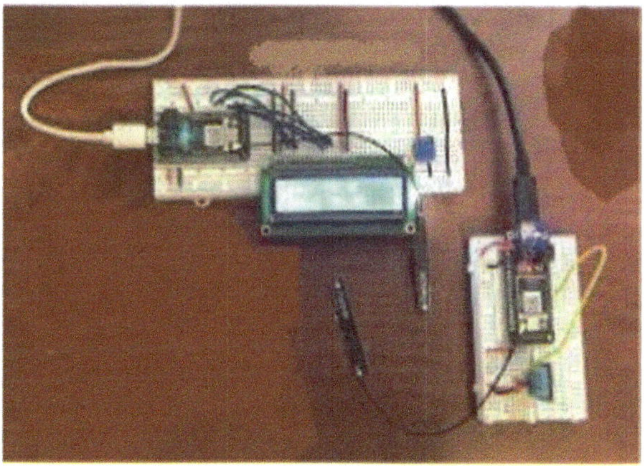

Fig. 6 IoT sensor device for climate predictions

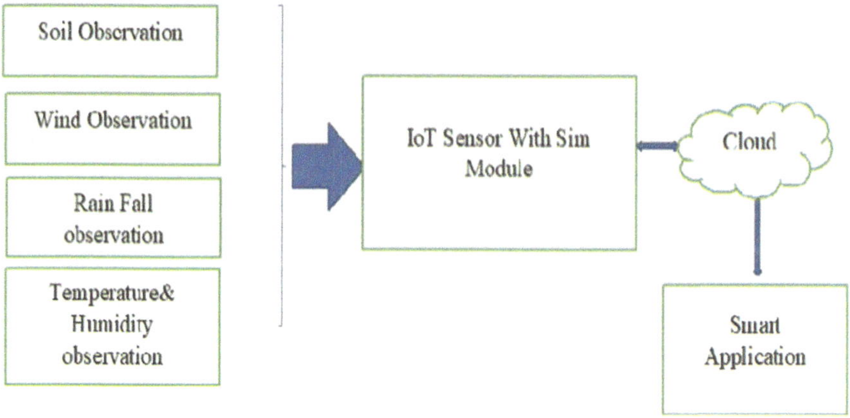

Fig. 7 IoT monitoring sensor device process for climate predictions

1.8.2 Optimized Monitoring Approach Using IoT Device and Sensors for Water Resources

The optimized water level monitoring sensor measures water level in the field in Fig. 9 in various forms such as water source from well and rainfall and from river irrigation sources. The smart IoT devices fixed in the field keenly monitor the essential water resource level needed for organic farming process and send report to the client smart application on an hour-based activity [8].

In Fig. 8, the IoT sensor device helps in monitoring the water resource level and sends the data to cloud platform and intern it report to smart client board for organic farming essentials Fig. 9.

1.8.3 Optimized Monitoring Approach Optical Sensor for Organic Farming

In smart organic farming, optical sensors play an important role by observing, measuring and recording the information about organic crops and the soil field by using optical sensor processes. Optical sensor is the process apply to smart farming system by emitting of light of shining to a specified wavelength at crops to get the condition of crops through the reflected light to the sensors. It helps in monitoring the crop condition from the beginning stage to harvesting stage of organic crop cultivation process.

The above is the optical monitoring device (Fig. 10) that help fully in observing and predicating the crop condition at every stage of the development process.

In Fig. 11, optical sensor used to capture and monitor the safety level of crop cultivation in an optimized manner with maximized efficiency.

1.9 Proposed Smart Application for Organic Smart Farming

Smart applications are those which help in sensing the process with actuation and also able to monitor and control for any condition to make a decision that can take place with the data obtained and derived for prediction process. In this research work, the smart application proposed for organic smart farming to the end user encompasses with the following process:

1. Supervised data for organic farming
2. Smart data from IoT devices and sensors
3. Machine learning predictions for organic smart farming

Fig. 8 IoT water level monitors

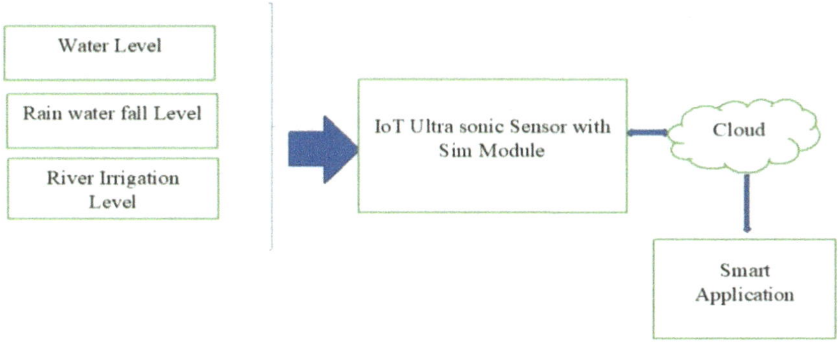

Fig. 9 IoT monitoring sensor device process for water level monitoring

Fig. 10 IoT optical sensors

1.9.1 Supervised Data for Organic Farming

The supervised data composition contains the basic essential process needed for organic farming. Let us take an example for organic farming method for rice cultivation; the essential data noted for cultivation are as follows: the average crop height of rice is in the range about 100–120 cm, and its mature period is ranging about 105–115 days for 1 acre rice yielding organic smart farming.

These data are learned and inputted to the application environment that needed for end users. Apart from the essential data quantity of organic fertilizer, water and natural pesticide data prerequisites present in the smart application computational process. Hence all the above essential data is labelled to use for computing process.

1.9.2 Supervised Organic Data Farming in South Indian-Based Farming System

The traditional South Indian farming system follows drilling method. It is a process of ploughing the land and sowing seed to cultivate crops in an organic manner. In that the smart application supervised learning process follows logistic regression algorithm for supervised learning method of organic farming. The supervised

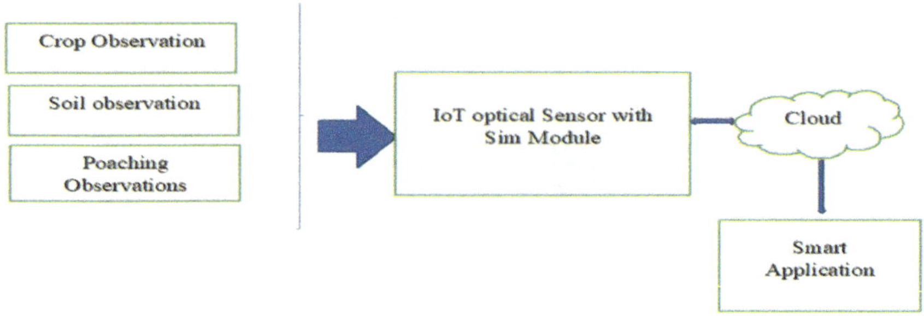

Fig. 11 IoT optical monitoring sensors for organic farming

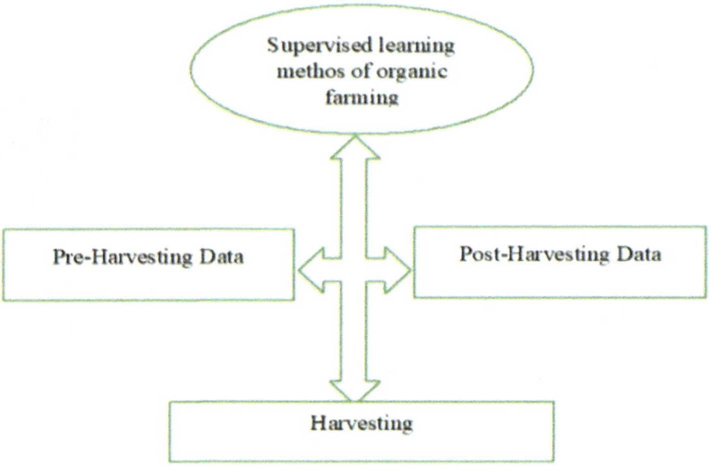

Fig. 12 Supervised learning data for smart farming

organic drilling method using smart application comprises three different stages; they are as follows and shown in Fig. 12:

1. Pre-harvesting method
2. Post-harvesting method
3. Harvesting method

2 Smart Data from IoT Devices and Sensors

IoT devices and sensor data are called as smart data collected from the sensors of IoT devices to monitor and control the environmental condition such as irrigation field-related real-time conditions, and all are connected over Internet of Things (IoT). The smart data collected are usually sent to cloud storage for analysis process to make decision-making. The smart data is classified into three different types such as:

- Device data
- Utility data
- Environment data

1. Device Data

   It mainly dealt with status of IoT devices that works for real time. Moreover, the device data obtained are centralized to get maximum predictions.

2. Utility Data

   It collects the data referred to utility basis on the ground of device data process; it helps in forecasting the earlier predictions for data processing.

3. Environment Data

   It monitors and measures the environmental condition used for agricultural process.

2.1 Machine Learning Predictions for Organic Smart Farming

The emergence of artificial intelligence in worldwide scenario also conquers the agricultural sector to promote the agricultural production in an effective way by implementing machine learning techniques to optimize result to the end user [9]. This article facilitates machine learning algorithm to support smart application used for organic farming methods.

This research article describes an optimized decision tree algorithm for machine learning process that is applied on the data from the sensor devices to predict the result for smart application efficiently to farming process.

2.1.1 Optimized Decision Tree Algorithm for Machine Learning Predictions

It is a supervised learning technique based on a data set that tend to find the result on the basis of finding all possible solutions using a decision tree. Figure 13 illustrates the flow of data for optimized smart applications.

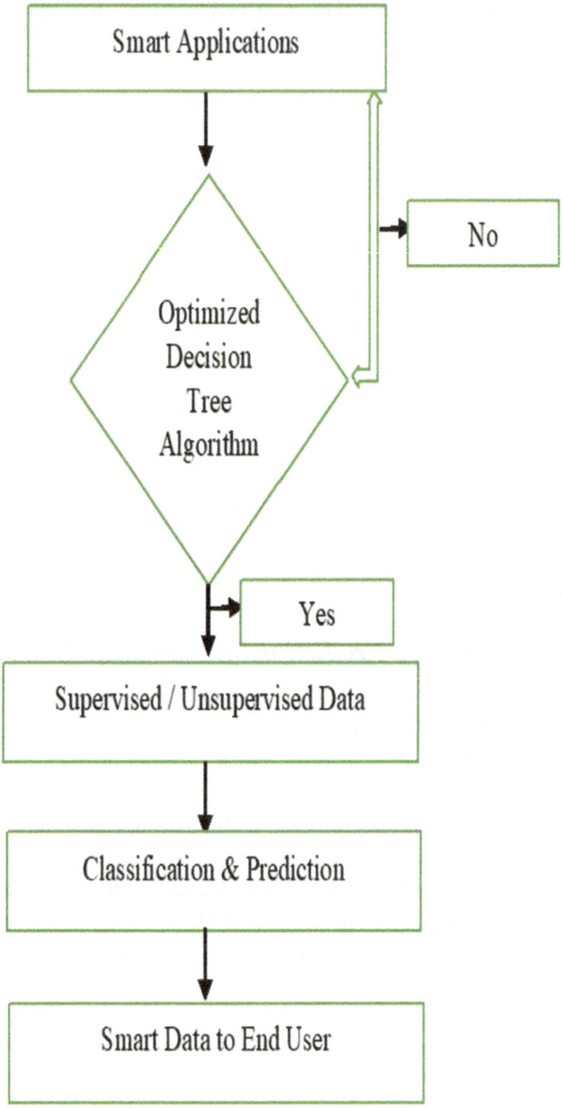

Fig. 13 Smart application environment

2.2 *Optimized Smart Organic Agriculture in South Indian-Based Farming System*

Figure 14 shows optimal farming system proposed for organic agricultural process in South India. The optimized approach works with four major modules as follows:

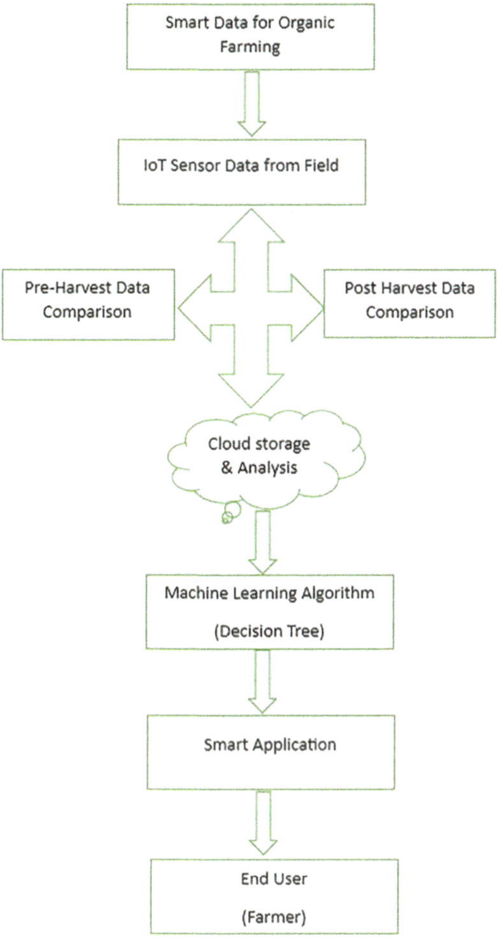

Fig. 14 Optimized farming system

- Smart data for agriculture
- IoT data
- Machine learned data
- Smart application

The process of smart farming system starts with smart information of organic farming flows to get and map IoT devices data by comparing pre- and post harvesting data in cloud platform, and it makes a flow towards smart application with machine learning algorithm for the end user [10].

2.3 Conclusion

This research focus towards smart farming that helps for agricultural sector to get technology facilities for harvesting process. It brings an agricultural revolution to automatize the entire process of organic agricultural activity in South Indian-based farming system. The procedural activity described in this research work really helps the farmer especially South Indian-based farmers to do their agricultural work in an optimized manner. Hence, this adaption of new advanced technologies to farming really helps South Indian-based farmer to promote their organic-based agricultural product globally.

References

1. W. Pedrycz, S.-M. Chen, *Deep Learning: Concepts and Architectures* (Springer Nature Switzerland AG 2020, 978–3–030-31756-0, 2019). https://doi.org/10.1007/978-3-030-31756-0
2. L. Alzubaidi, J. Zhang, A.J. Humaidi, A. Al-Dujaili, Y. Duan, O. Al-Shamma, J. Santamaría, M.A. Fadhel, Muthana Al-Amidie4 and Laith Farhan, Review of deep learning: Concepts, CNN architectures, challenges, applications, future directions. J. Big Data **8**(53), 1–74 (2021). https://doi.org/10.1186/s40537-021-00444-8
3. J. Wang, Y. Ma, L. Zhang, R.X. Gao, W. Dazhong, Deep learning for smart manufacturing: Methods and applications. J. Manuf. Syst. **48**, 144–156 (2018). https://doi.org/10.1016/j.jmsy.2018.01.003
4. M.-Y. Chen, E.D. Lughofer, E. Egrioglu, Deep learning and intelligent system towards smart manufacturing. Enterpr. Inf. Syst **16**, 189–192 (2022). https://doi.org/10.1080/17517575.2021.1898050
5. T. Kotsiopoulos, P. Sarigiannidis, D. Ioannidisb, D. Tzovaras, Machine learning and deep learning in smart manufacturing: The smart grid paradigm. Comput. Sci. Rev **40** (2021). https://doi.org/10.1016/j.cosrev.2020.100341
6. R. Rai, M.K. Tiwari, D. Ivanov, A. Dolgui, Machine learning in manufacturing and industry 4.0 applications. Int. J. Prod. Res. **50**, 4773–4778 (2021). https://doi.org/10.1080/00207543.2021.1956675
7. A. Jamwal, R. Agrawal, M. Sharma, Deep learning for manufacturing sustainability: Models, applications in Industry 4.0 and implications. Int. J. Inf. Manag. Data Insights **2**(2) (2022). https://doi.org/10.1016/j.jjimei.2022.100107
8. P.P. Kovac, B. Savković, D. Rodic, I. Maňková, Artificial inteligence approache to modeling of cutting force and tool wear relationships during dry machining. J. Product. Eng (2018). https://doi.org/10.24867/JPE-2018-02-013
9. N. Gnanasankaran, G. Rakesh, T. Manikumar (eds.), Chapter 3: "Multidisciplinary applications of machine learning", in *Machine Learning, Block Chain and Cyber Security in Smart Environments*, (Scopus SCIE, Taylor and Francis/CRC Press, Chapman and Hall, Copyrights 2022, Book ISBN: 9781003240310), pp. 41–57. https://www.taylorfrancis.com/books/mono/10.1201/9781003240310
10. G. Rakesh, Hybridized gradient descent spectral graph and local – Global Louvain based clustering of temporal relational data. Int. J. Eng. Adv. Technol **9**(03), 3515–3521 (2020) https://stm.bookpi.org/RPST-V3/article/view/9326

Smart Farming with Cloud Supported Data Management Enabling Real-Time Monitoring and Prediction for Better Yield

Robin Cyriac and Jayarani Thomas

1 Introduction to Cloud Computing and Smart Farming

A new era of agricultural progress is arrived, and it's all because of the integration of cloud computing and "smart farming." The foundation of contemporary precision agriculture can be built on cloud computing, which can remotely store, process, and analyze massive volumes of data. Farmers can now take data driven decisons due to the real-time data collection and processing facilitated by sensors, drones, smart agricultural machinery, and Cloud servers. Because of this interplay, conventional farming is transformed into an activity that is more productive, environmentally friendly, and linked. As we explore the details of this intersection of technologies, we find a world in which farmers are given the tools to increase their yields while decreasing their use of resources and their adverse effects on the environment.

1.1 Background and Motivation

The necessity of resolving issues troubling the agriculture sector and the desire to make use of promising new technology to increase crop yields while reducing environmental impact are the driving forces behind the adoption of cloud computing in smart farming. The circumstances that have led to this convergence all point to the ways in which cloud computing can revolutionize conventional farming.

R. Cyriac (✉)
Department of IT, Federation University, Brisbane, QLD, Australia

J. Thomas
Brisbane, QLD, Australia

S. Balasubaramanian et al. (eds.), *Intelligent Robots and Drones for Precision Agriculture*, Signals and Communication Technology,
https://doi.org/10.1007/978-3-031-51195-0_14

Manual labor, intuition, and broad strategies are common in traditional farming's approach to crop, soil, and resource management [1]. This method often results in wasted effort and materials and a slower adaptation to changing environmental circumstances. There are a number of problems that have prompted researchers to look at using cloud computing in agriculture.

1. Resource Scarcity: With a growing population comes a greater need for food, which puts more stress on the world's limited supplies of farmland, water, and electricity.
2. Environmental Concerns: Soil erosion, water contamination, and an overabundance of chemical use are just a few of the environmental problems that can result from conventional agricultural methods. To lessen negative effects on the environment, sustainable behaviors are crucial.
3. Climate Change: This change has brought extremely variable weather patterns, making it impossible to accurately forecast and manage agricultural production.
4. Food Security: Ensuring a stable and secure food supply is a significant concern, particularly in regions prone to food shortages.
5. Labor Shortages: Labor-intensive farming methods are becoming increasingly difficult to implement due to a lack of available workers and rising wages.

The reason to implement cloud computing in smart farming comes from the fact that smart farming has the ability to provide game-changing solutions to the problems hampering conventional farming. Cloud computing's ability to alter the way farming is done by fostering more efficient, sustainable, and productive methods is a major reason for its growing significance in this sector. Integrating cloud computing has significant implications for farmers and the global food supply chain because of the many advantages it delivers and the difficulties it solves in the agriculture business.

In cloud computing environment, massive volumes of data from the agriculture sector can be collected, stored, and analyzed. This data may tell you a lot about the soil, the weather, and the overall health of your crops. When this information is analyzed in real time, farmers can make better judgments about watering, fertilizing, and controlling pests. Precision agriculture, enabled by the cloud, allows farmers to adapt their methods to the unique requirements of individual plants or localized regions of a field. By administering water, fertilizer, and insecticides just when and where they are required can reduce wastage of resources. This results in less wasted resources and more productivity. Fields may be remotely monitored using cloud computing and technology like sensors, drones, and the Internet of Things. Anywhere there is an Internet connection, farmers can get up-to-the-minute information about things like soil conditions, temperature, humidity, and more. As a result, field workers aren't required to be on-site all the time, and emergencies may be addressed promptly.

Based on past and present data, cloud-based systems may create prediction models. Crop yields, disease outbreaks, and market movements are all within the scope of these models' predictions. These findings can help farmers anticipate problems and prepare for them in advance. Smart farming, enabled by the cloud, makes the most efficient use of water and fertilizer. Farmers may reduce their negative effects on the environment by applying the results of scientific analyses of soil and meteorological data. Farmers now have a single location from which to oversee and manage all facets of their farms, thanks to cloud-based farm management software. These systems allow for effective communication and cooperation between agricultural workers by integrating data from many sources. Farmers, agriculturists, and academics may now work together on a worldwide scale by collecting and maintaining data in cloud computing. Better results for the whole industry may be achieved via the sharing of knowledge, best practices, and new solutions.

Cloud services are extremely scalable, so farms of any size may take use of cutting-edge tools without having to make substantial initial monetary investments. This opens up state-of-the-art equipment and techniques to both large and small farms. Through the provision of safe data storage and backup, cloud computing increases farm resilience. Data that is vital to business operations can be restored quickly after a disaster or system failure. Cloud-based smart farming helps agriculture become more sustainable by maximizing efficiency and lessening the impact on the environment. This is essential in order to minimize the negative effects on the environment that the industry has. The quality and safety of food is protected by cloud-based technologies because of their enhanced capacity to track products from farm to fork. Customers now have a better idea of where their food comes from.

Cloud computing's value in agriculture arises from its potential to upgrade obsolete methods of farming to ones that are more modern, data-driven, and environmentally friendly. Farms may benefit from cloud computing because it allows for more accurate resource management, real-time monitoring, predictive analytics, and global cooperation, all of which help farmers overcome obstacles, increase output, and strengthen the global food supply chain. The agricultural sector may improve its capabilities and respond to the needs of a shifting global market by tapping into the potential of the cloud.

1.2 Overview of Cloud Computing

The term "cloud computing" is used to describe the practice of providing data storage, processing power, and software applications through the Internet. Users no longer need to rely entirely on locally available resources; instead, they may access and use these services remotely, without setting up a complex physical infrastructure. Cloud services are very flexible, so users may easily add or remove capacity to meet their specific requirements. This is especially helpful in agriculture, where planting, harvesting, and weather conditions all have a significant impact on demand.

With cloud computing, expensive infrastructure like servers and programs are no longer required. By charging users per unit of resource used, financial risks are mitigated, and smaller farms are given the opportunity to use cutting-edge equipment. As long as they have an Internet connection, farmers and agricultural professionals may use any number of online tools and databases. This allows for cross-locational collaboration, instantaneous decision-making, and remote monitoring. The ability to quickly handle and analyze data is essential for making good judgments in precision agriculture, and this is where cloud computing comes in. Insights on crop health, soil conditions, weather patterns, and more may be gleaned from massive databases using sophisticated algorithms. Many cloud providers claim of having many backups and servers. In the case of a hardware breakdown or natural disaster, this guarantees that vital agricultural data will still be preserved and accessible.

1.3 Concept of Smart Farming

Smart farming, often known as precision agriculture, is the practice of utilizing a wide range of technological advancements to enhance productivity in agricultural operations. The deployment of smart agricultural methods is greatly aided by cloud computing.

1. IoT and Sensor Integration: In order to monitor soil moisture, temperature, humidity, and other environmental factors, IoT devices and sensors are installed in fields [2]. In turn, this helps farmers make better decisions regarding issues like watering, fertilizing, and insect management in real time.
2. Predictive Analytics: Cloud-based solutions may use both historical and real-time data to build predictive models (predictive analytics). These models allow farmers to properly prepare for future crop yields, disease outbreaks, and market trends.
3. Resource Management: Cloud-based smart farming allows for more efficient use of available resources. Farmers may reduce the amount of fertilizers and pesticides used and their negative effects on the environment by analyzing data on soil conditions, weather forecasts, and crop health.
4. Crop Monitoring: High-resolution aerial photographs of fields are taken by drones fitted with cameras and sensors for crop monitoring. Then, regions with vulnerable crops, illnesses, or pests are pinpointed with the use of cloud-based picture processing.
5. Supply Chain Efficiency: Increased openness and traceability in the agricultural supply chain is one of the many ways in which cloud computing improves supply chain efficiency. Food safety and quality may be monitored across the whole supply chain, from farm to fork.

Smart farming can profit greatly from cloud computing, but there are drawbacks as well. Sensitive data, including personal information and proprietary algorithms, are involved in farming. It is crucial to guarantee data security and adherence to privacy laws. For access to cloud services and real-time data transfer, consistent Internet access is necessary, but this might be difficult in rural regions. Initial infrastructure, training, and system integration investments may be necessary for the successful implementation of cloud-based solutions. Particularly when there are several parties involved, it is important to establish explicit agreements on data ownership and usage rights.

In the world of agriculture, cloud computing has changed the game and brought in the era of smart farming. Productivity, sustainability, and profitability are increased in cloud-powered smart farming through real-time data analysis, predictive modeling, and resource optimization. The use of cloud computing to agriculture has the potential to completely transform the world's food supply chain and make it more adaptable to the problems of the twenty-first century, as long as technology keeps developing and connectivity keeps getting better.

2 Fundamentals of Cloud Computing

A technological paradigm known as cloud computing refers to the distribution of computer resources, services, and applications through the Internet. Users can access and utilize software, storage, and processing power without having to maintain or own physical infrastructure. With cloud computing, businesses can instantly scale up or down, pay for just the resources they use, and concentrate on their main goals instead of worrying about maintaining hardware and software. Cloud computing offers on-demand access to a wide range of resources and services. To accommodate diverse user requirements, it includes a range of deployment options (public, private, hybrid, multi-cloud) and service models (IaaS, PaaS, SaaS). Flexibility, scalability, affordability, and the capacity to develop and implement apps quickly are all available through cloud computing.

2.1 Cloud Deployment Models

Several cloud deployment models are intended to explain the deployment and management of cloud resources and services inside an enterprise [3]. The four primary cloud deployment models are as follows:

1. *Public Cloud*: A third-party cloud service provider provides resources and services in a public cloud deployment, which are then made accessible to the general public over the Internet. Pay-as-you-go access and usage of these resources is available to organizations, who share them with other consumers. Google Cloud Platform (GCP), Microsoft Azure, and Amazon Web Services (AWS) are a few examples of public cloud providers.
2. *Private Cloud*: Building a cloud infrastructure inside a company's on-premises or data center is known as a private cloud deployment. This infrastructure is exclusive to one firm and isn't used by any other businesses. More control over resource allocation, security, and customization is possible with private clouds. Organizations with certain regulatory or compliance needs frequently use them.

3. *Hybrid Cloud*: A hybrid cloud deployment integrates aspects of public and private cloud infrastructure. It enables businesses to combine public cloud resources with their on-premises infrastructure. This method offers flexibility and scalability by allowing data and workloads to be transferred between the two environments as needed. Organizations looking to keep sensitive data on-site while utilizing the public cloud's scalability and agility for other purposes might benefit from hybrid clouds.
4. *Community Cloud*: A cloud shared by two or more businesses with similar cloud needs is called a community cloud. A community cloud will be managed and utilized by a group of businesses with similar goals or particular security needs.

Every deployment model has benefits and drawbacks of its own, and an organization's needs, objectives, and factors like security, compliance, performance, and cost all influence in the model selection process.

2.2 Cloud Service Models

When it comes to administering and maintaining various levels of a cloud-based application or infrastructure, cloud service models specify the degree of control and accountability that a cloud service provider and a client share [4]. There are three primary models of cloud services:

1. *Infrastructure as a Service (IaaS)*: The cloud provider provides virtualized computer resources through the Internet under an IaaS paradigm. This covers networking, storage, virtual machines, and other essential infrastructure parts. Clients can create, install, and maintain their own software, operating systems, and apps with the help of these resources. Customers are in charge of administering and maintaining the virtual machines, OS upgrades, security patches, and other infrastructure-related activities, despite the fact that they have control over their apps and customizations.
2. *Platform as a Service (PaaS)*: In an IaaS model, the cloud provider makes virtualized computer resources available online. This includes virtual machines, networking, storage, and other crucial infrastructure components. With the use of these tools, clients may develop, install, and manage their own operating systems, applications, and software. Although they retain control over their programs and customizations, customers are responsible for managing and maintaining the virtual machines, OS updates, security patches, and other infrastructure-related tasks.
3. *Software as a Service (SaaS)*: The most customer-focused cloud service paradigm is SaaS. Under this strategy, software programs that are fully functional are delivered via the Internet by the cloud provider. Clients do not need to install or manage the program locally in order to access and utilize it through a web browser. The supplier takes care of infrastructure, security, upgrades, and maintenance in its entirety. From email and collaboration tools to customer relationship

management (CRM) and enterprise resource planning (ERP) systems, SaaS solutions address a broad spectrum of corporate demands.

Organizations may select the cloud service model that most closely matches their requirements, capabilities, and experience by selecting from a variety of levels of abstraction and administrative responsibilities offered by each model. A number of variables, including cost, scalability, control, and development speed, can be impacted by the service model selection.

2.3 Virtualization and Resource Allocation in Cloud

Two fundamental ideas in cloud computing are virtualization and resource allocation, which allow for the flexible creation and management of virtual environments for services and applications as well as the effective use of physical resources.

The act of turning real resources like computer hardware, storage, and networking components into a virtual form is known as virtualization. It essentially abstracts the underlying hardware and provides isolation between these virtual instances by enabling the operation of numerous virtual instances or environments on a single physical system. Virtualization is essential for the creation and administration of virtual machines (VMs), also known as containers, which are separated environments that resemble actual computers in the context of cloud computing. These virtual instances offer a degree of flexibility, scalability, and resource efficiency as they may run many operating systems and applications. Virtualization makes it possible for several users or tenants to share hardware resources effectively, which eliminates the need for separate physical hardware for each task.

In cloud computing, resource allocation is the process of allocating computer resources—like CPU, memory, storage, and network bandwidth—to different virtual instances in accordance with their needs and priorities. Cloud service providers allocate resources to their clients' workloads in a way that maximizes efficiency, scalability, and economy.

2.4 Scalability, Elasticity, and On-Demand Services

The capacity of a system, application, or infrastructure to accommodate growing volumes of work, data, or users without sacrificing responsiveness or performance is referred to as scalability. It involves constructing a system with the flexibility to grow or decrease to meet shifting demand. To guarantee that the system maintains ideal performance levels, scalability can be accomplished by adding or eliminating resources, such as processing power, storage, or network capacity.

There are two forms of scalability: horizontal scalability (also known as scale out) and vertical scalability (also known as scale up). The former refers to adding extra instances or servers to spread the burden over numerous computers, whereas the former is restricted by the capacity of a single system. This method has a better chance of managing heavier workloads and rising demand. The latter speaks about boosting a single instance's or server's capacity in terms of CPU, memory, or storage.

An extension of scalability, elasticity, concentrates on a system's capacity to dynamically and automatically modify its resources in response to demand. Elasticity allows resources to be automatically scaled up or down to accommodate variations in workload. This keeps the system operating at peak efficiency and reduces expenses when demand is low.

In the context of cloud computing, "on-demand services" refers to the flexibility to access and utilize resources, apps, and services whenever required, without the need for lengthy lead times or manual involvement. On-demand resources and services are provided by cloud service providers, enabling customers to rapidly provision and use them in accordance with their needs. On-demand services include several important features, such as flexibility, pay-as-you-go, instant provisioning, and self-service. demand.

On-demand services in the context of cloud computing refer to the ability to access and use resources, applications, and services whenever they are needed, without requiring manual intervention. Cloud service providers offer various resources and services on-demand, allowing users to quickly provision and utilize them according to their requirements. Key characteristics of on-demand services include self-service, pay-as-you-go, instant provisioning, and self-service.

2.5 Land Suitability Assessment Based on Soil Vegetation Indices from Satellite Data

Organizations must take security and privacy concerns seriously while utilizing cloud computing in order to safeguard their resources, data, and apps. Although cloud computing has many advantages, like flexibility and scalability, there are some disadvantages as well [3]. Data security, network security, vulnerability management, identity and access management, compliance, shared responsibility models, and incident response are among the security factors to be taken into account.

Data privacy, location and residency of data, ownership and control of data, portability of data, vendor lock-in, transparency, and consent management are among the privacy considerations. Organizations should do a thorough risk assessment, embrace best practices for cloud security, and collaborate closely with the cloud service providers of their choice to build efficient security controls and privacy safeguards in order to handle these factors. Regular evaluations and audits are necessary to guarantee continued compliance and the security of private data.

3 Applications of Cloud Computing in Agriculture

Because cloud computing provides scalable, economical, and effective solutions to a range of problems encountered by farmers and agribusinesses, it has had a major impact on the agriculture industry. Precision farming and data-driven farming, crop monitoring and yield prediction, livestock management and health monitoring, irrigation control and water management, supply chain optimization, and traceability are a few of the most prevalent applications of cloud computing in agriculture.

3.1 Precision Agriculture and Data-Driven Farming

Precision agriculture is a kind of agricultural management that adapts and optimizes farming techniques on a field-by-field or even plant-by-plant level via the use of technology and data. In order to evaluate variables including soil health, moisture levels, temperature, and crop health, data is first collected from a variety of sources, including weather stations, GPS, drones, satellites, and remote sensing. After the data has been gathered, analysis is done to produce suggestions and insights for certain agricultural practices. Data analytics and machine learning methods are frequently used in this. Based on the data-driven insights, inputs such as herbicides, fertilizers, and irrigation water are subsequently administered at different rates. This minimizes waste and its negative effects on the environment by ensuring that resources are used exactly where and when they are required [5]. To increase productivity and save labor costs, precision agriculture uses automation and robots for planting, harvesting, and other farming operations. With the assistance of GIS (Geographic Information Systems) software, farmers can precisely plan their planting, irrigation, and crop management strategies by creating comprehensive maps of their fields. Farmers may monitor and manage their activities from a single dashboard created by integrating data from many sources using farm management software. Through mobile applications and online platforms, farmers may remotely monitor crop conditions, equipment status, and environmental factors in real time, enabling them to make timely modifications as needed.

Precision agriculture is included in the larger term of "data-driven farming," which also refers to the general agricultural philosophy of using data to guide decisions rather than tradition or gut feeling. Farmers are able to make well-informed decisions on pest management, fertilizer, irrigation, planting schedules, and harvesting by using data analysis. These choices are customized for individual plants or even for each unique field. Throughout the growth season, ongoing observation and feedback are helpful in optimizing results and adapting to shifting circumstances. Although adopting a data-driven strategy in agriculture might help achieve sustainable agriculture, there are several obstacles in its way, including social, legal, and technological issues as well as practical limitations. The aforementioned obstacles hinder the exchange of data in order to obtain substantial advantages from it [5].

3.2 Crop Monitoring and Yield Prediction

Modern agriculture is dependent on yield prediction and crop monitoring, which are made possible by data analytics and technological breakthroughs. All during the growing season, farmers can closely monitor the health and growth of their crops, thanks to the combination of sensors, drones, satellite photography, and data-driven software powered through cloud. With the help of this real-time monitoring, irrigation, fertilizer, and pest control can be precisely adjusted to maximize resource efficiency and minimize waste. Furthermore, farmers can anticipate crop yields with exceptional precision using historical data and predictive modeling, which helps them make better decisions about harvesting schedules and market planning. Crop monitoring and yield prediction help to ensure food security in a world, minimize environmental impact, and boost agricultural productivity—all while contributing to sustainable farming practices.

3.3 Livestock Management and Health Monitoring

Since technology and data-driven solutions have been integrated, livestock management and health monitoring have seen a revolutionary transformation. Thanks to automated data gathering systems, wearable technology, and sensors, farmers can now keep a close eye on the health of their livestock. These technologies allow for the early diagnosis of disease and stress by providing real-time data on important factors including dietary habits, temperature, and heart rate. Furthermore, this data is processed by sophisticated analytics and machine learning algorithms to produce actionable insights that enable prompt intervention and individualized treatment for every animal. An Internet of Things (IoT) tool called animal health monitoring (AHM) packaging prevents medication abuse by guaranteeing pharmaceutical compliance [6]. This promotes overall farm efficiency and increases animal welfare while guaranteeing the production of high-quality, safe livestock products. In contemporary agriculture, livestock management and health monitoring are essential for fostering animal health, sustainability, and the prudent management of livestock resources.

3.4 Irrigation Control and Water Management

Modern agriculture now cannot function without irrigation control and water management, especially in areas where water is scarce and weather patterns are irregular. Precision sprinklers, drip irrigation, soil moisture monitors, and other advanced irrigation technology enable farmers to apply water precisely where and when it is required. These devices' data are processed by cloud-based data analytics platforms, which provide real-time weather forecasts and insights into soil moisture levels. Farmers may reduce water waste, maximize crop growth, and optimize irrigation schedules by utilizing this data. Additionally, by preserving water supplies and lessening the negative effects of agriculture on the environment, these methods support sustainability. Irrigation control and water management are essential to ensure food security and the prudent use of our planet's most valuable resource, water, as climate change exacerbates water difficulties.

3.5 Supply Chain Optimization and Traceability

Utilizing cutting-edge technology such as drones, data analytics, and Internet of Things sensors, smart farming optimizes the whole agricultural supply chain. Optimizing the supply chain guarantees effective management and delivery of resources, such as machinery, fertilizers, and seeds, to the farm. Furthermore, it makes it easier to monitor crop health and environmental conditions in real time, allowing farmers to make accurate, data-driven decisions at every stage of the farming process. At the same time, tracking technologies offer a digital history of every agricultural product, starting from the farm and ending at the market. This traceability minimizes customer worries about the origin of their food by guaranteeing food safety, transparency, and quality control. When combined, supply chain optimization and traceability improve smart farming methods' dependability, sustainability, and efficiency, which eventually advances the agriculture sector.

4 Building Blocks of Cloud-Based Smart Farming Systems

The essential elements that provide data-driven intelligence and efficiency to contemporary agriculture are the building blocks of cloud-based smart farming systems. These systems are primarily dependent on reliable sensors and Internet of Things (IoT) devices that are positioned strategically around the farm. These devices are able to gather an abundance of real-time data on crop health, weather patterns, soil conditions, and livestock status. After that, these data streams are sent to the cloud, where sophisticated machine learning and data analytics algorithms are applied. By acting as a single repository for data processing, analysis, and storage, the cloud helps farmers understand their operations better. Efficient resource allocation and decision-making are possible for farmers thanks to easily navigable dashboards and mobile applications that offer remote monitoring and control capabilities. In addition, cloud-based solutions frequently encourage cooperation and knowledge exchange among farmers, transforming our understanding of profitable and sustainable farming methods in the digital era.

4.1 Sensor Networks and Internet of Things (IoT) Devices

Agriculture is being revolutionized by sensor networks and Internet of Things (IoT) devices that offer real-time data and remote control capabilities. With the use of these tools, farmers may increase production, encourage sustainability in smart farming practices, and make data-driven choices.

With its ability to provide real-time data on a variety of agricultural factors, sensors form the core of smart farming. They gather information on animals, weather, crops, soil, and other topics. A variety of sensors, such as GPS trackers, cameras, motion sensors, temperature and humidity sensors, and soil moisture sensors, are utilized in agriculture. In order to make wise judgments in farming operations, sensors continually collect data. For analysis, this data is sent to central systems. A lot of sensors send data to central hubs or the cloud using wireless technologies like LoRaWAN, Zigbee, or cellular networks. This makes remote control and observation possible.

Connecting physical things and gadgets to the Internet is known as the Internet of Things (IoT), and it is a major influence on agriculture. IoT gadgets used in smart farming include drones, smart machinery, actuators, and sensors. These gadgets contain sensors to gather data, and they frequently have the capacity to act on the data. IoT devices assist in integrating data from several sources to present a comprehensive picture of the farm. To make judgments about irrigation, for instance, meteorological data from a weather station can be linked with information about soil moisture. IoT devices give farmers the ability to oversee and manage farm activities from a distance. For example, people may use their cellphones to control irrigation systems or keep an eye on the well-being of cattle. Precision agriculture is made possible by IoT devices because they provide data that facilitates efficient resource allocation. This may lead to lower waste, higher yields, and lower costs. IoT devices have many advantages, but there are drawbacks as well, such as the need for regular maintenance, connectivity problems, and data security.

4.2 Data Collection, Transmission, and Storage

The gathering, sharing, and storing of data are essential elements of smart farming. In the end, they increase agricultural sustainability and efficiency by empowering farmers to make data-driven decisions, automate procedures, and obtain real-time information [7].

Data Collection in Smart Farming

Data from a variety of sources, including as sensors, drones, GPS units, and even satellite pictures, is used in smart farming. Throughout the farm, sensors gather information on crop health (NDVI from drones), livestock factors (health, location), meteorological conditions (temperature, humidity, and pH), and soil conditions (moisture, pH, and temperature). Continuous data collecting yields real-time information essential for prompt decision-making. Soil moisture sensors, for instance, can tell you when to water.

Data Transmission in Smart Farming
Sensor and device data is wirelessly sent to cloud servers or central hubs. Wireless technologies that are often utilized include cellular networks, Wi-Fi, LoRaWAN, and Zigbee. Data processing may occasionally take place at the edge, or close to the data source, facilitating rapid decision-making. Only pertinent data may be filtered and sent to the cloud by edge devices. Real-time data may be accessed remotely using wireless data transmission. Even when they are not physically there, farmers may use computers or cellphones to keep an eye on their crops. One of the challenges is in guaranteeing dependable access, particularly in remote regions, and managing the security risks associated with wireless data transfer.

Data Storage in Smart Farming
Usually, sensor data transmissions are kept on the cloud. Redundancy, scalability, and accessibility are provided via cloud storage. Sensitive agricultural data must be protected. To protect data saved in the cloud, encryption, access restrictions, and frequent security upgrades are required. Utilizing strong algorithms and machine learning models, data analysis is made easier with cloud-based storage. Crop management, resource optimization, and predictive maintenance can all benefit from this approach. Farmers may use trend analysis on historical data saved in the cloud to make well-informed decisions based on previous seasons and patterns. Agribusiness data storage may need to abide by local laws, rules, and regulations, including those pertaining to data privacy.

4.3 Data Analytics and Machine Learning for Insights

Smart farming relies heavily on data analytics and machine learning, which use data's potential to provide insightful information. These technologies are able to detect patterns, abnormalities, and correlations that would be missed by human observation alone by analyzing enormous datasets gathered from sensors, satellites, drones, and different IoT devices [8]. In addition to predicting agricultural production, machine learning algorithms can identify plant diseases, improve irrigation schedules, and even recommend methods for managing animals. These insights enable farmers to decrease resource waste, boost production, make data-driven decisions, and advance agricultural sustainability [7]. Additionally, as these systems learn and adapt over time, they get better at optimizing agricultural techniques, which helps to produce food in a way that is both ecologically friendly and efficient.

4.4 Decision Support Systems and Real-Time Monitoring

Real-time monitoring and decision support systems are essential elements of smart farming, providing farmers with the resources they need to maximize their farming techniques. The continuous gathering of data from several sensors and Internet of Things (IoT) devices positioned around the farm is known as real-time monitoring [9]. These gadgets offer current data on crop health, animal status, weather trends, and soil characteristics. After that, decision support systems examine this data and provide useful insights based on past trends and advanced algorithms. Farmers may utilize the intuitive computer or smartphone interfaces to acquire this information. This helps them to make well-informed judgments about pest management, fertilization, irrigation, and other crucial farming decisions. In the end, this boosts productivity, lowers expenses, and increases crop yields while guaranteeing sustainable farming practices. Traditional agriculture is evolving into a precision- and data-driven sector and thanks to smart farming's capacity to deliver real-time data and decision assistance.

4.5 Integration of Cloud and Edge Computing

The management and optimization of agricultural operations have undergone a radical change because of the combination of edge and cloud computing in smart farming [10]. Farmers are able to access and analyze massive volumes of data from sensors and Internet of Things devices in real time thanks to cloud computing, which acts as the centralized hub for data processing, storage, and analysis. It facilitates the use of cutting-edge machine learning algorithms for predictive analytics, supporting choices about crop management and resource allocation. Conversely, edge computing lowers latency and speeds up decision-making by bringing computer resources closer to the data source [7]. Critical data may be processed locally by edge devices, enabling quick decisions like modifying irrigation schedules or sending out drones for in-the-field surveillance. The integration of edge and cloud computing in smart farming facilitates a smooth data flow, integrating the advantages of real-time responsiveness and data analytics capabilities. This eventually results in enhanced agricultural sustainability, less resource waste, and higher efficiency.

5 Cloud-Based Data Management and Storage

Managing and storing data on the cloud is the foundation of contemporary smart farming techniques. A centralized, scalable platform for gathering, storing, and analyzing massive volumes of agricultural data is offered by the cloud. Cloud storage guarantees that data on soil conditions, weather patterns, crop health, and animal metrics is effectively and securely saved, since sensors, drones, and Internet of Things devices are constantly producing data on these topics. Cloud-based data management makes it simple for stakeholders to access and share information, which helps farmers make educated decisions, agricultural researchers do analysis, and even promotes cooperation within the farming community [11]. In addition, cloud platforms provide strong security features, data redundancy, and the processing capacity required to implement sophisticated data analytics and machine learning algorithms. These capabilities enable farmers to streamline their processes, cut down on resource waste, and strive toward more productive and sustainable farming methods.

5.1 Cloud-Based Databases for Agricultural Data

A central location to store numerous kinds of agricultural data, such as details on crop health, weather patterns, soil conditions, and animal metrics, is provided via cloud-based databases. On farms of various sizes, cloud databases are easily scalable to handle massive amounts of data provided by sensors, drones, and IoT devices. Remote monitoring and analysis are made possible by the data's accessibility to farmers, academics, and stakeholders from any location with an Internet connection. Cloud providers use strong security protocols, such as encryption, access restrictions, and adherence to industry rules, to safeguard agricultural data. Real-time data updates are supported by cloud databases, guaranteeing that users may make decisions based on the most recent information available. Because they enable several users to access and interact with the same data at once, hence cloud-based databases encourage cooperation. This is beneficial for the agricultural community's knowledge and best practice exchange. Cloud systems provide the computing capacity to execute machine learning and data analytics algorithms on agricultural data, enabling farmers to make decisions based on facts. Cloud service providers usually include disaster recovery and data backup options, guaranteeing data integrity even in the case of unanticipated occurrences or hardware malfunctions. For farmers and other companies, cloud-based databases are an economical option since they do not require on-premises hardware or ongoing maintenance. A comprehensive farming ecosystem may be created by integrating these datasets with additional cloud-based agricultural technologies like precision agriculture software and decision support systems. Cloud service providers frequently follow compliance guidelines unique to their sector, which might be crucial for farms having to comply with legal regulations.

5.2 Data Warehousing and Data Lakes

Data warehousing is the process of arranging and structuring data such that it is readily available for reporting and querying. On the other hand, data lakes provide greater freedom for data exploration and analysis since they keep data in its unstructured and raw form. By handling structured data such as crop yields, weather records, and inventory levels, data warehousing may be used in smart farming to help farmers make well-informed decisions based on both historical and current data. However, unstructured data sources like sensor readings, satellite images, and drone video may be accommodated by data lakes, which makes it possible to extract insightful information for precision agriculture using sophisticated analytics, machine learning, and data mining. By combining data lakes with data warehousing, farmers may better allocate resources, increase crop yields, and advance sustainable agricultural methods thanks to a comprehensive data management ecosystem.

5.3 Data Security, Backup, and Recovery Strategies

Data security, backup, and recovery plans are essential in farming since decision-making and operational effectiveness depend on the availability and integrity of agricultural data. Farmers and agricultural groups should put strong security measures in place, such as encryption, access limits, and frequent security audits, to protect this important data. Moreover, it is imperative to build secure backup protocols. This entails routinely making duplicate copies of your data and keeping it in many geographically separated places to ensure data recovery in the case of cyberattacks, natural catastrophes, or hardware malfunctions. An extra degree of security is typically provided by the backup and recovery capabilities that are integrated into cloud-based solutions. In order to guarantee quick data restoration in the case of an unexpected incident, farmers must have well-defined data recovery procedures and test them on a regular basis. The agriculture sector may reduce risks and ensure the continuation of farming activities even in the face of unforeseen problems by placing a high priority on data protection, backup, and recovery.

5.4 Data Governance and Compliance in Agriculture

As farms and agricultural organizations handle and manage enormous volumes of data, data governance and compliance are becoming more and more crucial in the agriculture industry. Setting up guidelines, practices, and standards for data gathering, storing, access, and use is essential to effective data governance. It encourages appropriate data handling procedures while guaranteeing data security, correctness, and integrity. In agriculture, compliance frequently refers to following industry-specific rules and guidelines on data privacy, environmental protection, and food safety. Data governance, for example, is essential to complying with standards such as the General Data Protection Regulation (GDPR) for personal data or the traceability laws for food goods. The agriculture industry can use the power of data to improve farming operations' productivity and sustainability while navigating the complicated regulatory landscape and fostering trust via the implementation of strong data governance and compliance mechanisms.

6 Cloud-Enabled Agricultural Services

The agriculture sector is undergoing a radical change because of cloud-enabled technologies. These services provide farmers access to a variety of tools and capacities by utilizing cloud computing technologies. Through the cloud, farmers may get weather predictions, data-driven insights, remote monitoring, and precision agricultural solutions. Real-time decision-making, cost-effectiveness, and resource optimization are all made possible by it. Moreover, cloud-enabled agricultural services promote knowledge sharing and best practices by facilitating data sharing and cooperation between farmers and agricultural specialists. With the use of this technology, conventional agricultural practices are being completely transformed. Agriculture is becoming more productive, sustainable, and climate-adaptive, which will ultimately lead to higher yields and better food security.

6.1 Weather Forecasting and Climate Modeling

Climate modeling and weather forecasting are essential elements of smart farming, serving as critical decision-making and resource-optimization tools. Farmers may take timely measures like planting, watering, and pest management by using weather forecasting, which gives them real-time insights into the short-term weather. It protects agricultural productivity by lessening the effects of abrupt weather occurrences. On the other hand, climate modeling provides a long-term view, enabling farmers to predict and adjust to small changes in climatic patterns. Farmers are better equipped to choose crops, arrange planting dates, and implement sustainable agricultural techniques when they have a clear picture of future climatic patterns. By combining these modeling and forecasting technologies, agricultural operations become more resilient, resource-efficient, and environmentally sustainable.

6.2 Pest and Disease Prediction and Management

Predicting and managing pests and diseases is a proactive, data-driven strategy used in smart farming to protect agricultural harvests. Smart farming monitors fields and crops for early indicators of pest infestations or illnesses using a variety of technology, such as sensors, drones, and AI-powered algorithms. Through the analysis of various plant health, temperature, and humidity data, these systems are able to predict possible dangers with impressive precision. Afterward, farmers may reduce the need for broad-spectrum insecticides and lessen their negative effects on the environment by using focused treatments, such as precision pesticide applications or crop rotations. In addition to improving crop protection, this strategy encourages resource conservation and sustainable farming methods, which raise agricultural production.

6.3 Farm Management and Planning Software

The effectiveness of smart farming methods is largely dependent on the use of farm management and planning software. Farmers can now improve resource allocation, streamline operations, and make data-driven choices. This gives farmers the ability to keep an eye on and oversee a variety of agricultural operations, such as scheduling irrigation and planting crops, tracking animals, and maintaining farm equipment. These software programs combine sensor data, weather forecasts, and historical records to deliver insightful information about crop health, yield forecasts, and resource efficiency. They also make long-term planning easier, assisting farmers in adjusting to shifting environmental and market situations. In the upcoming years, it is projected that the worldwide market for farm management and planning software would increase at an exponential rate. The world's increasing need for food and agricultural by-products is the cause of this considerable growth [12]. In the end, farm management and planning software is crucial to contemporary agriculture as it fosters profitability, productivity, and sustainability.

6.4 Agricultural Marketplaces and Trading Platforms

In the age of smart farming, agricultural markets and trade platforms are transforming the way farmers purchase, sell, and advertise their goods. These digital platforms provide an effective and transparent marketplace for agricultural commodities by bringing farmers and buyers, suppliers, and consumers together on a worldwide scale. Farmers are able to increase their client base, make educated price decisions, and obtain up-to-date market data. These platforms also frequently use smart contract technology, which makes transactions safe and automated. Through use of these technologies, farmers are enabled through agricultural markets and trading platforms to enhance their income, minimize wastage, and establish a more robust and competitive agricultural ecosystem, therefore playing a role in the industry's sustainability and expansion.

6.5 Remote Equipment Monitoring and Maintenance

A key component of smart farming is remote equipment monitoring and maintenance, which helps farmers effectively manage and take care of their assets and machines. Farmers can monitor the operation and state of their equipment remotely in real time by utilizing IoT sensors and linked devices. With the use of these technologies, predictive maintenance is made possible by learning about the condition of the equipment, consumption trends, and possible problems. Alerts are delivered to farmers or service providers when abnormalities or maintenance needs are found, guaranteeing prompt repairs or upkeep. In addition to minimizing downtime, this lowers operating costs, increases the longevity of farming equipment, and encourages sustainable farming methods. Modern agriculture depends heavily on remote equipment repair and monitoring, which boost production and efficiency while guaranteeing trouble-free agricultural operations.

7 Challenges and Future Directions

There are potential for future growth and challenges specific to smart farming. The digital gap, which occurs when certain farmers lack access to the tools and technology needed for precision agriculture and result in differences in revenue and productivity, is one of the biggest obstacles to smart farming. Furthermore, there are issues with data management, analysis, and privacy arising from the massive volume of data created by smart agricultural systems. The variability of smart agricultural methods is further challenged by climate change and its unexpected effects on weather patterns. In the future, agriculture's sustainability and efficiency can be further improved by combining blockchain technology, sophisticated robots, and artificial intelligence. Moreover, as smart farming is vital to fulfilling the world's growing food demand while reducing its environmental impact, resolving environmental issues and guaranteeing security measures will be critical to the industry's future expansion.

7.1 Connectivity and Network Infrastructure

At the core of smart farming's issues is connectivity and network infrastructure. The importance of connection is only going to increase in the dynamic field of agriculture. Low-power wide-area networks (LPWANs), satellite-based Internet, and 5G and beyond will all be integrated into smart farming in the future to provide ubiquitous, fast, low-latency access to even the most distant agricultural locations. The extraordinary growth of autonomous machinery, remote-controlled operations, and real-time data analytics will be made possible by these breakthroughs. But as we move forward, issues like cybersecurity risks and data privacy issues will only become worse. Ensuring network infrastructure security and preserving the accuracy of agricultural data will be crucial. It will also continue to be difficult to close the digital gap and guarantee that rural agricultural communities in the United States have fair access to these cutting-edge networks. Overall, the future of smart farming will be shaped by the development of connection and network infrastructure, offering the agricultural sector both challenging and exciting potential.

7.2 Data Interoperability and Standardization

Two key elements influencing how smart farming develops in the future are data interoperability and standards. Agriculture is becoming more and more dependent on data-driven decision-making; thus being able to integrate and communicate data from many sources in an effortless way is essential. Future developments in smart farming will see the creation of common data formats and communication protocols that provide effective interaction across various platforms, sensors, and equipment. Farmers will be able to obtain a comprehensive understanding of their operations by combining data from many sources, including weather stations, soil sensors, and machinery. Establishing and upholding these norms in a quickly changing technology environment will be difficult, though. There will be constant problems in ensuring interoperability between new technology and existing systems while resolving privacy and data security issues. Furthermore, because agriculture is a worldwide sector, international collaboration will be necessary to unify standards internationally. In conclusion, achieving data standardization and interoperability will be essential to maximizing the benefits of smart farming, but overcoming the difficulties involved in creating and upholding these standards will continue to be a major obstacle.

7.3 Cost Considerations and Return on Investment

Smart farming's future is heavily influenced by cost and return on investment (ROI) factors. Even though using smart farming technology has a lot of potential benefits, farmers still need to consider the upfront expenditures and continuing operating expenses. The ability to lower the cost and increase the accessibility of these technologies for farmers of all sizes will determine the future course of smart farming. Technological developments in hardware, software, and networking might assist lower initial costs, but there are still issues in making sure that the return on investment is worthwhile. In order to assess the financial advantages of smart farming practices—such as higher yields, less resource use, and enhanced operational efficiency—farmers will require precise and measurable measurements. For smart farming to be widely adopted and expand sustainably, it will be imperative to address these financial concerns and show a favorable return on investment, especially for small- and medium-sized farms. One of the biggest challenges facing the business going ahead will be balancing the initial costs with the long-term benefits.

7.4 Ethical and Legal Issues in Data Ownership

The future paths and problems of smart farming raise important ethical and legal questions around data ownership. The amount of data produced in agriculture is increasing, raising more and more urgent concerns regarding data ownership and governance. There are complex discussions around data rights, privacy, and security because farmers, technology suppliers, and data aggregators all have vested interests. To solve these challenges, smart farming in the future will need well-defined legal and ethical frameworks. The difficulty is striking a balance between protecting sensitive information and farmers' rights while weighing the possible advantages of data analysis and sharing for better farming methods [13]. The growth of data markets, data-sharing agreements, and data governance systems in the agriculture sector is expected to be influenced by the ownership and management of data. Finding the ideal balance between data security and accessibility will be crucial to ensuring data is used responsibly and fairly in the rapidly changing field of smart farming.

7.5 Emerging Trends in Cloud Computing for Smart Farming

Smart farming is undergoing a transformation because of new developments in cloud computing. The growing use of edge computing in agriculture is one of the most noticeable developments. Real-time decision-making in the field is made possible by edge computing, which lowers latency by moving data processing closer to the source. This is especially helpful for jobs like autonomous machinery and precision farming. The incorporation of machine learning (ML) and artificial intelligence (AI) into cloud-based systems is another noteworthy trend. With the help of these technologies, farmers may optimize crop management and resource allocation by taking use of actionable insights from more sophisticated data analytics, predictive modeling, and automation. A unified and comprehensive picture of a farmer's activities may be created by smoothly connecting different equipment and sensors with cloud-based systems, which are also becoming increasingly interoperable. These new developments in cloud computing hold the potential to improve agricultural production, sustainability, and efficiency as smart farming develops further.

8 Collaborative Initiatives and Partnerships

Partnerships and cooperative projects are essential to the development of smart farming. To promote innovation and sustainable agriculture in this fast-paced industry, farmers, technology suppliers, academic institutions, and government agencies are collaborating more and more. These collaborations make it easier to share technology, exchange ideas, and co-create solutions to challenging agricultural problems. Collaborative approaches may include collaborations between technology enterprises and agricultural cooperatives to deliver affordable technological solutions to small-scale farmers or farmers exchanging data with researchers to create more effective crop management practices. These programs, which combine knowledge and resources, hasten the implementation of intelligent agricultural techniques and enhance the robustness and sustainability of the world's food supply chain.

8.1 Envisioning the Future of Smart Farming with Cloud Computing

Using cloud computing to envision the future of smart farming opens up a world of previously unimaginable possibilities. The foundation of smart farming will remain cloud computing, which makes it easier to integrate data from several sources, such as sensors, drones, satellites, and agricultural equipment. In order to make well-informed decisions about crop management, resource allocation, and pest control, farmers will have access to real-time data analytics and machine learning algorithms through the power of the cloud. Edge computing is going to proliferate, making quick decisions possible right there in the field. Furthermore, cloud-based technologies that optimize water consumption, minimize the environmental effect of agriculture, and reduce chemical inputs will be an integral component of the future of smart farming, which will incorporate sustainable practices. Global farmer connections will be made possible via collaborative data-sharing systems, which will promote information exchange and aid in meeting the world's expanding food needs. Future agricultural practices will become more intelligent, effective, and ecologically sensitive as cloud computing develops, guaranteeing a lucrative and sustainable future for both farmers and customers.

9 Conclusion

Cloud computing for smart farming sheds light on how technology might revolutionize the agricultural industry. Farmers may fully use precision agriculture, data-driven decision-making, and sustainable practices by integrating cloud-based technologies. Real-time data processing, predictive analytics, and remote monitoring are made possible by the cloud, which gives farmers the ability to maximize resource allocation, raise yields, and lessen their impact on the environment. Cloud computing will continue to lead smart farming in the future by encouraging cooperation, creativity, and scalability. It does, however, also present issues with privacy, data security, and equal access. Stakeholders must collaborate to overcome these problems in order to fully fulfill the potential of cloud computing in agriculture and guarantee that the advantages of smart farming are promising to address the global food crises, accessible, and sustainable. In the end, cloud computing for smart farming offers a bright future where agriculture and technology come together to create a more sustainable world.

References

1. C. Prakash, L.P. Singh, A. Gupta, S.K. Lohan, Advancements in smart farming: A comprehensive review of IoT, wireless communication, sensors, and hardware for agricultural automation. Sensors Actuators A Phys. **362**, 114605 (2023). https://doi.org/10.1016/j.sna.2023.114605
2. E. Navarro, N. Costa, A. Pereira, A systematic review of IoT solutions for smart farming. Sensors (2020). https://doi.org/10.3390/s20154231
3. Australian Cyber Security Centre, Cloud computing security considerations (2021, 6 Oct). Retrieved from: https://www.cyber.gov.au/resources-business-and-government/maintaining-devices-and-systems/cloud-security-guidance/cloud-computing-security-considerations
4. Amazon Web Services, Types of cloud computing (2023). Retrieved from amazon.com website: https://aws.amazon.com/types-of-cloud-computing/
5. Rozenstein. et al., Data-driven agriculture and sustainable farming: Friends or foes? Precis. Agric (2023) Open Access. Retrieved from: https://link.springer.com/article/10.1007/s11119-023-10061-5
6. M.S. Farooq et al., A survey on the role of IoT in agriculture for the implementation of smart livestock environment. IEEE Access (2022) Retrieved from: https://ieeexplore.ieee.org/stamp/stamp.jsp?tp=&arnumber=9681084
7. M. Amiri-Zarandi et al., Big Data privacy in smart farming: A review. Sustainability (2022). https://doi.org/10.3390/su14159120
8. R. Cyriac et al., *LMH-RPL: A Load Balancing and Mobility Aware Secure Hybrid Routing Protocol for Low Power Lossy Network*, vol 17 (Inderscience Publishers, 2022), pp. 224–269

9. R. Cyriac et al., RPL enhancement with mobility-aware two-stage objective function for improving network lifetime in IoT by Robin Cyriac; M.A. Saleem Durai. Int. J. Electron. Bus. (IJEB), **17**(3), (2022)
10. V. Balamurugan et al., Enhanced Elman spike neural network based fractional order discrete Tchebyshev encryption fostered big data analytical method for enhancing cloud data security. Wirel. Netw **29**. Springer, 523–537 (2023)
11. R. Karthikeyan et al., COSCO2: AI-augmented evolutionary algorithm based workload prediction framework for sustainable cloud data centers. Trans. Emerg. Telecommun. Technol. **34**, e4652 (2023) Wiley
12. I. Hristoski et al., Farm management software for increased productivity and competitiveness, in *5th International Balkan and Near Eastern Social Sciences Congress Series (IBANESS)*, (2017, Sept), pp. 375–392
13. S. Leonelli, H.F. Willianson, *Towards Responsible Plant Data Linkage: Data Challenges for Agricultural Research and Development* (Springer, Cham, 2022. ISBN 978-3-031-13275-9). https://doi.org/10.1007/978-3-031-13276-6

Applications of UAV-AD (Unmanned Aerial Vehicle-Agricultural Drones) in Precision Farming

Sandhya Soman, Rakesh Gnanasekaran, Gnanasankaran Natarajan, and Fatema Khalifa Said ALSaidi

1 Introduction

Farming has undergone a considerable transition from hunting and gathering to settled agriculture. As [1] humankind evolved, there has been a continuous effort to understand and observe the plants and their domestication to transform them into crops. Every revolution unfolded techniques to increase productivity in agriculture, unveiled new crop breeds, and introduced new agricultural technologies.

Farming is witnessing another revolution, popularly termed precision farming, aka smart agriculture. It has been defined as [2] "*the usage of techniques and tools which facilitate the farmers to enhance the quality of soil and increase the productivity through technological intervention.*"

With advanced technological techniques, crop-specific treatment is possible, and the exact intervention time can be gauged and utilized to enhance crop productivity. This technological advancement has given farmers [3] the insight to understand specific crop needs and adapt their farming practices accordingly. Studies have revealed that precision farming has significantly reduced the environmental risk.

S. Soman (✉)
GITAM (Deemed-to-be) University, Bangalore, Karnataka, India
e-mail: ssoman@gitam.edu

R. Gnanasekaran · G. Natarajan
Department of Computer Science, Thiagarajar College, Madurai, Tamil Nadu, India

F. K. S. ALSaidi
Department of Information Technology, University of Technology and Applied Sciences-Al Mussanah, Muladdah, Oman
e-mail: fatema.alsaidi@act.edu.om

S. Balasubaramanian et al. (eds.), *Intelligent Robots and Drones for Precision Agriculture*, Signals and Communication Technology,
https://doi.org/10.1007/978-3-031-51195-0_15

1.1 What Led to Precision Farming

The population around the world is increasing rapidly, with an expectation of reaching ten million by 2050 [4], which has led to an increased demand for agricultural production. The involvement of the technology can facilitate the production of such scale, which can boost the farm output. Precision farming is, thus, very crucial for the farmers and agricultural sector. With the growth seen in the past few years in the field of Big Data and data analytics, the power of these techniques and methodologies can be leveraged in smart farming. It can help alleviate the problems related to the degradation of soil. Since farmers can get crop-specific information, it can be used to reduce the chemical application for the production of crops.

Efficient water utilization and quality improvement are other benefits that can be derived from this. Most importantly, in an agro-nation like ours, it can help improve the farmers' social and economic conditions. Figure 1 depicts the timeline of developments in agriculture.

1.2 Data Points of Precision Farming

In this section, we shall discuss some of the fundamental data points used in smart farming.

Precision farming is crop-specific. The characteristics of the crop, i.e., the type of nutrients required, the growth cycle of the crop, diseases affecting the yield, and stages of the life cycle, are pivotal data points.

The main steps of precision farming include gathering data, mapping variability, determining the nature of the soil and suitable crop type, and making decisions based on the above information.

The qualitative aspects of the soil include the properties, nutrient content, temperature, and toxicity.

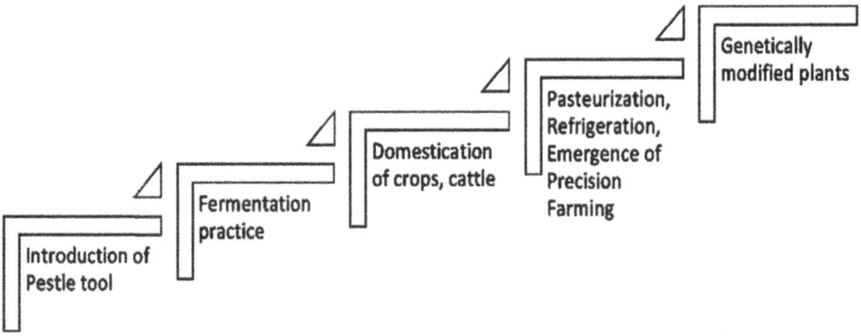

Fig. 1 Some of the significant events in the timeline of agriculture. (Own image)

The crop's seasonal and daily climate data are needed at the micro- and macro-levels.

Water requirements for the plant.

Technological equipment and sensors that facilitate:

Application of fertilizer, pesticides, etc., in the right amount in the right place.

Detection of variation in the color of the field due to changes in the type of soil, boundaries of the area, etc.

Recording of yield across a field

Detect soil properties

Detecting acidic areas, eroded soils, water-logged areas, dry areas

- Detection of soil roughness and moisture content
- Identification of insects harmful to crops
- Finding those areas in the field where nitrogen is more

2 Application of Drones in Precision Agriculture

India's agriculture, forestry, and fishery contribute slightly over 20% of the national GDP. This sector offers employment to the highest number of people, valued at over 152 million. Precision agriculture in India can significantly boost productivity and yield in this sector.

Drones are unmanned aerial vehicles. Typically, a drone consists of sensors, actuators, a communication module, and a ground control station. The algorithms govern the drone's positioning, the speed at which it flies, and its pitch, roll, and yaw [5].

One of the primary advantages of these devices is that [6] they can be customized according to the present needs. All that needs to be done is to install the required hardware and the algorithms for the task.

The union government has approved the usage of drones for activities in agriculture, which has paved the way for a new revolution. Farmers hire drones from entrepreneurs to perform tasks like pesticide spraying on their fields [7]. This is more cost-effective for them when compared to the usage of satellites or other aircraft.

Per the literature published, drones have been primarily classified as fixed-wing and multi-rotor. The fixed-wing cuts the air at a specific angle, while the multi-rotor uses the motor's direction and speed to propel. Based on the number of engines, they can be further classified as single, quad (4), hex (6), and Oct (8)-copters. The classification has been depicted in Fig. 2.

From the multiple applications of drones in the agriculture sector, in the present chapter, we shall be focusing only on the five most important application areas, i.e., crop spraying, crop monitoring, mapping, and soil analysis, livestock monitoring, and seed planting, as shown in Fig. 3.

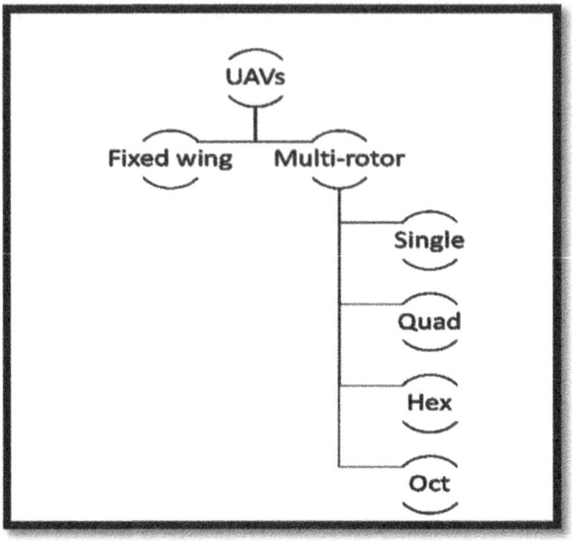

Fig. 2 General classification of drones

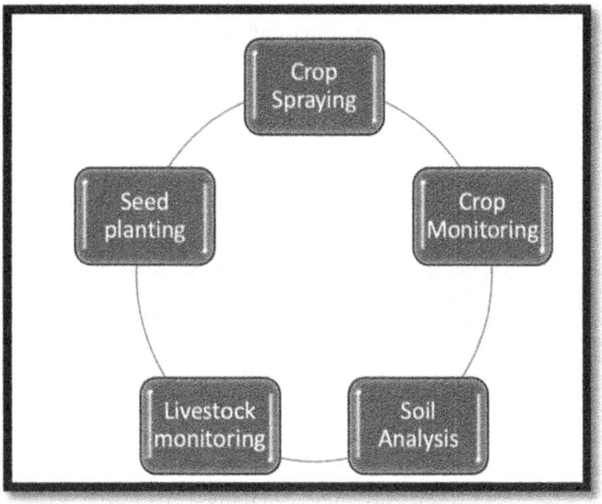

Fig. 3 Common applications of UAV-AD

2.1 Monitoring of Crops [8]

Drones and UAVs enable capturing images of high resolution both in spatial and temporal front since they can fly at low altitudes, making them helpful in capturing cereal images. They also assist in crop phenotyping, i.e., it enables us to know the condition of crops under variable/changing environmental conditions [2]. Other than this, it also reveals information about the weeds, pests, nutrient quantity, and other diseases of the soil/farm.

Weed management requires high-resolution cameras and a topography survey to differentiate the plants. AI and computer-based algorithms can help in this segmentation task.

Another area where crop monitoring is beneficial is the identification of pests and disease conditions in plants. The biomass content in a plant is measured, which can reveal any nutrient deficiency. This permits us to take corrective measures, and any deficiency can be treated well before the harvest. This is also beneficial in identifying areas with nutrient deficiency, like less nitrogen content. Drones can also help spray once such deficient zones/regions are identified. The drones can be equipped with sensors that can deviate the drone path based on the speed of the wind and its intensity.

Monitoring can benefit phenotyping, where the plant features can be monitored against different growth stages and generations. Manual observations for the same are error-prone and quite time-consuming. These are beneficial in determining those plant species that can adapt/survive well compared to others.

The crop images can also be used to determine vegetative indexes. The biomass contents, canopy structure, etc. are selected from red/infrared bands. Visible bands help determine nitrogen contents.

Usually, multi-rotor, helicopter, or fixed-winged types are used for monitoring. However, each style has its limitations. Multi-rotors are suitable for smaller areas because of their lower speed and capacity. Helicopters and fixed-winged can cover larger areas and hence are preferred.

2.2 Crop Spraying

Drones have helped spray pesticides and nutrient solutions. Once the monitoring has revealed significant results, the areas to be operated upon can be decided. A drone used for spraying [9] typically follows this mechanism. Once the joystick position of the drone is adjusted, the gyroscope adjusts the calibration, and based on the movements in the throttle and the signals from the gyroscope, the UAV moves to the specified location. Without throttle movement, the drone returns to its saved start location.

Such drones must be small, have significant payload capabilities, be easy to use, deal with RGB and other spectral images, and be lightweight. However, the existing drones cannot function well with images captured in poor weather conditions, making it an open area to be worked on. Multi-copters are a preferred choice in this area because of their stability. The significant challenges in this area include:

- The cost involved
- Battery capacity
- Vision limitation of drones
- Lack of accurate algorithms for detection and identification

2.3 Soil Analysis

Soil is the primary medium for the nutrients required for crop growth. The health of the soil determines the quality of the yield of the crop. Traditionally, soil analysis was done manually, which was a cumbersome, error-prone, and time-consuming process. The sample was collected from one area and sent for analysis to a third party, which, in turn, took some processing time before the results would reach the farmer. Also, this process is quite cumbersome and nightmarish for big fields.

A remote monitoring system for analyzing the condition of the soil turned out to be a blessing as the manual approach is tedious. Farmers (without third-party support) usually need more expertise to do the same.

Drones have been particularly beneficial in analyzing the condition of the soil as they can operate on a predefined path in a pre-specified pattern. They have cameras that capture information from the EM spectrum, capturing light in near-infrared, infrared, and UV light. The wavelengths reflected by different elements are collected and analyzed by software to get information about the composition of the soil. The land images of the path are taken, which are processed algorithms like inverse modeling, interpolation methods, etc. The data collected by the drones in this way is compared to the ground truth values. This information can be further used to estimate the yield of the crop. These techniques enable covering large acres of land in significantly less time.

Drones have been advantageous over satellite images as the latter lack visibility during cloudy weather conditions, and images might not be available for the current season on time. Drones also facilitate getting higher-resolution images at a cm-level (centimeter) compared to m-level (meter) in satellites. An example of such a drone has been quoted in [10]. DJI Phantom 4 Pro [11] observes the conditions around the field, captures images, processes them, and sends them to the connected AR glasses, which the farmer can wear in real time to observe and understand the field conditions. The information gathered can be used to determine the soil's seeding pattern, watering need and pattern, fertilizer requirement, etc.

2.4 Monitoring Livestock

Agricultural drones can be used for livestock management, i.e., they can be used to detect, count, and track animals [12]. Drones can be used to provide live feed about the location of the cattle and can help farmers act accordingly. Studies have revealed the usage of quadcopters for monitoring live stocks. Such drones have ML algorithms for detection, movement tracking, and counting. For detection and counting, earlier thresholding, morphological operations, and masks were used, which could separate the cattle images from the background and remove the noises. The resultant blobs were counted for the final result. Owing to their primitive nature, the accuracy is low. To boost the accuracy, CNN-based architecture with RPN (regional proposal networks) created regions of interest as boxes around the cattle, which were further fed to an R-CNN for count and SVM for classification. Other techniques adopted were YOLO and segmentation using U-Net/Mask R-CNN with residual networks.

The usage of drones effectively in this segment poses several challenges. The first one includes the selection of the correct drone. The features of drones, like their wing type, maneuver capability, payload, and application domain, play a significant role in their selection. Another area is sensor selection, often limited by battery life, load, and price. The drone's capabilities are also determined by its flight time and the areas it can cover. This can be solved by fitting larger batteries in the drones. Operational issues like working around rough terrains can lead to problems. Reducing the payload and autopilot working with GPS can help in this regard.

Economic factors could hinder the adoption of drones on a broader scale. Other challenges include atmospheric conditions and cattle behavior like grazing in groups, which can challenge effective segmentation.

Some solutions that can address the mentioned challenges include using energy-aware and low-power networks and deep learning architectures that could function with limited UAV resources.

2.5 *Seed Planting*

Drones can be used for planting seeds, especially in areas where manual accessibility is low. Sites that are hindered by natural obstacles like hills and other landscape features have low accessibility and also increase the risk for the workers involved. Drones can help to mitigate this problem. One of the advantages drones offer, besides accessibility, is the ability to save time. The drones initially gauge the land on which the seeds must be planted and are equipped with canisters to hold the seeds. The funnel facilitates the dropping of sources at the selected location. Generally, the seed containers/pods have a nutrient solution that helps in seed germination. The seed dispersal mechanism might vary and may be controlled with the help of IoT. Since these drones need to carry more payload, a hex-copter configuration is preferred to a quad [13]. A larger compartment for releasing seeds could help the drone have more smooth and precise rotations and reduce the hindrance in planting seeds.

3 Challenges Faced

Agricultural drones present a plethora of opportunities in farming [14]. However, its adoption introduces several challenges as well [15].

A summary of the challenges faced have been listed below and are also depicted in Fig. 4. Some of the prominent ones have been explained below.

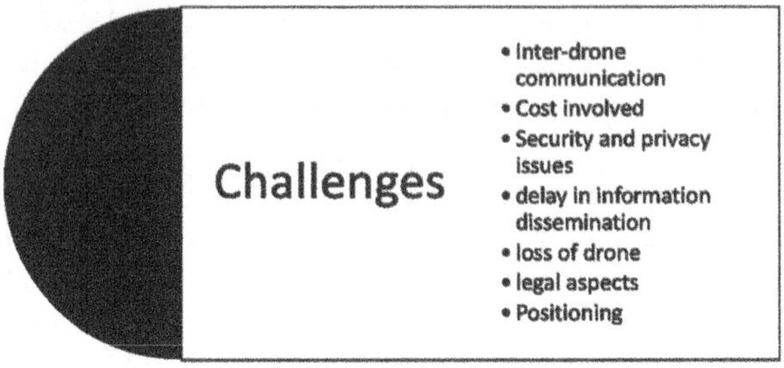

Fig. 4 Challenges with drones

3.1 Inter-drone Communication

Drones require a wireless medium for communication. The cellular network can be adopted for this purpose. But, since these networks are designed for terrestrial devices, they might be overloaded, which may result in a collision, primarily since these drones operate in data above LOS.

A possible solution to the above could be to have a combination of cellular and non-cellular drones. The cellular drones would connect with the base station through the network. The non-cellular can have a mutual connection with each other and other cellular nodes. This can significantly bring down the cost of communication as well.

3.2 Associated Costs

Implementing drones requires specific hardware and software installations to be in place. This includes systems for communication channels, devices with battery backups, and devices having ample storage for capturing data and images. This shoots up the installation charges and hence becomes unaffordable for most farmers. The technical know-how required for these systems can also hinder their adoption.

3.3 Security and Privacy Issues

Since drones use the cellular channel for communication, they pose a vulnerability threat to data privacy. Attacks like the localization attack may provide falsified farm images to the farmer, and they may be misled about the quality and quantity of farm produce. The data about the farm, crop, and other sensitive information may be exposed to malicious intruders. One of the most accessible possible solutions to combat the former situation would be incorporating authentication protocols. Literature has also shown the usage of blockchain for storing and transmitting this data.

3.4 Delay in Information Dissemination

The agricultural drones are designed [16] with limited capabilities to reduce their price. The results should be processed in real time for the entire scheme to be effective. The latency in processing can be significantly reduced if the processing happens in the drone itself. If the processing occurs in the base station, this will bring up the processing time, and the delay could also cause harm in some cases, like an intruder attack on the farm.

3.5 Loss of Drone

There is a threat of losing the drones because of malicious attacks by intruders. The functioning of the drones may also be affected by the loss of network and changes in weather conditions.

3.6 Legal Aspects

To ensure hassle-free operation, appropriate licensing needs to be procured. The drones used for spraying and other agricultural purposes must be certified by the Director General of Civil Aviation in India. In other countries, the respective certifying agencies must be approached to secure all legal permissions and ensure hassle-free operations.

3.7 Positioning

The operations of drones are relatively autonomous with the help of GPS and networks, but the attack on the drone's trajectories might hinder the maneuvering capabilities. The positioning of drones faces challenges, especially in heavily populated areas.

4 Sensors Used in Precision Agriculture

A sensor can be defined as a device [2] that accepts specific inputs and transforms them into signals that we can further process to understand different phenomena. For example, the sensors can capture the moisture level in the soil, and the farmer can use it to take corrective measures. A wide range of sensors have been used in agriculture. This section summarizes a few of them and their functionalities through Fig. 5.

5 Embedded Systems in Smart Farming

To obtain the full benefits of precision agriculture, drones or other devices must be able to perform the calculations/processing in the device itself instead of in the base stations. If the information is processed only after it reaches the base station, it would increase the response time. It would also delay timely responses, which may incur huge losses in some situations.

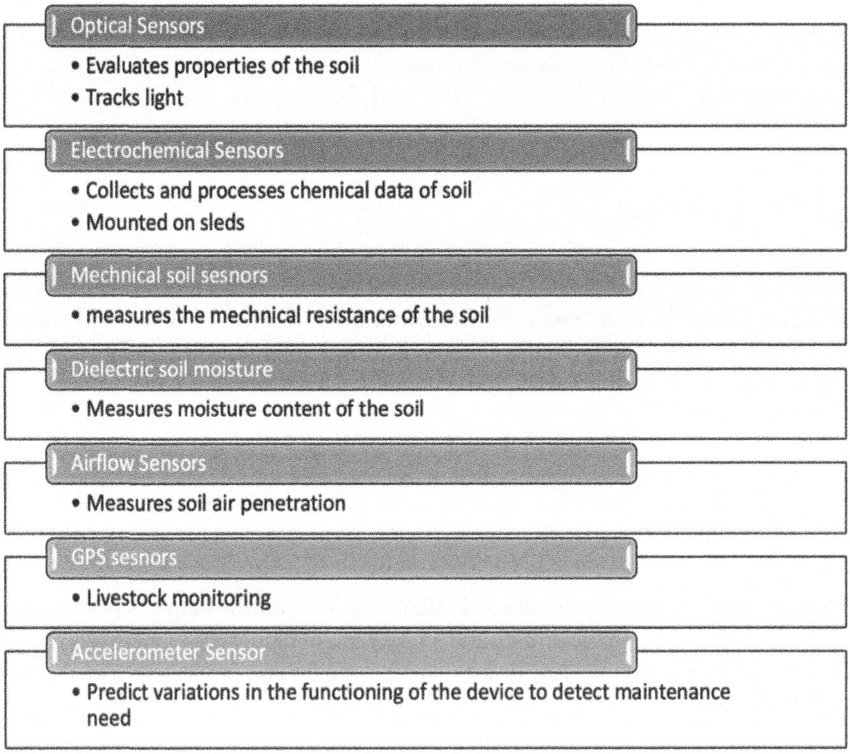

Fig. 5 Sensors in precision agriculture

One possible solution to the above problem is to have a design approach [3] that combines hardware and software and allows the inclusion of algorithms capable of processing the data collected by the sensors.

Various works in the literature have used embedded systems. The following is a list of applications in which embedded systems have been used.

- Systems have been used for the detection of plant diseases in real time. Although the system cannot adapt to the data flow from the camera, it still offers real-time data processing.
- The binary patterns captured in the images have been used to classify weeds from the plants.
- Embedded algorithms can monitor honey bees' activities, especially in fruit farms.
- CNN-based algorithms to detect the germination of seeds.
- FCN-based systems for management of greenhouses.
- Semantic and CNN for detection of early symptoms of diseases.
- R-CNN algorithms for counting the cattle.
- CNN-based systems for tracking animals that may potentially attack the farms.
- Estimating the water level in the field using neural networks.
- Microcontroller-based systems for evaluating the various states of the plants.

These algorithms have been developed in different types of environments. The environments are either homogenous, e.g., CPU-CPU, GPU-GPU, or maybe heterogenous, e.g., CPU-GPUs. Figure 6 summarizes the other application areas.

The algorithms heavily depend on the following factors:

1. Power required for processing
2. The dimensions of the device
3. The computing capability

A significant constraint for these algorithms is the size of the field they operate upon. Operating larger areas would require more consumption of energy and battery power. Hence, energy-efficient algorithms can be an area that requires further exploration.

Another area that can significantly improve their performance includes DSS. The existence of autonomous decision-making systems for this real-time embedded system is an open area of research and requires further attention.

6 Role of ML and AI in Smart Farming

Most authors have classified the various agricultural tasks into three main categories. The *tasks* before harvesting are grouped under one category; the tasks done during harvesting form the next category, while the tasks done after harvesting fall under the third category.

Fig. 6 Application of ES in precision agriculture

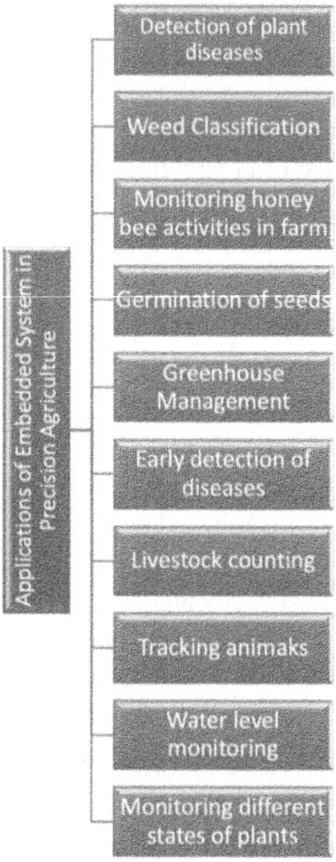

Each of the categories has specific essential parameters that are considered significant during each task. Various machine learning algorithms have been employed to capture these parameters and use them to derive crucial decisions. This section provides a list of ML algorithms that have been engaged in the past to capture these parameters at various stages.

6.1 Supervised Algorithms

The following figure (Fig. 7) gives a summary of the unsupervised algorithms that have been used in intelligent farming.

6.2 Unsupervised Algorithms

The following figure (Fig. 8) gives a summary of the unsupervised algorithms which have been used in smart farming.

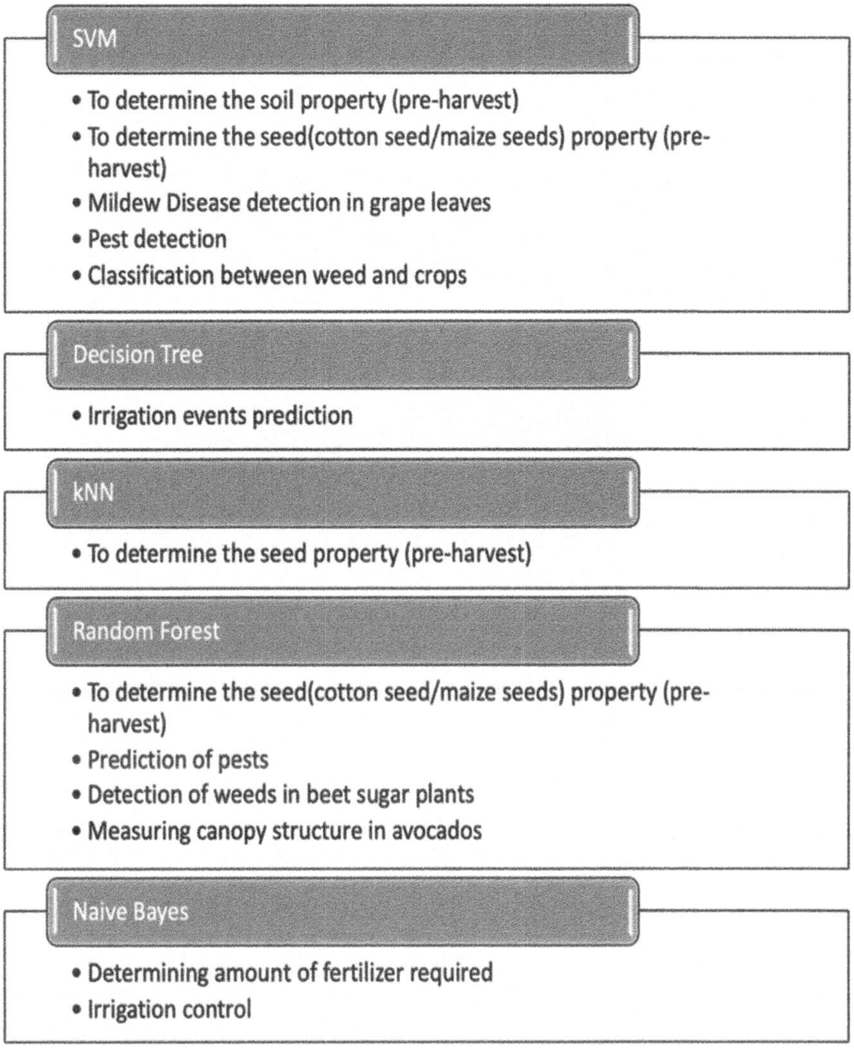

Fig. 7 Supervised algorithms

7 Proposition of the Components for an Intelligent System for Precision Agriculture (ISPA)

The following section proposes the components of an Intelligent System for Precision Agriculture (ISPA). Figure 9 depicts the main modules. The textual description of the modules for an ISPA can be explained as follows:

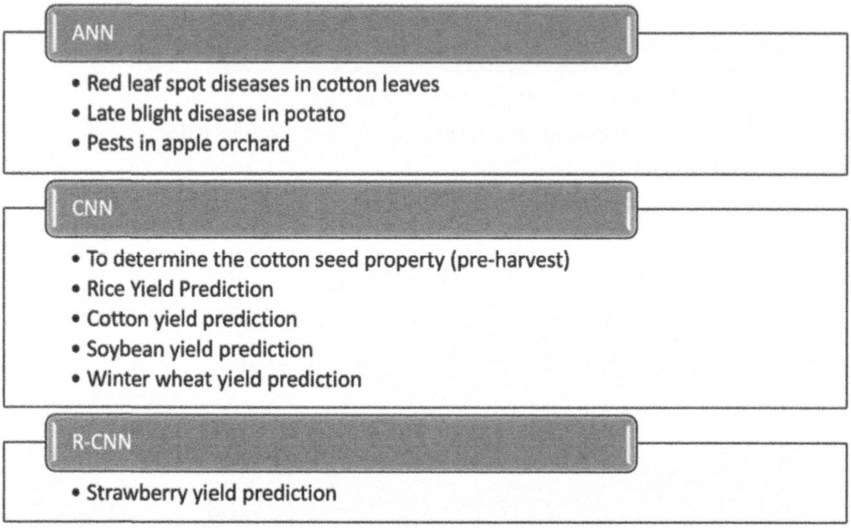

Fig. 8 Unsupervised algorithms

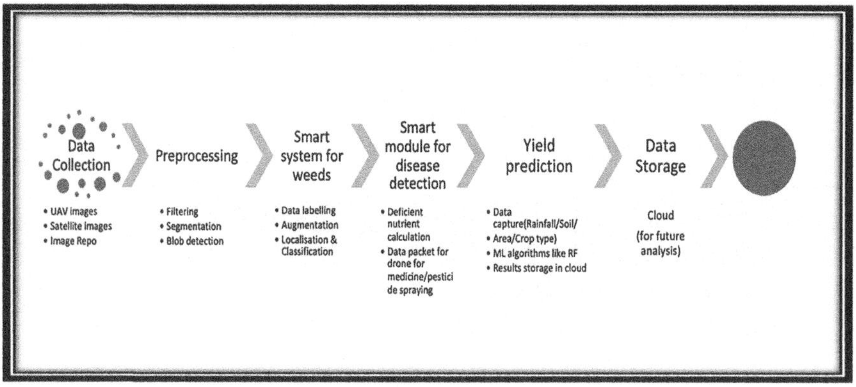

Fig. 9 Main components of ISPA

7.1 Data Collection

Data for the IS may be collected from the captured drone or satellite images. Some standard pictures can also be considered from public data repositories for training the ML models.

7.2 Preprocessing

The acquired data may not be in a form which can be utilized for analysis. Hence, preprocessing steps make it fit for analysis. Applying filters like Decor stretch has been advocated in the literature as beneficial. Segmenting the portions from the image and identifying blobs are the primary tasks at this stage to make it suitable for weed detection.

7.3 Weed Detection

Some properties of the image that can be utilized for differentiating weeds from crops include the brightness determining the medium red and medium green. The classifier uses these properties to localize the weed locations and label them.

7.4 Disease Detection

The drone images can be compared with the pre-trained model from standard datasets; blobs can be identified, and affected areas of the crops can be localized. The above process can help detect discolored plant parts, which can signify nutrient deficiency. The data can be sent to the drone analyzer subsystem to demarcate the areas where the nutrient solution needs to be applied.

7.5 Yield Prediction

From the drone images and data from sensors, properties of the area, like rainfall, soil type, humidity, and other location-specific features, can be captured to determine yield for a crop. Literature has shown algorithms using random forest to predict the yield of maize crops.

7.6 Data Storage

Instead of being stored locally, the data can be stored in the cloud and used later for further and future analysis.

8 The Way Forward: Agriculture 4.0 [17]

Smart farming introduced the usage of technology in agriculture. A new revolution in agriculture is being unveiled—popularly termed Agriculture 4.0. This represents more efficient usage of techniques and technology to enhance farm yields. The various operations of agriculture generate tremendous data, [18] which so far had been overlooked. Agriculture 4.0, with the aid of Big Data, AI, IoT, and AR, brings together digital equipment that can gather and process volumes of data and produces results and insights that can be used for timely processing and conduct of actions.

A list of application areas and advantages of the involvement of technology in agriculture has been described in this section and illustrated in Fig. 10.

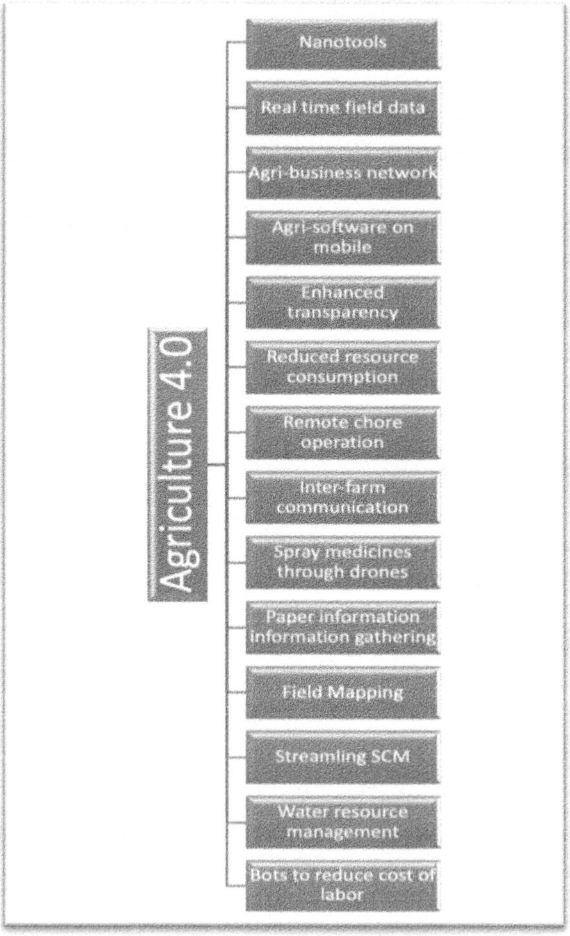

Fig. 10 Application of Agriculture 4.0

- Applying nanotools like nano-biosensors, nano-pesticides, and sensor data enables intelligent and timely decisions. The usage of robots helps in the reduction of regular farming chores.
- Agricultural drones and satellites can be used to capture real-time field data, which assists the farmers in understanding the nature of the soil, its nutrient level, and its health. This data can also be consolidated and stored as a blockchain record.
- The formation of a network connecting agriculture and business, also termed the farmers, can use the agri-business network to streamline prices according to the market demand and trends. The distributors can also use this information to utilize the market advantage.
- The introduction of mobile software can easily collect data about the farm and geography and can be used for trusted interaction through [19] techniques like blockchain technology.
- Technology usage will enhance the traceability and transparency of the entire system.
- It can bring down resource consumption to a significant level as the farmers can handle many chores remotely and without mainstream involvement. This can further strengthen the socioeconomic condition of our farmers.
- The farmers can access information from other farmers working with similar crops and climatic conditions. This will be particularly helpful in preventing the spread of infectious diseases in crops and cattle and thus prevent severe losses incurred due to them.
- Agricultural drones can be helpful while dealing with crop diseases. They can be used to spray medicines across the field. This is helpful because[20] it would reduce the need for direct human contact and provide enhanced field coverage.
- Time-sensitive data can be maintained in a paperless fashion. This digitization prevents unnecessary data tampering and speeds up the transmission time.
- As real-time data is available to the farmers, they can analyze and make accurate decisions about farm issues like crop selection and the amount of pesticide dosage required.
- Drones can be used for field mapping.
- The data generated from each field is massive and can be collectively used to streamline the farms' entire supply chain management.

9 Research Opportunities

The study has unveiled a couple of opportunities for open research, which can be listed as follows:

- Creation of a common set of protocols and standards that could be used for drone communication and information dissemination.[21]
- Protocols to reduce the delay in information transfer.

- The drone can process information in real-time using local models rather than relying on base systems to do the processing.
- Applications are reachable to the commoner to educate them and are relatively easy to operate upon. This can significantly scale up the adoption of agricultural drones.

10 Conclusion

This book chapter attempts to summarize the myriad areas where drones aid in improving the techniques used in present-day agriculture. With the population increasing rapidly, it is the need of the hour to have measures to increase agricultural productivity as well.

The chapter also proposes the "components of an intelligent system" for precision agriculture, which includes subsystems for data collection, weed, disease, yield detection, and a mechanism to store the data for future use.

References

1. M.A. Alanezi, M.S. Shahriar, M.B. Hasan, S. Ahmed, Y.A. Sha'aban, H.R.E.H. Bouchekara, Livestock management with unmanned aerial vehicles: A review. IEEE Access **10**, 45001–45028 (2022). https://doi.org/10.1109/ACCESS.2022.3168295
2. M, What types of sensors are used in precision agriculture? GeoPard Agriculture, (2022), https://geopard.tech/blog/what-are-the-types-of-sensors-used-in-agriculture/. Accessed 16 Dec 2022
3. Wikipedia Contributors, Precision agriculture, Wikipedia, (2023), https://en.wikipedia.org/wiki/Precision_agriculture. Accessed 16 July 2023
4. ED, Seed Sowing or Tree Planting Drone, (2022), https://www.skyfilabs.com/project-ideas/seed-sowing-tree-planting-drone. Accessed Nov 2022
5. M. Javaid, A. Haleem, R.P. Singh, R. Suman, Enhancing smart farming through the applications of Agriculture 4.0 technologies. Int. J. Intell. Netw. **3**, 150–164 (2022). https://doi.org/10.1016/j.ijin.2022.09.004
6. M.Z. Ghazali, A. Azmin, W. Rahiman, Drone implementation in precision agriculture – A survey. Int. J. Emerg. Technol. Adv. Eng. **12**(4), 67–77 (2022). https://doi.org/10.46338/ijetae0422_10
7. P. Biswas, Saves chemicals, time and money, in *Sangli District, Drone Use for Spraying Fields Takes Off*, (The Indian Express, 2022) https://indianexpress.com/article/cities/pune/saves-chemicals-time-money-sangli-drone-spraying-fields-maharashtra-8228003/. Accessed 12 May 2023
8. J. Cuaran, J.I. Leon, Crop Monitoring using Unmanned Aerial Vehicles: A Review. Agric. Rev. (2021). https://doi.org/10.18805/ag.r-180
9. A. Hafeez, M.A. Husain, S.K. Singh, A. Chauhan, M.S. Khan, N. Kumar, A. Chauhan, S. Soni, Implementation of drone technology for farm monitoring & pesticide spraying: A review. Inf. Process. Agric. (2022). https://doi.org/10.1016/j.inpa.2022.02.002
10. R.S. Rolle, G. Mrema, P. Soni, G. Agriculture, *A Regional Strategy for Sustainable Agricultural Mechanization. Sustainable Mechanization Across Agri-Food Chains in Asia*

and the Pacific region (RAP Publication, 2015) http://www.fao.org/publications/card/en/c/78c1b49f-b5c2-43b5-abdf-e63bb6955f4f/
11. Z. Ünal, Smart farming becomes even smarter with deep learning – A bibliographical analysis. IEEE Access **8**, 105587–105609 (2020). https://doi.org/10.1109/access.2020.3000175
12. Wikipedia Contributors, Agricultural revolution, Wikipedia, (2022), https://en.wikipedia.org/wiki/Agricultural_revolution. Accessed 1 Apr 2023
13. McCormick, Precision farming: What is it and what benefits does it offer, (2021), https://wwfw.mccormick.it/as/precision-farming/. Accessed 23 Nov 2022
14. R. Shakeri, M.A. Al-Garadi, A. Badawy, A. Mohamed, T. Khattab, A. Al-Ali, K.A. Harras, M. Guizani, Design challenges of multi-UAV systems in cyber-physical applications: A comprehensive survey and future directions. IEEE Commun. Surv. Tutor. **21**(4), 3340–3385 (2019). https://doi.org/10.1109/comst.2019.2924143
15. C.P. Singh, R. Mishra, H.P. Gupta, P. Kumari, The internet of drones in precision agriculture: Challenges, solutions, and research opportunities. IEEE Internet Things Mag. **5**(1), 180–184 (2022). https://doi.org/10.1109/iotm.006.2100100
16. M. Usama, Soil sampling with drones for precision agriculture, Drone Below, (2018), https://dronebelow.com/2018/09/13/soil-sampling-with-drones-for-precision-agriculture/. Accessed 26 Jan 2023
17. Wikipedia Contributors, Unmanned aerial vehicle, Wikipedia, (2023), https://en.wikipedia.org/wiki/Unmanned_aerial_vehicle. Accessed 16 Apr 2023
18. A. Wright, Drone Soil Analysis: Multispectral Remote Sensing for Soil Mapping. Mapware – 3D Mapping | Photogrammetry | Flight Planning, (2022), https://mapware.com/blog/drone-soil-analysis-multispectral-remote-sensing-for-soil-mapping/. Accessed 26 Mar 2023
19. A. Saddik, R. Latif, A. Elouardi, A. Singh, A. Khelifi, Computer development based embedded systems in precision agriculture: Tools and application. Acta Agriculturae Scandinavica Sect. B Soil Plant Sci. **72**(1), 589–611 (2022). https://doi.org/10.1080/09064710.2021.2024874
20. T.A. Shaikh, T. Rasool, F.R. Lone, Towards leveraging the role of machine learning and artificial intelligence in precision agriculture and smart farming. Comput. Electron. Agric **198**, 107119 (2022). https://doi.org/10.1016/j.compag.2022.107119
21. Precision Agriculture- Revolutionizing Agriculture, (n.d.), https://www.cropin.com/precision-agriculture. Accessed 20 Apr 2023

Crop and Fertiliser Recommendation System for Sustainable Agricultural Development

K. Sankareswari and G. Sujatha

1 Introduction

In India, agriculture is vital to the country's economy. In spite of notable improvements in the service industry, agriculture continues to be India's biggest employer and source of income. An efficient system is required to address various agricultural problems and increase agricultural production. With so many possibilities for Internet searches available nowadays, it might be challenging to decide what we actually need. Indian farmers' economic circumstances are unstable, and they are finding it difficult to decide which crop to plant in their fields. The entity that keeps life alive and responsible for sustainable agriculture on earth is the soil, also known as the soul of infinite life. A portion of the world's land area is covered by soil, a dynamic, three-dimensional substance. It changes depending on the location. The parent rock beneath the surface, time, organisms, geography, and climate are the five elements that make it up. Four primary tasks are carried out by soil: It acts as a habitat for burrowing mammals, bacterial, insects, fungi and other species; it recycles raw components; it cleans water; it provides the structural support for engineering projects like bridges, buildings and roads; and it also serves as a growing environment for plants.

Additionally, soil contributes to the production of food and preserves the harmony of regional, national and international environmental standards. Farmers in

K. Sankareswari (✉)
Department of Computer Science, Sri Meenakshi Govt. Arts College for Women (A), Affiliated to Madurai Kamaraj University, Madurai, India

Department of Computer Science, The American College, Madurai, India

G. Sujatha
Department of Computer Science, Sri Meenakshi Govt. Arts College for Women (A), Affiliated to Madurai Kamaraj University, Madurai, India

S. Balasubaramanian et al. (eds.), *Intelligent Robots and Drones for Precision Agriculture*, Signals and Communication Technology,
https://doi.org/10.1007/978-3-031-51195-0_16

India have been using traditional agricultural practices for generations to ensure crop yield and production. They controlled the soil fertility maintenance. However, the use of some types of seeds, herbicides and chemical fertilisers upsets the equilibrium between agricultural yield production and fertility [1]. The population has been expanding significantly lately. Crop production is insufficient to meet the needs of the current population. Nutrient-rich soil has become increasingly significant in the production of food, agriculture and industry. We focused on the soil nutrients in this study and used time-series data obtained from the previous farming season.

Crop selection is usually based on their land soil parameters, i.e. nitrogen, phosphorous, potassium and pH level [2]. The fundamental introduction to soil is shown in Sect. 1. A relevant information matrix and a description of an existing recommender system are presented in Sect. 2 of the literature study. It was also considered how crop recommendation research has changed over time. Crop selection categorisation algorithms are presented in Sect. 3. The method used for crop selection is defined in Sect. 3 in addition to the results and an explanation of the study in Sect. 4. In Sect. 5, the publication has finally summarised the study's findings while illuminating several research perspectives.

1.1 Physical Properties of Soil

Soil is made up of minerals, organic materials, water and air. These combinations determine the properties of the soil (Fig. 1). The most obvious aspects of soil are its physical characteristics, which may be seen without the aid of instruments like scanners or microscopes. They serve as a reflection of the placement of the sand, silt and clay solid soil particles. They can be used to categorise different horizons and soil types. Additionally, they work fantastically well for both laboratory and outdoor demonstrations.

1.1.1 Texture

The fineness or coarseness of the mineral particles in the soil is referred to as the soil texture. Silt, clay and sand proportions affect the texture of the soil. Loam is the sum of all three of them (Fig. 2). Sand and silt have no importance to the soil because they don't help the soil's capacity to retain water or nutrients in the soil. Due to its small particle size, high surface area per mass and assistance in storing ions and water, clay is an active component of soil texture. The drainage, compressibility, nutrient fixation, water holding capacity and aeration of the soil are all impacted by the soil texture.

Minerals and biological materials are both present in the soil. Soil texture can be used to determine its mineral content. Soil texture is defined by the ratios of silt, sand and clay [3].

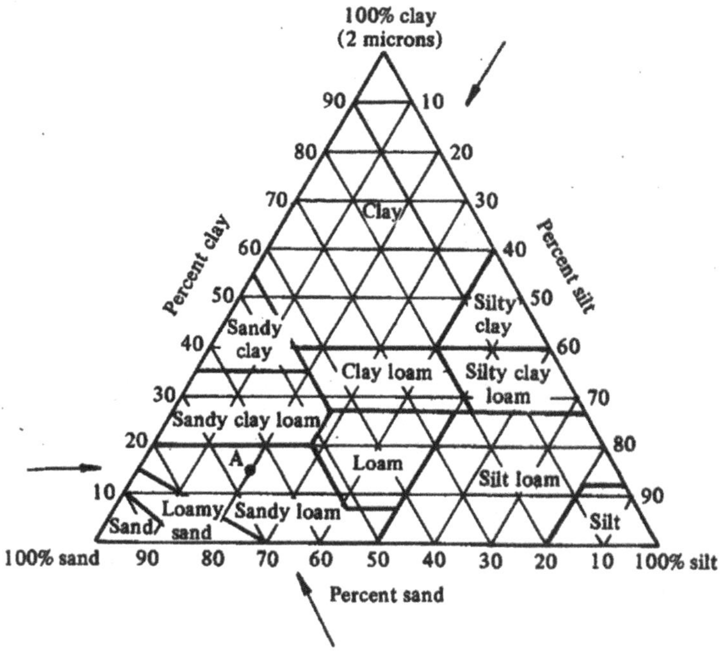

Fig. 1 Soil properties

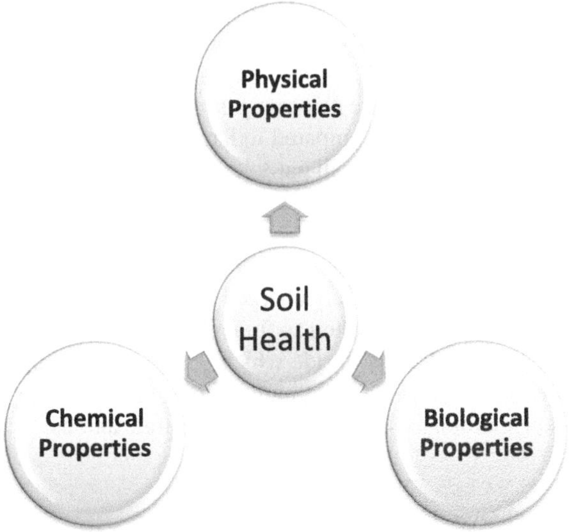

Fig. 2 Texture of soil. (Source: European Soil Data Centre)

1.1.2 Structure

The arrangement of sand, silt and clay particles is referred to as the soil structure. Soil organisms like bacteria and earthworms and organic materials (decaying plants and animals) also have an effect on soil structure. Silt and clay are usually always clumped together into bigger units called aggregates, but sand is frequently found in soil as individual particles. The structure of soil is determined by how this aggregation occurs.

1.1.3 Pore Space

The gaps between soil particles, known as pores, are rarely firmly packed together. Pore space is the portion of the bulk volume of soil that is neither covered by mineral nor organic materials but rather is an open space that is either filled with gases or water.

1.2 Chemical Properties of Soil

In comparison to the rocks and minerals from which they were formed, soils are chemically different in that they have a higher concentration of relatively insoluble elements like iron and aluminium and a lower concentration of soluble weathering products like sodium (Na), magnesium (Mg), calcium (Ca) and potassium (K). Iron oxides and aluminium are typically found in high concentrations in old, severely worn soils.

Because chemical reactions take place on particle surfaces, soil chemical activity and particle size are connected. Compared to large particles, small particles have a substantially larger surface area. Small particles involve a significant contribution in two chemical processes, management of soil pH and support of the soil capacity to store nutrients (CEC).

First of all, it's crucial to understand that fertilisers are salts. A positively charged ion and an anion which is a negatively charged ion are formed when salts dissolve in the soil solution. For instance, when sodium chloride dissolves in water, sodium which is positively charged and chloride which is negatively charged ions are produced. When we apply fertiliser containing sodium nitrate to the soil, it disintegrates into the soil solution as sodium cations and nitrate anions.

1.2.1 Cation Exchange Capacity (CEC)

The phrase "cation exchange capacity" describes how well soil can store and exchange cations. Constantly, ions are exchanged between plant roots, CEC sites on clay and humus particles and the soil solution. The electron charge affects this process, which is not random [3].

Humus and clay both display high CECs because of their numerous negative sites, high surface-to-volume ratios and small particle sizes. Sand exhibited extremely low CEC due to its huge particle size, low surface-to-volume ratio and consequently lower number of negative sites.

When growing in a lump of clay or humus, a gardener can less frequently use higher rates of fertiliser as opposed to sandy soil because the soil's particles can hold cations. It is preferable to fertilise in sandy soil more regularly with lower amounts of fertiliser since sandy soil cannot contain the same quantity of cations.

1.2.2 pH

The pH of the soil describes how acidic and alkaline the soil reacts. The pH scale has numbers 0 through 14. The soils typically have a pH range of 4.0 to 8.0. For soil, a pH of 7 is regarded as neutral. A soil's pH level greater than 7 suggests an alkaline soil, whereas one less than 7 indicates an acidic soil.

The pH level of the soil is very important factor because it has a significant influence on plant growth and it controls how readily each nutrient is available to plants in the soil. It affects the types of microorganisms, quantity and activity of soil. These microorganisms in turn affect how quickly agricultural wastes, manures, sludge and other organic materials decompose. Additionally, it has an impact on other nutrient conversions, solubility, plant and several crucial plant nutrients. Phosphorus is most readily available in soils that are slightly acidic to slightly alkaline soils, but all other vital micronutrients except molybdenum become more readily available when pH decreases. Even aluminium, manganese and even iron can become soluble if pH is less than 5.5 and become hazardous to plants. In general, bacteria that are important in a variety of nutrient transformation pathways under soils tend to be most active in conditions that range from slightly acidic to slightly alkaline.

1.3 Biological Properties

The direct and indirect effects of the living things that inhabit a given soil are represented by the soil's biological properties. The biological properties of soil represent how well the soil supports life [3].

The biological characteristics of soil are influenced by the microorganisms that reside there. Organic matter in soil contains metabolites, waste and residue from plants and animals that decompose well and serve as fertilisers.

The organic matter in the soil is transformed by bacteria into usable forms such as ammonia, phosphate, sulphate, etc. All living organisms and soil microorganisms including bacteria, viruses, fungi, protozoa, nematodes, rodents, etc. contribute to keeping the soil's ecosystem in balance.

1.4 Plant Nutrients

Oxygen, carbon and hydrogen which are supplied by water and air are the three important elements for plant growth. Plant nutrients are obtained from the soil or applied as fertilisers which enter plants through their roots. Plant nutrients can be classified into two types, i.e. macronutrients and micronutrients. Nitrogen, potassium, phosphorus, magnesium and calcium are those that plants need in high quantities. On the other hand, micronutrients that are required in small amounts for plant growth include boron, cobalt, copper, iron, zinc, sodium and so on. Micro- and macronutrients are needed in different amounts; the regular growth and development of plants depend on both macronutrients and micronutrients and those required in different amounts.

Numerous goods and procedures can deliver nutrients. When deciding on the right fertiliser and application technique for each circumstance, factors such as cost, accessibility, usability, required tools, time and philosophy should be taken into account. Some micronutrients may occasionally be sprayed onto crop foliage in cases of extreme nutrient deficit, but the majority are applied to the soil and absorbed by plant roots. Nutrients are dissolved in water and applied to the visible roots of plants in hydroponic production systems.

Most soils still contain some nutrients. Only a soil analysis can determine this. Spending money and resources on fertiliser without considering the results of a soil test can exacerbate an already-existing nutrient imbalance. Additionally, nutrients may occasionally be in adequate supply but unavailable due to an improper pH balance. This can be discovered with a soil test, and experts in soil labs and crop consultants can suggest solutions.

1.5 Types of Soil

The word “soil” is a term to describe the loose coating of earth that covers the surface of the globe. The soil contains humus, crushed rock as well as both inorganic and organic components. The growth of soil from rocks often takes 500 years or longer. Normally, soil is created when rocks break down into their constituent elements. The soil is formed when the rocks are broken up into tiny bits as a result of several forces, such as the effect of wind, water and salt reactions. There are three soil stages: (i) soil that is firm, (ii) soil that has air pockets and (iii) soil with pores filled with water.

Different environmental influences are applied to diverse soil types. The texture, ratios and different mineral and organic compositions of soil make up its main properties. Four categories of soil exist: there are sandy soil, silty soil, loamy soil and clay soil shown in Fig. 3.

Fig. 3 Four types of soil

1.5.1 Sandy Soil

Sandy soils are among the worst forms of soil for growing plants because they have a low level of nutrients and a poor ability to hold water. It is challenging for plant roots to absorb water. Sand-like soil is typically produced by the breakdown or fragmentation of rocks like granite, limestone and quartz.

1.5.2 Silty Soil

Silt is composed of smaller particles than sand and larger mineral particles than clay. Due to its smoothness and fineness, it retains water better than sand. The third type of soil, silt, is the most fertile and is frequently found next to rivers, lakes and other bodies of water. Additionally, it is used in agricultural practices to improve soil fertility.

1.5.3 Clay Soil

Clay has the smallest particles, which are closely packed together and devoid of any airspace, as compared to the other two forms of soil. Due to this soil's substantial ability to retain water, it is impervious to both air and moisture. When wet, it feels somewhat sticky to the touch; after drying, it feels smooth. The densest and heavy form of soil is clay, which does not drain properly or allow plant roots to spread out very far.

1.5.4 Loamy Soil

The fourth type of soil is loam. Sand, silt and clay are mixed together to highlight each material's advantages. It is more suitable for agricultural growing and cultivation since it can keep nutrients and moisture. Given that all three types of soil materials are present and in harmony, it is sometimes referred to as agricultural soil. Additionally, the calcium and pH levels are higher.

1.6 Soil Testing

Soil testing is a useful tool for assessing soil's characteristics and nutrient level, selecting the right crops to be cultivated and aids in calculating nutrients to be needed to a particular soil depending on its soil fertility and crop requirements [5]. The soil test reports are used to categorise several important soil fertility indices like soil pH, nitrogen (N), phosphorus (P), potassium (K), organic carbon (OC) and boron (B). The classification and forecasting of soil parameters at the village level help to reduce the excessive use of fertiliser, save the time of experts in chemical soil analysis, boost up soil fertility and profitability and enhance the health of the environment.

Smart agriculture and precision agriculture is a modern technique that is helpful to address various problems that have affected agricultural production. Different algorithms based on machine learning and deep learning have been proposed to suggest the suitable crop based on the location-specific data such as soil types, soil characteristics, soil nutrients at the micro and macro level and so on. Different algorithms based on the soil properties have been used to recommend the crop for the farmers.

2 Related Work

Nowadays, crop and fertiliser recommending systems have developed as a result of various research using various methods relating to agriculture crops in the agriculture field. This section describes different researchers' contributions to the domain.

Motwani et al. and Panchamurthi [1, 2] proposed a system for crop recommendation that employs on random forest and convolutional neural network (CNN) to suggest a suitable crop based on various characteristics including the soil type, region, crop yield, selling price and so on. The accuracy of the random forest algorithm was 75%, whereas CNN was 95.21%. The primary challenge for Indian farmers is that they usually fail to select the right crop for their soil.

Bharath et al. [3] proposed a data mining technique that uses the naive Bayes algorithm to recommend the crop based on nutritional features obtained from soil samples collected by the farmer which was tested in a soil testing lab. The proposed model's accuracy was 75%.

Bouni et al. [4] presented a crop classification system using the deep reinforcement learning (DRL) method for supporting precision agriculture to recommend suitable crops for the farmers. The proposed system for crop recommendation compared with various machine learning algorithms such as KNN, random forest and naive Bayes for high accuracy and efficiency to suggest a site-specific crop.

Suchithra et al. [5] developed a model to classify soil based on soil fertility indices and pH values. The system might help to decide system to overcome the soil nutrient deficiency problems. The results of the machine learning classifier and decision system showed that optimized extreme learning machine (ELM) parameters help to develop a model for classifying soil based on soil fertility index. Classification issues have been solved using the fast learning classification method termed as extreme learning machine (ELM) with various activation functions such as sine-squared, Gaussian radial basis, triangular, hyperbolic tangent and sine squared. Gaussian radial basis functions outperformed with better performance in classification.

Precision agriculture which is a modern farming technique has been used to solve this issue for farmers. In order to solve this issue, Pudumalar et al. (2017) proposed system for precision agriculture that uses an ensemble model using different classifiers random tree, naive Bayes, KNN and CHAID to suggest a crop based on location-specific parameters such as soil type, characteristics and crop yield.

In order to recommend the crop and crop productivity, [7] built a system for crop recommendation which combines the predictions together from different machine learning classifiers to recommend the suitable crop based on the soil characteristics and its type. Random forest, naive Bayes and linear SVM are used in this ensemble model which classifies the soil input dataset for Kharif and Rabi crop types. The dataset includes information such as chemical and physical characteristics of the soil as well as climatic parameters including temperature and rainfall. The average classification accuracy achieved by using this model is 99.91%.

Doshi et al. [8] designed an intelligent system that uses Big Data and machine learning together to suggest the right crop for their farmer land based on environmental factors, physical and chemical characteristics of soil and geographical location.

Pruthviraj et al. [9] proposed machine learning (ML)-based model intends to categorise the sample soil datasets into four different groups, namely, very high fertile, high fertile, moderately fertile and low fertile soil using the support vector machine (SVM). Additionally, it offers fertilisers that can be used to further increase the soil fertility of the soil and forecasts the acceptable crops that can be grown based on the class to which the soil sample belongs. Farmers can choose which crop to produce based on the soil classification and determine the appropriate nitrogen (N), phosphorous (P) and potassium (K) fertiliser ratio by using the proposed model. SVM performed more accurately when compared to decision tree (DT), K-nearest neighbour (KNN) and other algorithms.

Kumar et al. [10] presented a model for predicting the suitable crop and pest detection and suggests pest control techniques. In this study, they use the logistic regression, SVM and decision tree method. They proved that the SVM classification model provides greater accuracy than other algorithms.

Rajak et al. [11] proposed a model that uses learners such as SVM, random forest, naive Bayes and ANN to recommend a suitable crop with higher accuracy and efficiency for location-specific values.

To suggest suitable crops and forecast production value, Venugopal et al. [12] proposed and compared three machine learning algorithms such as random forest, logistic regression and naive Bayes. They gathered historical information on the climate, temperature and a variety of other variables to predict the crops. The random forest algorithm outperformed with the higher accuracy than the other two algorithms.

Mahendra et al. [13] presented a technique that predicted the best crop based on soil characteristics such as composition, soil PH, weather and rainfall to predict rainfall. The SVM algorithm was applied for rainfall prediction whereas decision tree was used to predict the crops. For the benefit of farmers, we used machine learning techniques to create the system. Additionally, the system offers details on the type and amount of fertilisers and seeds needed for production. Therefore, farmers can grow a new variety of crops to raise their profit margin and prevent soil contamination.

Pande et al. [14] proposed a crop prediction system which is a mobile application for the farmers. GPS assist the farmers to find their location. Machine learning algorithms help the farmers to choose the right crop which is more profitable and predict its yield. Machine learning algorithms, including random forest (RF), support vector machine (SVM), multivariate linear regression (MLR), artificial neural network (ANN) and K-nearest neighbour (KNN), are used to estimate crop productivity. The random forest has outperformed the best accuracy of 95% than others. The algorithm also makes recommendations for fertilisers to achieve high crop yield.

Chakraborty et al. [15] suggested a system focuses on site-specific crop management and help farmers to choose right crop by taking into account all the factors such as soil type, sowing season and geographic location.

Goel et al. [16] To reach the final goal of advising farmers on the best crop based on a variety of regional and farm-related characteristics, a softmax classifier and a nature-inspired algorithm are used in the computational intelligence-based expert system that has been presented. It incorporates the creation of a hybrid technique employing two nature-inspired optimisation algorithms, namely, optimisation based on biogeography and optimisation based on plate tectonics. Later, a hybrid PBO/Adam algorithm is produced by combining PBO and the Adam optimisation technique. The weights of the classifier are then maximised using the proposed hybrid technique.

Priya and Ramesh [17] have recognised that stability of crop's productivity cannot be stretched without the prudent application of macro- and micronutrients to address any deficits in soil nutrients. The suggested study focuses on applying the Adaboost.RT methods to precisely determine the N-P-K content of the designated land that are also helpful for the farmers to determine how much amount of fertiliser has been used for their land. The discussed method is helpful for increasing the crop yield. A comparison of the nutritional intake using the suggested strategy and traditional approaches is also made. Experimental results show that the proposed algorithm outperformed than other current approaches.

Priya et al. [18] proposed random forest algorithm to predict the crop yield. The proposed model consists of parameters such as temperature, production season and rainfall.

Khan and Ghosh [19] proposed a regression model based on neural network to predict the rainfall in the specific geographical location and crop yield. They used Meteorological Data of Chhattisgarh for the last 10 years which includes rainfall, temperature and humidity values, and they also collected the data from Ahmednagar, India, weather station. They used support vector machine (SVM) as a machine learning classifier and obtained an accuracy of 97%.

Suresh et al. [20] proposed a framework based on supervised machine learning approach to suggest appropriate harvests and recommend crop yield based on the predicted values.

Sharma et al. [21] proposed a supervised learning approach to predict the uncertain rainfall in uncertain volume in regions and compared the accuracy using the ROC curve of machine learning classifiers such as naive Bayes, multilayer perceptron, random forest and SMO. This model is used to classify low, medium and heavy rainfall, and it is practically implemented in regions which have different uncertain rainfall in uncertain volume.

Rajak et al. [22] proposed a crop recommendation system that uses an ensemble model with a majority voting technique using different classifiers random forest, artificial neural network (ANN), support vector machine (SVM) and naive Bayes to suggest a crop based on soil features such as pH, moisture, water density, etc., which are collected from universities and soil testing labs.

Champaneri et al. [23] used random forest algorithm to predict the crop yield for the specific state of Maharashtra. They collected climatic parameters including temperature, vapour pressure, precipitation and cloud cover from different government websites at a monthly basis which was used to train the model.

Reddy et al. [24] suggested a two-step model suggest the crop for the farmers. Soil classification can be done during the first step, and crop suggestion can be done during the second step. In the first phase, chemical characteristics of the soil, such as potassium, magnesium, moisture, etc., were used to predict the type of soil. To suggest crops in the second step, classification algorithms including KNN, support vector machine and bagging were used.

Crop selection is one of the major problems for the farmers at the beginning state of the agricultural process which would produce a high yield. An efficient system is required for monitoring agricultural problems and supporting farmers to identify the crops to be cultivated [25–27]. Farmers are depending on soil for growing crops which is a warehouse of minerals. Soil is of different types in India, and soil properties and nature of the soil can vary from one location to another location. Each soil can have different levels of minerals, nutrients and organic matter and can have different characteristics based on the location. So, farmers need to know the soil types and features of various kinds of soils to understand the right crop to be cultivated in that particular soil type in different climate conditions and what kind of pesticides and fertilisers can be used for better crop yield. This process requires farmers' experience. To assist this process, Table 1 shows the information matrix describing the detailed study of related work to address the problem based on different approaches and techniques.

3 Input Sources and Methods

3.1 Dataset

Machine learning and deep learning algorithms will not achieve the desired results without the contribution of high-quality and large number of training data. When algorithms for machine learning and deep learning are trained on insufficient or irrelevant data, they are useless. The publications under study offered a variety of suggestions for inputs. The dataset provided in reviewed articles includes characteristics like nitrogen (N), phosphorus (P), potassium (K), temperature, rainfall, humidity and pH level of soil. Table 2 provides a list of the attributes that are present in the dataset.

3.2 Dataset Pre-processing

Data pre-processing uses raw data to generate clean data. When the data are acquired from various sources in raw form, analysis of data is not possible. We can change data into a coherent format by using various techniques, such as substituting missing values and null values. The final phase in the data pre-processing process is the separation of training and testing data. Data pre-processing was done in the reviewed articles due to the fact that the model training usually requires as much high-quality data as possible.

3.3 Classifiers for Crop and Fertiliser Recommendation

In this study, we investigated various articles provided by the researchers in the same domain. Precision agriculture-enabled farming methods would provide the best results for the farmers with the least amount of input. The most widely used machine learning algorithms such as linear regression, naive Bayes, SVM, decision tree, random forest, logistic regression, Ada Boost and XG Boost were quickly introduced in this section.

Table 1 Related work information matrix

Reference	Methodology/ technique	Dataset	Strength	Weakness	Accuracy
[1, 2]	CNN and random forest	Image dataset comprises the 15–200 images of clay, black, red and alluvial soil for phase 1, and phase 2 comprises the features such as crop, area of land, season, district, state, etc. which was collected from kaggle	CNN gave better results than random forest model for image classification-based soil dataset	Adapt to mixture of soil types and dataset in less number	Accuracy of CNN, 95.21%; and the random forest, 75%
[3]	Naive Bayes algorithm	Soil input from soil testing lab	To predict the suitable crop in right time to grow using data mining techniques based on soil features	A proposed model is not compared with other ensemble techniques	75%
[4]	Deep reinforcement learning method	Soil parameters obtained from soil testing lab	Proposed system is evaluated against different machine learning algorithms including random forest, naive Bayes and KNN to order to recommend a right crop with high level of accuracy for the specific location	Previous record should be included through LSTM to improve the outcome	98%
[5]	Extreme machine learning with various activation functions	Soil input dataset collected from farmers and soil testing lab	Classify soil and predict soil pH value and soil fertility indices to make a decision system to overcome the soil nutrient deficiency problems	Result obtained from only particular geographical location	The Gaussian radial basis functions obtained better performance in classification (nearly 90%)

(continued)

Table 1 (continued)

Reference	Methodology/ technique	Dataset	Strength	Weakness	Accuracy
[6]	KNN, naive Bayes, random tree and CHAID	The soil dataset comprises the features including colour, pH, texture, depth, permeability, drainage, water holding and erosion	Recommend a crop with high level of accuracy and efficiency to increase crop productivity and profit	Improved dataset collection with many attributes is required	88%
[7]	Naive Bayes, linear SVM and random forest	Soil input dataset for Kharif and Rabi crop types. The dataset contains climatic parameters such as temperature, rainfall, the chemical and physical of the soil	Predictions from various machine learning models combined to predict the crop selection using soil-specific parameters and characteristics with high accuracy	Input dataset is classified into crop type Kharif and Rabi only	The classification accuracy was 99.91%
[8]	Neural network and linear regression	Crop yield, soil and climatic parameters	Uses machine learning and big data together to help the farmers to recommend the right crop for their land using environmental factors, physical and chemical characteristics of soil and geographical location	Improved dataset collection with many attributes is required	Neural network, 89.88% Linear regression, 88.26%
[9]	SVM	Soil features such as nitrogen-phosphorous-potassium (NPK)	The proposed system compared the results of SVM with KNN and decision tree	Fewer details are provided for the proposed model and its performance	SVM outperformed with higher accuracy

(continued)

Table 1 (continued)

Reference	Methodology/ technique	Dataset	Strength	Weakness	Accuracy
[10]	SVM, the decision tree and the logistic regression	The dataset consists of soil integral attributes and information about the general crops	The proposed model for predicting the suitable crop, pest detection and suggest pest control techniques	Fewer soil attributes and a small dataset	SVM – 89.66% KNN – 88% Random forest – 88% Naive Bayes – 88% Decision tree – 86.8% Logistic regression – 86.4%
[11]	SVM, random forest, naive Bayes and ANN	Dataset comprises the data obtained from sensors such as NPK, temperature, soil moisture and soil pH level	Sensors for soil testing integrated into the proposed ML and IoT-based system. The proposed system was useful maintaining crop health and reducing the likelihood of soil deterioration	No evaluation metrics are used	Not mentioned
[12]	Logistic regression, naive Bayes and random forest	The dataset comprised the factors like temperature, rainfall, area, etc.	Accurate prediction of crop and calculation of crop yield	Soil parameters were not added for crop prediction	Random forest – 92.81% Naive Bayes – 91.5% Logistic regression – 87.8%

(continued)

Table 1 (continued)

Reference	Methodology/ technique	Dataset	Strength	Weakness	Accuracy
[13]	SVM – To predict rainfall Decision tree – To suggest crop	Data of soil and weather features such as temperature, humidity, soil pH and rainfall collected from different resources like VC farm Mandya, Govt. websites, APMC website etc.	GUI-based farmer-friendly system that predict the right crop for particular farmer's land Also provide information about the required nutrient level, required amount of fertilisers, required seeds, expected crop yield and its market price using GPS location. Therefore, farmers can grow a new variety of crop to raise their profit margin and prevent soil contamination	GPS location was given manually to predict the best crop	Not mentioned
[14]	Random forest, SVM, multivariate linear regression, ANN and K-nearest neighbour (KNN)	Soil and climatic parameters	GPS-enabled mobile applications assist farmers to find their location and choose the more profitable crop. Additionally, the proposed work also suggests to farmers when to use fertilisers for the crops to increase the crop yield	Fewer details are provided for the proposed model and its performance	The random forest has outperformed higher accuracy of 95% than others

(continued)

Table 1 (continued)

Reference	Methodology/ technique	Dataset	Strength	Weakness	Accuracy
[15]	Decision tree, KNN, linear regression, naive Bayes, neural network, SVM	GPS and soil parameters	Modern agricultural technology is used to implement precision agriculture	Fewer details are provided for the proposed model and its performance	SVM – 78% Linear regression – 88.26% Decision tree – 81% Naive Bayes – 82% KNN – 85% Cross validation – 88% Neural network – 89.88%
[16]	Hybrid PBO/ Adam algorithm	Soil-based features obtained from Landsat images	Crop suggestion based on farm and geographical parameters	No evaluation metrics are used	Not mentioned
[17]	Adaboost.RT method	The dataset comprises the features of soil parameters	Accurate prediction of N-P-K content	Less details provided for the proposed model and its performance	Proposed method achieved higher accuracy than traditional methods
[18]	Random forest	The dataset consists of rainfall, karif and rabi seasons of every district, max and min temperature and crop production in tonnes sourced for the years 1997 to 2013	Accurate crop yield prediction using the random forest algorithm	Proposed model's efficiency is not compared with other ensemble techniques	Not mentioned
[19]	SVM	Meteorological data of Chhattisgarh include rainfall, humidity and temperature for the last 10 years	Proposed neural network regression model was built to predict the value of crop yield and rainfall in the specific geographical location	Fewer details are provided for the proposed model and its performance	97%

(continued)

Table 1 (continued)

Reference	Methodology/ technique	Dataset	Strength	Weakness	Accuracy
[20]	Data mining techniques and machine learning algorithm	NPK and soil pH	A supervised machine learning-based framework was proposed to suggest appropriate harvests and recommending crop yield based on the predicted values	No evaluation metrics are used	Not mentioned
[21]	Multilayer perceptron, naive Bayes, random forest and SMO	Dataset consists of rainfall data volume hourly, weekly and monthly	The proposed method is used to classify the rainfall into low, medium and high	Fewer details are provided for the proposed model and its performance	–
[22]	Naive Bayes, random forest, support vector machine (SVM) and artificial neural network(ANN)	Soil features such as pH level, water holding capacity, water density, etc. collected from universities and soil testing labs	Recommend a crop with high level of accuracy and efficiency to increase crop productivity and profit	Proposed method can be tested with rich set of dataset collection with many attributes	SVM and ANN provided high accuracy and efficiency than others
[23]	Random forest	Climatic parameters including temperature, vapour pressure, precipitation and cloud cover	In all crops, the usability and accuracy are greater than 75%	Proposed model's efficiency is not compared with other ensemble techniques	>75%
[24]	KNN, support vector machine and bagging	Soil dataset and crop dataset	Forecast suitable crop for a farmer's specific land	The accuracy values were very poor	Not mentioned

3.3.1 Decision Tree

Machine learning, which is used for both classification and prediction, has benefited from the decision tree's expanded application. The objective is to create a model that uses a tree-like structure and predicts the value of the target variable. A decision tree's single node is the starting point, and it branches out to possible decisions. Each choice opens up new nodes that branch off into other options [32].

Table 2 Features of soil dataset

1	Nitrogen (N)
2	Phosphorus (P)
3	Potassium (K)
4	Temperature
5	Humidity
6	pH
7	Rainfall

3.3.2 Linear Regression

Linear regression is a supervised machine learning algorithm and uses independent variables to design a model to predict a value. It performs a regression task and used to establish the relationship between variables.

3.3.3 Logistic Regression

The supervised machine learning classification approach known as logistic regression uses the sigmoid function to forecast the data and is based on the idea of probability (Eq. 1).

$$g(x) = \frac{1}{1 + e - x} \tag{1}$$

The real value is mapped with another value between 0 and 1 by a sigmoid function.

3.3.4 Random Forest

Random forest is a well-known machine learning algorithm can be applied for both classification and regression. It is based on the principle of ensemble learning and the supervised learning technique, which uses a number of decision trees on various subsets of the provided dataset. Each tree prediction is used to predict the final output. When it has a greater number of trees, it improves accuracy and avoids the overfitting issue [32].

3.3.5 Naive Bayes

The naive Bayes is a statistical classifier model developed based on the Bayes theorem to predict the probability of occurrence that a given sample or set of data points will fall into a specific class. The Bayes theorem's equation (Eq. 2) is as follows:

$$P(H \mid X) = \frac{P(H \mid X)P(H)}{P(X)} \tag{2}$$

where $P(H)$ is the prior probability of class (target) and $P(H|X)$ is the posterior probability, whereas $P(X|H)$ and $P(X)$ are the likelihood and prior probabilities of the predictor, respectively [32].

3.3.6 Support Vector Machine (SVM)

It is a machine learning technique that applies supervised learning for both classification and regression. In machine learning, classification is mostly accomplished by using it. Linear SVM and non-linear SVM are the two categories into which it can be divided. Non-linear SVM is utilised for datasets which cannot be classified by using a straight line, i.e. non-linearly separable data, while linear SVM is utilised for datasets which can be divided into two classes by using a straight line, i.e. linearly separable data.

3.3.7 *K*-Nearest Neighbours

KNN is a non-parametric technique that applies supervised learning for classification and regression problems. The KNN categorises test samples based on the majority of its *K*-nearest neighbours with shortest distance representing the characteristics that are shared by most test samples [32].

3.3.8 Ada Boost

Ada Boost algorithm is also referred to as adaptive boosting. The weights are redistributed to each instance, giving instances that were incorrectly recognised additional weights; this is why the method is known as "adaptive boosting". Boosting is used to reduce bias and variation in supervised learning. It operates on the premise that learners develop in stages. Except for the first, every student after that is created from a previous learner. Simply put, weak students become strong students [32].

3.3.9 XGBoost

The acronym for XGBoost is Extreme Gradient Boosting, and it uses gradient boosting framework. The ensemble machine learning algorithm is made to be very effective, adaptable and portable for classification, regression and ranking issues [28–31].

4 Results and Discussion

Several evaluation metrics have been employed in reviewed articles. Formulation of metrics used for evaluation in the reviewed articles is provided in Table 3. Equation 3, which employs numerical information of correctly categorised classes from all of the soil samples in the dataset, has been used to calculate the accuracy of the soil classification and crop recommendation system.

$$\text{Accuracy} = \frac{\text{no.of identified samples}}{\text{total no.of samples}} * 100 \tag{3}$$

For system evaluation, the following formulas are used to calculate the precision, recall and F1 score, which are all important factors [32]:

$$\text{Precision} = \frac{\sum \text{True Positives}(\text{TP})}{\sum(\text{True Positives}(\text{TP}) + \text{False Positives}(\text{FP}))} * 100 \tag{4}$$

$$\text{Recall} = \frac{\sum \text{True Positives}(\text{TP})}{\sum(\text{True Positives}(\text{TP}) + \text{False Negatives}(\text{FP}))} * 100 \tag{5}$$

$$\text{F1 Score} = 2 * \frac{\text{Precision} * \text{Recall}}{\text{Precision} + \text{Recall}} \tag{6}$$

5 Conclusion

Despite the fact that for the vast majority of Indians, agriculture is their main source of income and a key component of the national economy, precision agriculture and smart farming were significant advancements in the agricultural field and were utilised by the corporates involved in the agricultural field. Precision farming and smart farming have not yet attracted the Indian farmers. In this study, we have provided a thorough review of 25 articles that were published in the recent 5 years from 2017 to 2022. The aim of this study was to provide insights and propose new methods to propose the right crop for the farmers. Such insights are useful for the farmers and

Table 3 Performance metrics used for evaluation in reviewed articles

S.No.	Metrics for evaluation
1	Accuracy
2	Precision
3	Recall
4	F1 score

provide a good understanding of recent research trends in the agricultural field. We look forward that the detailed study inspired by the researchers in the same domain and that would be helpful to propose new methods to tackle the specific challenges faced by the farmers and build efficient crop and fertiliser recommender systems.

References

1. A. Motwani, P. Patil, V. Nagaria, Verma, S. Ghane, *Soil Analysis and Crop Recommendation Using Machine Learning* (International Conference for Advancement in Technology (ICONAT), 2022), pp. 1–7. https://doi.org/10.1109/ICONAT53423.2022.9725901
2. S. Panchamurthi, Soil analysis and prediction of suitable crop for agriculture using machine learning. Int. J. Res. Appl. Sci. Eng. Technol. **7**, 2328–2335 (2019). https://doi.org/10.22214/ijraset.2019.3427
3. K.R. Bharath, K. Balakrishna, C.A. Bency, M. Siddesha, R. Sushmitha, Crop recommendation system for precision agriculture. Int. J. Comput. Sci. Eng. **7**(5), 1277–1282 (2019)
4. M. Bouni, B. Hssina, K. Douzi, S. Douzi, Towards an efficient recommender systems in smart agriculture, a deep reinforcement learning approach. Procedia Computer Science **203**, 825–830 (2022). https://doi.org/10.1016/j.procs.2022.07.124
5. M.S. Suchithra, M. Pai, Improving the prediction accuracy of soil nutrient classification by optimizing extreme learning machine parameters. Inf. Process. Agric. **7**, 72 (2019). https://doi.org/10.1016/j.inpa.2019.05.003
6. S. Pudumalar, E. Ramanujam, R.H. Rajashree, C. Kavya, T. Kiruthika, J. Nisha, *Crop Recommendation System for Precision Agriculture* (Eighth International Conference on Advanced Computing (ICoAC), 2017), pp. 32–36. https://doi.org/10.1109/ICoAC.2017.7951740
7. N.H. Kulkarni, G.N. Srinivasan, B.M. Sagar, N.K. Cauvery, *Improving Crop Productivity Through A Crop Recommendation System Using Ensembling Technique* (3rd International Conference on Computational Systems and Information Technology for Sustainable Solutions (CSITSS), 2018), pp. 114–119. https://doi.org/10.1109/CSITSS.2018.8768790
8. Z. Doshi, S. Nadkarni, R. Agrawal, N. Shah, *Agro Consultant: Intelligent Crop Recommendation System Using Machine Learning Algorithms* (Fourth International Conference on Computing Communication Control and Automation (ICCUBEA), 2018), pp. 1–6. https://doi.org/10.1109/ICCUBEA.2018.8697349
9. A. Pruthviraj, G.C. Shastry, K.A. Nagaraj, Nikhil, Crop and fertilizer recommendation system based on soil classification, in *Recent Advances in Artificial Intelligence and Data Engineering. Advances in Intelligent Systems and Computing*, ed. by D.P. Shetty, S. Shetty, vol. 1386, (Springer, Singapore, 2022). https://doi.org/10.1007/978-981-16-3342-3_3
10. A. Kumar, S. Sarkar, C. Pradhan, *Recommendation System for Crop Identification and Pest Control Technique in Agriculture* (International Conference on Communication and Signal Processing (ICCSP), 2019), pp. 0185–0189. https://doi.org/10.1109/ICCSP.2019.8698099
11. R.K. Rajak, A. Pawar, M. Pendke, P. Shinde, S. Rathod, A. Devare, *Crop Recommendation System to maximize Crop yield using Machine Learning* (Issue 12 IRJET, 2017)
12. A. Venugopal, S. Aparna, J. Mani, R. Matthew, V. Williams, Crop yield prediction using machine learning algorithms. Int. J. of Eng. Res. and Technol. **9**(13), 87–91 (2021)
13. N. Mahendra, V. Dhanush, K. Nischitha, Ashwini, M.R. Manjuraju, Crop prediction using machine learning approaches. Int. J. of Eng. Res and Technol. **9**, 23–26 (2020)
14. S.M. Pande, P.K. Ramesh, A. Anmol, B.R. Aishwarya, K. Rohilla, K. Shaurya, *Crop Recommender System Using Machine Learning Approach* (5th International Conference on Computing Methodologies and Communication (ICCMC), 2021), pp. 1066–1071. https://doi.org/10.1109/ICCMC51019.2021.9418351

15. P.A.S. Chakraborty, A. Kumar, O.R. Pooniwala, *Intelligent Crop Recommendation System using Machine Learning* (2021 5th International Conference on Computing Methodologies and Communication (ICCMC), 2021), pp. 843–848. https://doi.org/10.1109/ICCMC51019.2021.9418375
16. L. Goel, A. Jindal, S. Mathur, Chapter 4 – Design and implementation of a crop recommendation system using nature-inspired intelligence for Rajasthan, India, in *Cognitive Data Science in Sustainable Computing, Deep Learning for Sustainable Agriculture*, (Academic Press, 2022), pp. 109–128), ISBN 9780323852142. https://doi.org/10.1016/B978-0-323-85214-2.00005-7
17. R. Priya, D. Ramesh, Adaboost. RT Based Soil N-P-K Prediction Model for Soil and Crop Specific Data: A Predictive Modelling Approach, in *Big Data Analytics. BDA 2018. Lecture Notes in Computer Science*, ed. by A. Mondal, H. Gupta, J. Srivastava, P. Reddy, D. Somayajulu, vol. 11297, (Springer, Cham, 2018). https://doi.org/10.1007/978-3-030-04780-1_22
18. P. Priya, U. Muthaiah, M. Balamurugan, Predicting yield of the crop using machine learning algorithm. Int. J. of Eng. Sci and Res. Technol. **29**(11), 1248–1255 (2020)
19. H. Khan, S.M. Ghosh, Crop yield prediction from meteorological data using efficient machine learning model, in *Proceedings of International Conference on Wireless Communication. Lecture Notes on Data Engineering and Communications Technologies*, ed. by H. Vasudevan, Z. Gajic, A. Deshmukh, vol. 36, (Springer, Singapore, 2020). https://doi.org/10.1007/978-981-15-1002-1_57
20. G. Suresh, A.S. Kumar, S. Lekashri, R. Manikandan, Efficient crop yield recommendation system using machine learning for digital farming. Int. J. Modern Agric. **10**(1), 906–914 (2021)
21. A.K. Sharma, S. Chaurasia, D.K. Srivastava, Supervised rainfall learning model using machine learning algorithms, in *The International Conference on Advanced Machine Learning Technologies and Applications (AMLTA 2018). AMLTA, Advances in Intelligent Systems and Computing*, ed. by A. Hassanien, M. Tolba, M. Elhoseny, M. Mostafa, vol. 723, (Springer, Cham, 2018). https://doi.org/10.1007/978-3-319-74690-6_27
22. R.K. Rajak et al., Crop recommendation system to maximize crop yield using machine learning technique. Int. Res. J. of Eng. and Technol. **4**(12), 950–953 (2017)
23. M. Champaneri, D. Chachpara, C. Chandvidkar, M. Rathod, Crop yield prediction using machine learning. Int. J. of Sci. Res. **10**(1), 1–3 (2018)
24. A.K.M. Reddy, S. Chithra, H.M. Hemashree, K. Thanu, Soil classification and crop suggestion using machine learning. Int. J. for Res. in Appl. Sci. Eng. Technol. **8**(7), 1625–1628 (2020)
25. N. Gnanasankaran, E. Ramaraj, A multiple linear regression model to predict rainfall using indian meteorological data. Int. J. Adv. Sci. Technol. (SCOPUS) **29**(8) ISSN: 2005-4238, 746–758 (2020)
26. N. Gnanasankaran, E. Ramaraj, The effective yield of paddy crop in Sivaganga district: an initiative for smart farming. Int. J. Sci. Technol. Res. **9**(2) ISSN: 2277-8616, 6553–6556 (2020)
27. N. Gnanasankaran, E. Ramaraj, T. Manikumar, An intelligent framework for Rice yield prediction using machine learning based models. International Journal of Scientific and Engineering Research **12**(1) ISSN: 2229-5518, 422–431 (2021)
28. https://extension.umaine.edu/gardening/manual/soils/soil-and-plant-nutrition/
29. https://soilhealthnexus.org/resources/soil-properties/
30. https://towardsdatascience.com/logistic-regression-classifier-8583e0c3cf9
31. https://www.kaggle.com/datasets/atharvaingle/crop-recommendation-dataset
32. G. Sujatha, K. Sankareswari, A comparative study on machine learning based classifier model for wheat seed classification, in *Mining Intelligence and Knowledge Exploration. MIKE 2021. Lecture Notes in Computer Science*, ed. by R. Chbeir, Y. Manolopoulos, R. Prasath, vol. 13119, (Springer, Cham, 2022). https://doi.org/10.1007/978-3-031-21517-9_12

The Revolution of Edge Computing in Smart Farming

D. Sathya, R. Thangamani, and B. Saravana Balaji

1 Introduction

Agriculture, the cornerstone of human civilization, has been on a remarkable journey of transformation throughout history. From the earliest days of manual cultivation to the mechanization of the Industrial Revolution, farming has consistently evolved to meet the demands of a growing global population. In the twenty-first century, a new and pivotal chapter is being written in the annals of agriculture, and at the heart of this transformation lies a technological marvel—edge computing. The story of agriculture has not been without its challenges. Traditional farming practices, deeply rooted in centuries-old traditions, have often struggled to keep pace with the complexities of modern society. Challenges such as labor-intensive practices, uncertainty in decision-making, environmental sustainability, and vulnerability to climate change have tested the resilience of the agriculture sector.

Traditional farming methods, which relied heavily on manual labor and were often guided by historical knowledge, have been challenged to keep up with the ever-increasing demands for food production. Farmers faced the daunting task of optimizing resource allocation while contending with the unpredictability of weather, soil conditions, and pest infestations. The environmental consequences of excessive water use, overuse of fertilizers, and pesticide application have raised concerns about sustainability. Furthermore, the specter of climate change, with its erratic weather patterns and extreme events, has added an element of

D. Sathya (✉) · R. Thangamani
Kongu Engineering College, Perundurai, Tamilnadu, India
e-mail: sathya.ai@kongu.edu; thangamani.ai@kongu.edu

B. S. Balaji
Lebanese French University, Erbil, Iraq
e-mail: saravanabalaji.b@lfu.edu.krd

S. Balasubaramanian et al. (eds.), *Intelligent Robots and Drones for Precision Agriculture*, Signals and Communication Technology,
https://doi.org/10.1007/978-3-031-51195-0_17

unpredictability to farming, making it increasingly difficult for farmers to safeguard their livelihoods. In response to these challenges, the agriculture sector has embarked on a technological revolution that is reshaping every facet of the industry. This revolution is powered by a convergence of cutting-edge technologies, including the Internet of Things (IoT), artificial intelligence (AI), data analytics, and automation. At the heart of this transformation is edge computing—an innovation that is redefining the way farming is conducted, known as "smart farming"(Bolan, S, et al., 2015).

Edge computing is a paradigm shift in the world of technology. It brings computational power closer to the data source, allowing real-time data processing and decision-making. In the context of agriculture, this means that data collected on the farm, whether it be soil moisture [71] levels, crop health indicators, or even livestock behavior, can be processed and acted upon immediately, without the need for data to travel to distant cloud servers. This immediacy has profound implications for farming operations. With the advent of edge computing, the agriculture sector has entered a new era—one characterized by data-driven, precision agriculture. IoT sensors and devices, strategically placed across farms, continuously collect data and transmit it to edge computing devices. These devices, equipped with AI and machine learning algorithms (Tiwari, S et al., 2018) [8], analyze the data and provide actionable insights to farmers in real-time. The implications of this technological leap are far-reaching:

Precision Agriculture: Edge computing empowers farmers to practice precision agriculture with unparalleled accuracy. Every action on the farm, from planting and irrigation to pest control and harvesting, is guided by precise data and analysis. This precision minimizes resource use, reduces waste, and maximizes crop yields, ultimately leading to higher profitability for farmers.

Resource Optimization: The ability to process data at the edge allows for precise resource allocation. Farmers can make real-time decisions on water usage, fertilizer application, and energy consumption. This optimization not only reduces operational costs but also contributes to sustainable farming practices.

Environmental Sustainability: Edge computing plays a pivotal role in promoting sustainable agriculture. By optimizing resource use, minimizing environmental impact, and reducing the reliance on chemicals, farmers can transition to eco-friendly and regenerative farming practices that are beneficial for both the environment and consumers.

Early Disease Detection: Edge devices equipped with AI models are capable of detecting signs of disease or pest infestations in crops or livestock at an early stage. This early detection enables farmers to take prompt and targeted action, reducing crop losses and the need for chemical interventions.

Risk Mitigation: The real-time data and predictive analytics made possible by edge computing empower farmers to better mitigate risks associated with climate change, disease outbreaks, and market fluctuations. This enhanced ability to adapt to changing conditions leads to more resilient and sustainable farming practices.

As we embark on this exploration of the revolution of edge computing in smart farming(Rai, P.K., 2009) [10], we will journey deeper into the applications, benefits, challenges, and future directions of this transformative technology. Edge computing is not merely a tool; it is a catalyst for change in an industry that sustains us all. It represents the embodiment of the farming adage, "Make hay while the sun shines," in a world where technology and data are the new suns that illuminate the path to a brighter, more sustainable future for agriculture and food production.

2 Smart Farming and Agricultural Transformation

The agricultural sector is undergoing a remarkable transformation, driven by a convergence of cutting-edge technologies. This transformation, often referred to as "smart farming" or "precision agriculture," is reshaping traditional farming practices and revolutionizing the way we produce food. In this era of smart farming, data-driven decision-making, automation, and sustainability are at the forefront of agricultural innovation.

Data-Driven Insights

One of the pillars of smart farming is data. From Fig. 1, modern farms are increasingly equipped with sensors and Internet of Things (IoT) devices that continuously collect data on soil conditions, weather patterns, crop health, and livestock behavior. This wealth of data is processed and analyzed using artificial intelligence (AI) and machine learning (Tiwari, S et al., 2018) [8] algorithms to provide farmers with real-time insights. Farmers can make informed decisions about resource allocation,

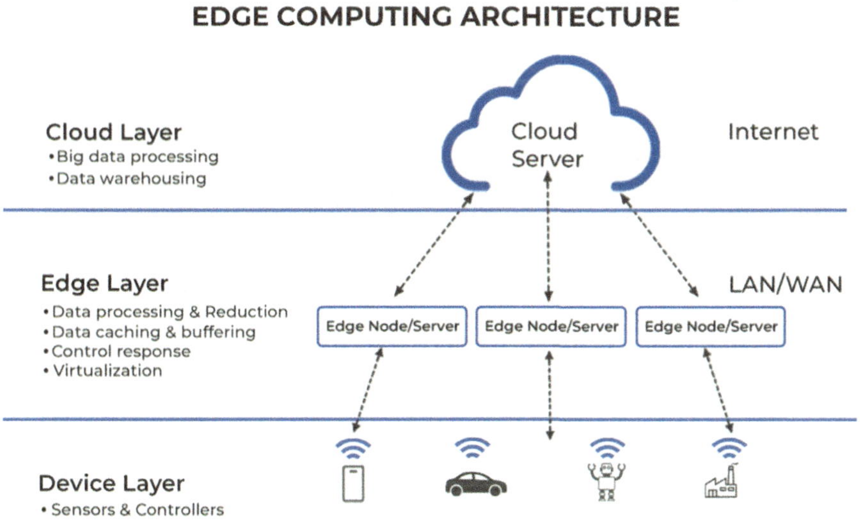

Fig. 1 Edge computing architecture

pest control, and irrigation, leading to optimized crop yields and reduced environmental impact.

Precision Agriculture
Precision agriculture is a hallmark of the smart farming revolution. It involves the precise management of farming practices based on data analytics. For example, GPS-guided tractors and machinery can plant seeds and apply fertilizers with centimeter-level accuracy, reducing resource wastage and enhancing productivity. This precision extends to all aspects of farming, from soil management to harvesting techniques.

Resource Efficiency
Smart farming (Kim, S.W, et al., 2015) is synonymous with resource efficiency. With the ability to monitor soil moisture levels [71], weather forecasts, and crop health in real-time, farmers can tailor their resource usage to specific needs. This results in reduced water consumption, minimized chemical use, and optimized energy usage, all of which contribute to cost savings and environmental sustainability.

Environmental Sustainability
Sustainability is a central focus of smart farming. The reduction of chemical inputs, the adoption of organic practices, and the promotion of regenerative agriculture are all made possible by data-driven insights. Farmers can implement practices that not only protect the environment but also regenerate soil health and enhance biodiversity.

Automation and Robotics
Automation and robotics are transforming labor-intensive farming tasks. Drones equipped with cameras and sensors can monitor large fields, identifying areas that require attention. Autonomous robots (Acosta, J.A, et al., 2011) [15] can perform tasks such as weeding and harvesting with precision and efficiency. These technologies reduce the need for manual labor, address labor shortages, and improve farm productivity.

Early Disease Detection
Smart farming technologies, including AI-powered image recognition systems, enable early disease detection in crops and livestock. These systems can identify signs of disease or pest infestations before they become widespread, allowing farmers to take targeted and timely action. This reduces crop losses and minimizes the use of chemical treatments.

Market Insights
Access to real-time data and market analysis tools empowers farmers to make strategic decisions regarding crop selection and timing. They can align their production with market demand, ensuring that their products reach consumers at the right time and price.

2.1 The Need for Data-Driven Agriculture

In a world where the global population (Arora, M, et al., 2008) [4] continues to burgeon, and the impacts of climate change pose new challenges to food production, the agricultural sector faces an imperative: to produce more food with fewer resources while mitigating environmental harm. In this context, the advent of data-driven agriculture emerges as an indispensable solution.

Growing Global Population
With the world's population [Arora, M, et al., 2008] [4] projected to surpass 9 billion by 2050, the demand for food is escalating. To meet this demand, agriculture must embrace innovation and efficiency. Data-driven agriculture leverages technology to optimize resource use, boost crop yields, and ensure food security for a growing populace.

Resource Scarcity
Scarce resources, notably arable land and freshwater, intensify the pressure on agriculture. Data-driven approaches enable precise resource allocation, ensuring that water, fertilizers, and other inputs are used judiciously. This minimizes waste and promotes resource efficiency, crucial in a world where resource scarcity looms.

Climate Change and Environmental Impact
Climate change exacerbates the unpredictability of weather patterns, making agriculture more vulnerable to extreme events. Data-driven agriculture equips farmers with tools for climate adaptation. Additionally, it promotes sustainable practices that reduce environmental harm, mitigating agriculture's contribution to climate change.

Optimizing Crop Yields
Data-driven insights enable farmers to make informed decisions at every stage of cultivation. Precision agriculture, guided by data analytics, ensures optimal planting densities, timely irrigation, and targeted pest control. This leads to maximized crop (Nagajyoti, P.C, et al., 2010) [5] yields while minimizing inputs, bolstering agricultural sustainability.

Minimizing Environmental Impact
The environmental consequences of traditional agriculture, such as soil degradation and water pollution, necessitate a paradigm shift. Data-driven agriculture minimizes these impacts by enabling the adoption of eco-friendly practices. Farmers can implement regenerative agriculture and reduce the reliance on harmful chemicals.

Resilience to Disease and Pest Outbreaks
Data-driven agriculture incorporates early warning systems through AI and sensor technology. This proactive approach enables farmers to detect disease outbreaks and pest infestations swiftly. By taking targeted measures, they can minimize crop losses(Nagajyoti, P.C, et al., 2010) [5] and reduce the need for chemical treatments.

Market Alignment
Data-driven agriculture empowers farmers with market insights. They can make informed decisions about crop selection and timing, aligning production with market demand. This reduces food waste and enhances the economic viability of farming operations.

Challenges and the Path Forward
While the need for data-driven agriculture is clear, it is not without challenges. Data privacy, connectivity in rural areas, and the digital literacy of farmers are notable hurdles. Nonetheless, collaborative efforts among stakeholders, investment in digital infrastructure, and education are key to realizing the full potential of data-driven agriculture.

2.2 *The Rise of IoT in Agriculture*

In recent years, the agricultural landscape has witnessed a profound transformation, catalyzed by the rise of the Internet of Things (IoT) as shown in Fig. 2. This transformative technology, which connects physical objects and devices to the Internet, has found fertile ground in agriculture, giving rise to what is now commonly referred to as "smart farming." In this section, we explore the ascent of IoT in agriculture and its far-reaching implications.

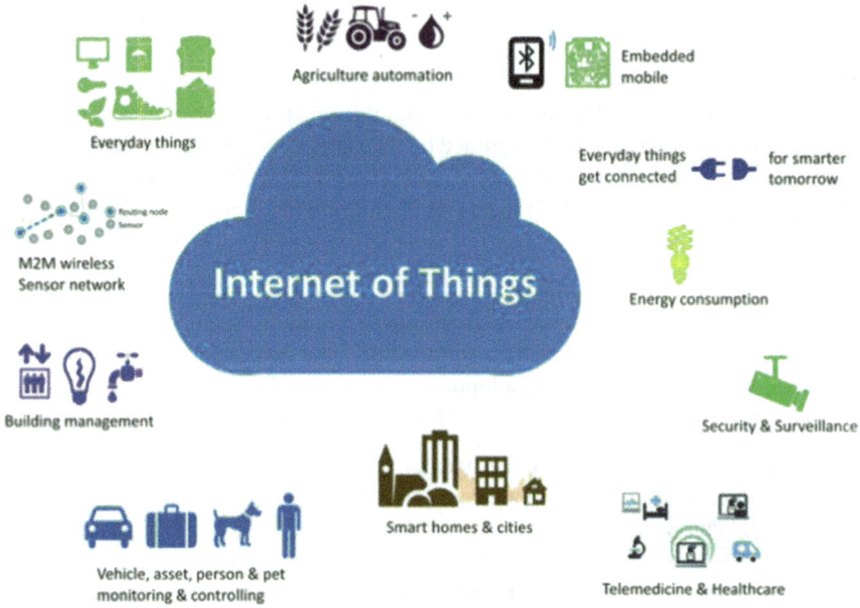

Fig. 2 Application of IoT in various fields

Sensors and Data Collection
At the heart of IoT in agriculture are sensors—tiny, yet powerful devices that can be deployed across fields, orchards, and livestock farms. These sensors are equipped to collect a vast array of data, including soil moisture levels [71], temperature, humidity, and even the health and behavior of livestock. The data collected by these sensors forms the foundation of data-driven agriculture.

Real-Time Monitoring and Insights
IoT sensors provide real-time monitoring capabilities, granting farmers unprecedented insights into their operations. They can track soil conditions, detect early signs of disease in crops (Nagajyoti, P.C, et al., 2010) [5], and monitor livestock health—all from a digital dashboard. This real-time visibility empowers farmers to make informed decisions promptly.

Precision Agriculture
IoT in agriculture has ushered in the era of precision agriculture. With the ability to precisely measure and analyze data, farmers can optimize their farming practices down to the square meter. GPS-guided tractors can plant seeds with millimeter accuracy, while automated irrigation systems adjust water flow based on real-time soil moisture [71] data. Precision agriculture maximizes resource efficiency and crop yields.

Resource Optimization
Resource scarcity is a pressing issue in agriculture, with water being a prime example. IoT-driven irrigation systems use data to determine precisely when and how much water is needed. This not only conserves water but also reduces energy usage and operational costs.

Environmental Sustainability
The environmental impact of agriculture has raised concerns worldwide. IoT empowers farmers to adopt sustainable practices that mitigate harm to the environment. By minimizing chemical inputs, reducing waste, and optimizing resource use, IoT plays a pivotal role in fostering environmentally friendly farming practices.

Early Detection of Issues
One of IoT's most compelling applications in agriculture is early issue detection. Through AI-powered image recognition and sensor data analysis, farmers can spot signs of disease outbreaks or pest infestations before they spread. Early intervention reduces crop losses and lessens the need for chemical treatments.

Data-Driven Decision-Making.
The data collected by IoT sensors and devices serve as the foundation for data-driven decision-making. Farmers can assess historical data, weather forecasts, and real-time field conditions to make informed choices about planting, harvesting, and resource allocation.

Challenges and Future Directions
Despite its immense potential, IoT in agriculture faces challenges such as data security and connectivity in rural areas. However, ongoing advancements in technology, coupled with investments in digital infrastructure and farmer education, promise to overcome these obstacles.

In conclusion, the rise of IoT in agriculture has unlocked a world of possibilities, transforming farming into a data-driven, efficient, and sustainable industry. With IoT sensors and devices as their allies, farmers are poised to address the pressing challenges of resource scarcity, climate change, and food security. The journey of IoT in agriculture is one of innovation, resilience, and a commitment to shaping a more sustainable and productive future for farming.

3 Edge Computing: Enabling Real-Time Intelligence

In the realm of technology, there is a constant pursuit of pushing boundaries, making systems smarter, and achieving greater efficiency. Edge computing represents one such leap in the evolution of computing paradigms. It is a concept that brings computation closer to data sources, and its impact on industries, including agriculture, is profound. In this section, we explore the essence of edge computing and its role in enabling real-time intelligence.

The Essence of Edge Computing
Edge computing represents a fundamental shift from the conventional cloud computing paradigm. Instead of relying on distant data centers, edge computing places computational power at the "edge" of the network, in close proximity to where data is generated and needed. This closeness to data sources reduces latency, enhances real-time processing, and enables faster decision-making.

Real-Time Data Processing
In the context of agriculture, where timeliness is critical, edge computing shines. With IoT sensors and devices deployed throughout farms, data is continuously generated, whether it's soil moisture [71] readings, weather data, or livestock health metrics. Edge computing devices, strategically placed on the farm, process this data locally and instantaneously.

Immediate Decision-Making
One of the primary benefits of edge computing in agriculture is immediate decision-making. Consider a scenario where a field's soil moisture [71] level drops unexpectedly due to a sudden heatwave. An edge computing device can analyze this data in real-time and trigger an irrigation system to ensure the crop [Nagajyoti, P.C, et al., 2010] [5] receives adequate water promptly. This real-time responsiveness optimizes resource use and safeguards crop health.

Enhanced Precision Agriculture
Edge computing elevates the concept of precision agriculture to new heights. With data analyzed at the edge, farming practices become highly precise and granular. Planting, irrigation, fertilization, and pest control can all be tailored to specific conditions within a field. This level of precision minimizes resource waste and maximizes crop yields.

Resource Efficiency
Resource efficiency is at the core of edge computing's impact on agriculture. With immediate access to data on soil conditions, weather forecasts, and crop health, farmers can optimize the use of water, fertilizers, and pesticides. This not only conserves resources but also reduces operational costs and environmental impact.

Environmental Sustainability
Sustainability is a pressing concern in agriculture. Edge computing supports sustainable practices by enabling the adoption of eco-friendly farming methods. By reducing chemical inputs, minimizing waste, and optimizing resource use, edge computing contributes to environmentally responsible farming.

Early Disease Detection
Edge computing is a crucial tool in early disease detection. AI models running on edge devices can analyze images of crops or livestock and detect signs of disease or pest infestations swiftly. Timely identification allows farmers to take targeted action, reducing crop losses and the need for chemical treatments.

3.1 Understanding Edge Computing

Edge computing is a crucial concept in the context of smart farming, where it plays a pivotal role in revolutionizing agricultural practices. Here, we'll delve into understanding edge computing with respect to smart farming:

1. Definition of Edge Computing
   - Edge computing refers to a decentralized computing paradigm where data processing and analysis occur closer to the data source, at the "edge" of the network, rather than relying solely on centralized cloud servers.
   - In smart farming, the "edge" often represents various on-farm devices and sensors, such as IoT (Internet of Things) (Routray et al., 2019) [1] sensors, drones, and machinery, which collect and generate a wealth of data.
2. Significance in Smart Farming
   - **Real-Time Data Processing:** Edge computing enables real-time data processing, which is critical in smart farming. It allows for immediate analysis of data generated by sensors in the field, providing timely insights for decision-making.

- **Reduced Latency:** By processing data at the edge, edge computing minimizes latency, ensuring that critical decisions, such as irrigation adjustments or pest control measures, can be implemented swiftly.
- **Bandwidth Optimization:** Edge computing reduces the need to transmit vast amounts of data to centralized servers, optimizing bandwidth usage and lowering data transmission costs for farmers.
- **Offline Capabilities:** Edge devices in smart farming can operate offline or with intermittent connectivity, ensuring that data collection and analysis continue even in remote areas.

3. Use Cases in Smart Farming

- **Precision Agriculture:** Edge devices and sensors on farming equipment collect data on soil conditions, weather, and crop health. This data is processed locally to provide real-time insights for optimizing planting, fertilization, and irrigation.
- **Livestock Monitoring:** Wearable sensors on livestock animals collect health and behavior data. Edge computing processes this data to detect anomalies, helping farmers identify and address health issues promptly.
- **Crop Monitoring:** Drones equipped with sensors capture high-resolution images of crops. Edge computing on the drone processes these images to identify areas requiring attention, such as disease outbreaks or nutrient deficiencies.
- **Farm Equipment Automation:** Edge computing is integral to autonomous farming equipment, allowing tractors and harvesters to make real-time decisions about field navigation and precision tasks.
- **Climate and Environmental Monitoring:** Edge devices can measure environmental parameters like temperature, humidity, and air quality. This data helps farmers manage their agricultural practices in response to changing environmental conditions.

4. Benefits in Smart Farming

- **Real-Time Decision-Making:** Edge computing empowers farmers with immediate insights, enabling them to make informed decisions on the spot, leading to improved crop yields and resource efficiency.
- **Cost Savings:** By reducing the need for constant data transmission to centralized servers, edge computing can lower data transmission costs and reduce the dependence on expensive, always-on Internet connections.
- **Scalability:** Smart farming systems can scale easily by adding more edge devices as needed, without overloading centralized servers.
- **Privacy and Security:** Local data processing at the edge enhances data privacy and security since sensitive agricultural data can be kept on-farm rather than transmitted to external servers.

In essence, edge computing is a cornerstone of smart farming, enabling farmers to harness the power of real-time data processing and analysis to optimize their agricultural practices, reduce costs, and enhance sustainability. It represents a significant leap forward in modernizing agriculture and addressing the growing challenges of food production in an increasingly complex world.

3.2 Edge Devices and Sensors in Agriculture

Edge devices and sensors are integral components of modern agriculture, playing a crucial role in collecting, monitoring, and analyzing data at the field level. These devices are deployed directly in the agricultural environment, often at the "edge" of the network, to gather a wide range of data that can inform decision-making and optimize farming operations. Here's an overview of edge devices and sensors in agriculture:

1. Types of Edge Devices and Sensors

   - **Sensors:** Internet of Things (IoT)[Rai, P.K. et al., 2019] [6] sensors are widely used in agriculture to collect data on various parameters. These sensors include soil moisture[71] sensors, weather stations, temperature and humidity sensors, and nutrient sensors. They provide real-time information about environmental conditions, helping farmers make informed decisions about irrigation, fertilization, and pest control.
   - **GPS and GNSS Receivers:** Global Positioning System (GPS) and Global Navigation Satellite System (GNSS) receivers are used for precision agriculture. They enable accurate mapping and tracking of field activities, allowing for precise planting, harvesting, and field navigation.
   - **Imaging and Camera Systems:** Drones and other imaging systems equipped with cameras and multispectral sensors capture high-resolution images of crops. This imagery can be used for crop health assessment, disease detection, and yield prediction.
   - **Livestock Monitoring Devices:** Wearable sensors and RFID tags are used for monitoring livestock health and behavior. They collect data on parameters like body temperature, activity levels, and feeding patterns, helping farmers detect health issues and optimize animal care.
   - **Automated Farm Equipment:** Modern farm equipment, such as tractors and harvesters, often come equipped with edge computing capabilities and sensors. These sensors monitor equipment performance, fuel efficiency, and field conditions, contributing to more efficient and precise farming operations.

2. Key Functions and Applications

   - **Data Collection:** Edge devices and sensors collect data on a wide range of agricultural parameters, including soil moisture [71], temperature, humidity, crop growth, livestock health, and equipment performance.
   - **Real-Time Monitoring:** These devices provide real-time monitoring of field conditions and livestock behavior. Farmers can access this data remotely

through mobile apps or web interfaces, allowing them to make timely decisions.

- **Decision Support:** The data gathered by edge devices and sensors serve as the foundation for decision support systems in precision agriculture. Farmers can use this data to optimize irrigation, fertilization, pest control, and harvesting schedules.
- **Predictive Analytics:** Data from sensors and edge devices can be analyzed using predictive analytics algorithms to forecast crop yields, disease outbreaks, and equipment maintenance needs.
- **Automation:** Some edge devices are integrated with automation systems that can control irrigation, machinery, and other farm operations based on real-time data. For example, automated irrigation systems can adjust water delivery based on soil moisture [71] readings.

3. Benefits and Impact

- **Increased Efficiency:** Edge devices and sensors improve the efficiency of farming operations by providing data-driven insights and automation capabilities.
- **Resource Optimization:** Farmers can optimize resource use, including water, fertilizers, and pesticides, based on real-time data, reducing waste and environmental impact.
- **Crop Yield Improvement:** Precision agriculture techniques enabled by these devices often lead to higher crop yields and better quality produce.
- **Livestock Welfare:** Livestock monitoring devices improve animal welfare by allowing early detection of health issues and optimizing feeding and care practices.
- **Cost Reduction:** Edge devices can help reduce operational costs by optimizing resource use and minimizing downtime through predictive maintenance.
- **Sustainability:** Sustainable farming practices are promoted through the efficient use of resources and reduced environmental impact.

In conclusion, edge devices and sensors have become indispensable tools in modern agriculture, enabling data-driven decision-making, automation, and resource optimization. They empower farmers to enhance productivity, reduce costs, and embrace more sustainable and environmentally friendly farming practices.

4 Real-Time Data Processing at the Field Edge

Real-time data processing at the field edge is a critical aspect of modern agriculture, revolutionizing how farmers collect, analyze, and act upon data to optimize their operations. This approach involves processing data directly on-site, near the source of data generation, rather than relying solely on centralized cloud-based solutions. Here's an exploration of the concept of real-time data processing at the field edge in agriculture:

1. Data Sources

   - **IoT Sensors:** Internet of Things (IoT) (Kohzadi, S, et al., 2019) [11] sensors placed throughout the field collect data on soil moisture, temperature, humidity, nutrient levels, and other environmental factors. Livestock wearables, such as health monitoring devices, also contribute data.
   - **Imaging Systems:** Drones equipped with cameras and multispectral sensors capture high-resolution images of crops. These images can be used for crop health assessment, disease detection, and yield prediction.
   - **Automated Farm Equipment:** Modern agricultural machinery, including tractors and harvesters, are equipped with sensors that monitor equipment performance, fuel consumption, and field conditions.

2. Key Components of Real-Time Data Processing

   - **Edge Devices:** These are computing devices deployed in the field, such as IoT gateways[Rai, P.K., et al., 2019] [6] or embedded systems on farm equipment, capable of processing data locally.
   - **Edge Computing:** Edge computing technology enables the processing and analysis of data at or near the data source. It involves running algorithms and applications on the edge devices to make sense of the collected data.
   - **Connectivity:** Edge devices are often connected to the Internet or private networks to facilitate data transmission and remote monitoring.

3. Importance and Benefits

   - **Immediate Insights:** Real-time data processing provides farmers with immediate insights into field conditions, livestock health, and equipment performance. This timely information empowers them to make informed decisions on the spot.
   - **Latency Reduction:** By processing data at the edge, latency is significantly reduced compared to sending data to a centralized cloud server. This is crucial for time-sensitive actions, such as adjusting irrigation or applying pest (Feng, W.et al., 2020) [13] control measures.
   - **Data Privacy:** Processing data at the field edge enhances data privacy and security since sensitive agricultural data can be kept on-site rather than transmitted to external servers.
   - **Bandwidth Optimization:** Edge processing minimizes the need to transmit large volumes of data over the Internet, reducing data transmission costs and reliance on high-speed connectivity.
   - **Offline Capabilities:** Edge devices can operate offline or with intermittent connectivity, ensuring data collection and analysis continue even in remote areas with limited network access.

4. Applications in Agriculture

   - **Precision Agriculture:** Real-time data processing informs precision agriculture practices, helping farmers optimize irrigation, fertilization, and pest control based on current field conditions.

- **Livestock Management:** Livestock wearables and sensors process data on animal health, behavior, and environmental conditions, allowing early disease detection and improved animal welfare.
- **Crop Monitoring:** Drones equipped with edge computing capabilities analyze crop images on-site to detect issues like nutrient deficiencies, pest [13] infestations, and drought stress.
- **Equipment Optimization:** Farm machinery with edge devices monitor equipment performance, enabling predictive maintenance and efficient operation.

In conclusion, real-time data processing at the field edge represents a fundamental shift in agricultural practices, empowering farmers with immediate insights, reducing latency, and enhancing data privacy. This approach contributes to increased efficiency, resource optimization, and sustainability in modern farming operations.

4.1 Immediate Benefits of Local Data Processing

The immediate benefits of local data processing in agriculture, particularly at the field edge, are profound and have a direct impact on farming practices and outcomes. Here are the key advantages:

1. **Real-Time Decision-Making:** Local data processing enables farmers to make informed decisions in real-time. By analyzing data collected from sensors and devices within the field, farmers can adjust various parameters like irrigation, fertilization, and pest (Feng, W.et al., 2020) [13] control instantly. This agility in decision-making minimizes response times, crucial for managing changing environmental conditions or unexpected issues.
2. **Reduced Latency:** Processing data at the field edge significantly reduces latency compared to sending data to centralized servers or the cloud for analysis. This reduction in delay is vital for critical operations such as automated irrigation, where immediate adjustments based on soil moisture levels can save water resources and improve crop health.
3. **Bandwidth Optimization:** Local data processing optimizes the utilization of limited network bandwidth in rural or remote farming areas. Instead of constantly transmitting large volumes of data to centralized servers, only relevant information or actionable insights are sent, reducing data transmission costs and potential network congestion.
4. **Offline Capabilities:** Edge devices can operate autonomously and continue to collect and process data even in areas with poor or intermittent connectivity. This capability ensures that data collection and analysis persist without interruption, maintaining the integrity of the farming operation.
5. **Data Privacy and Security:** Keeping sensitive agricultural data within the farm's local network enhances data privacy and security. Farmers have greater

control over who accesses their data, reducing the risk of unauthorized access, data breaches, or information leaks.

6. **Resource Optimization:** Local data processing supports the efficient use of resources such as water, fertilizers, and pesticides. By analyzing data on soil conditions, weather, and crop health at the field edge, farmers can precisely target resource application, minimizing waste, and environmental impact.
7. **Early Issue Detection:** Immediate data analysis allows for the early detection of issues like pest (Feng, W.et al., 2020) [13] infestations, diseases, or equipment malfunctions. Early intervention can prevent crop damage, reduce losses, and improve overall farm productivity.
8. **Scalability and Adaptability:** Local data processing systems can be easily scaled up or modified to accommodate changing needs or expanding farms. Farmers can add more edge devices or sensors as required without overburdening centralized data infrastructure.
9. **Compliance and Traceability:** For farms that need to comply with regulations or require traceability in their supply chain(Burhan, M. et al., 2018) [16], local data processing systems can store and manage necessary records efficiently while maintaining data integrity.
10. **Improved Crop Yields and Efficiency:** Ultimately, the immediate benefits of local data processing contribute to improved crop yields, resource efficiency, and the overall economic viability of farming operations.

In summary, local data processing in agriculture offers a range of immediate advantages that empower farmers to make data-driven decisions, optimize resources, enhance security, and respond quickly to changing conditions, all of which collectively lead to more sustainable and productive farming practices.

4.2 Case Studies: On-Field Sensor Data Analysis

Case studies showcasing on-field sensor data analysis in agriculture highlight the practical applications and benefits of local data processing. Here are two illustrative examples:

Case Study 1: Precision Irrigation in Vineyards

Background: A vineyard in California sought to improve water efficiency in their irrigation practices to conserve water resources while maintaining high-quality grape yields.

Solution: The vineyard deployed soil moisture sensors throughout the field. These sensors collected data on soil moisture levels at different depths. Edge computing devices installed in the vineyard processed this data locally.

Results:

1. **Real-Time Decision-Making:** Local data processing allowed vineyard managers to access up-to-the-minute soil moisture information, enabling them to adjust irrigation schedules as needed.

2. **Water Savings:** By optimizing irrigation based on real-time data, the vineyard reduced water usage by 30% while ensuring grapevines received the right amount of moisture for healthy growth.
3. **Improved Grape Quality:** Precise control over soil moisture levels led to improved grape quality, resulting in higher market prices for the vineyard's wine.

Case Study 2: Pest Management in Citrus Orchards

Background: A citrus orchard in Florida faced challenges with pest infestations, which were affecting crop yields (Feng, W.et al., 2020) [13] and quality. Traditional pest management (Feng, W.et al., 2020) [13] methods were costly and often resulted in pesticide overuse.

Solution: The orchard deployed insect monitoring traps equipped with sensors that tracked pest activity. Data from these traps, including pest counts and activity patterns, were processed locally using edge computing devices.

Results:

1. **Timely Pest Alerts:** Local data analysis allowed for immediate alerts when pest activity reached predetermined thresholds, enabling rapid response to potential infestations.
2. **Reduced Pesticide Usage:** By targeting pesticide application based on real-time data, the orchard reduced pesticide use by 40%, resulting in cost savings and reduced environmental impact.
3. **Enhanced Crop Health:** Early pest detection and targeted interventions improved the overall health of the citrus trees, leading to increased fruit yields and improved fruit quality.

These case studies highlight how on-field sensor data analysis, powered by edge computing, can transform agricultural practices by optimizing resource usage, improving crop quality, and reducing environmental impact. These benefits demonstrate the value of local data processing in precision agriculture.

5 Reducing Latency for Efficient Farming

Reducing latency in agriculture is a critical aspect of achieving efficient and data-driven farming practices. Latency refers to the delay or lag in data transmission and processing, and in the context of agriculture, minimizing latency is essential for timely decision-making and optimizing various farming operations. Here's an exploration of how reducing latency contributes to efficient farming:

1. **Real-Time Decision-Making:**
   - **Precision Agriculture:** Minimizing latency allows farmers to access real-time data from sensors and devices in the field. This immediate information empowers them to make timely decisions regarding irrigation, fertilization, pest control, and harvesting.

- **Crop Health Monitoring:** Low-latency data from imaging systems and drones provide instant insights into crop health. Detecting issues like nutrient deficiencies or disease outbreaks in real-time enables rapid corrective action.

2. **Automated Systems:**

- **Equipment Control:** Modern farm machinery, equipped with low-latency communication systems, can be remotely operated or guided with minimal delay. This feature enhances the efficiency of operations like planting, harvesting, and field navigation.
- **Irrigation Management:** Low-latency data from soil moisture sensors allows for immediate adjustments to irrigation systems. This precision prevents over- or under-irrigation, conserving water resources and optimizing crop growth.

3. **Environmental Response:**

- **Weather and Climate Monitoring:** Timely weather data with minimal latency enables farmers to respond swiftly to weather changes, protecting crops from adverse conditions like frost, storms, or excessive heat.
- **Pest and Disease Control:** Rapid alerts about pest or disease outbreaks through low-latency monitoring systems enable farmers to implement targeted interventions, reducing crop damage.

4. **Resource Efficiency:**

- **Resource Optimization:** Reduced latency contributes to efficient resource management. Farmers can precisely target the application of water, fertilizers, and pesticides based on real-time data, minimizing waste and environmental impact.
- **Energy Efficiency:** In livestock management, low-latency data from sensors on animal behavior and health can help optimize feeding schedules, leading to energy savings and improved livestock welfare.

5. **Crop Yield Enhancement:**

- **Yield Prediction:** Low-latency data processing enables the use of predictive analytics to forecast crop yields accurately. This information aids in planning harvesting and post-harvest processes for optimal efficiency.

6. **Data Transmission Efficiency:**

- **Bandwidth Optimization:** Reduced latency minimizes the need for constantly transmitting large volumes of data to centralized servers, optimizing bandwidth usage and lowering data transmission costs for farmers.

7. **Remote Farming:**

- **Remote Monitoring:** With low-latency connectivity and data analysis, farmers can remotely monitor their fields, livestock, and equipment. This capability is especially valuable for large-scale or geographically dispersed farms.

8. **Sustainability:**
    - **Water Conservation:** Efficient irrigation through reduced latency not only conserves water resources but also contributes to sustainable farming practices.
    - **Reduced Chemical Use:** Targeted pest and disease control based on real-time data lead to reduced pesticide and chemical usage, aligning with environmentally friendly agriculture.

In summary, reducing latency in agriculture is fundamental to achieving efficient, data-driven, and sustainable farming practices. It empowers farmers with timely information and enhances their ability to make informed decisions, ultimately leading to improved crop yields, resource conservation, and the economic viability of farming operations.

5.1 Challenges of Latency in Agriculture

Reducing latency in agriculture is crucial for efficient and data-driven farming practices, but it comes with its own set of challenges. These challenges can impact the adoption and implementation of low-latency solutions in agricultural operations. Here are some of the key challenges associated with latency in agriculture:

1. **Network Connectivity:**
    - **Limited Coverage:** In rural and remote farming areas, access to high-speed Internet or reliable network connectivity can be limited. This lack of infrastructure hinders the seamless transmission of real-time data from the field to the farm office or cloud-based systems.
    - **Intermittent Connectivity:** Even when connectivity is available, it may be intermittent. Fields with poor or unreliable network connections can experience delays in data transmission, affecting the timeliness of decision-making.
2. **Data Volume and Complexity:**
    - **Data Overload:** Agriculture generates vast amounts of data from various sensors and devices. Processing and transmitting this data in real-time can strain network bandwidth and lead to congestion.
    - **Data Processing:** Analyzing complex data, such as high-resolution images from drones or multispectral sensors, in real-time requires significant computing power and can introduce latency.
3. **Security and Privacy:**
    - **Data Security:** Real-time data transmission may expose sensitive agricultural information to potential cyber threats if not adequately secured. Ensuring data security and privacy is a critical concern.

- **Privacy Regulations:** Compliance with data privacy regulations, such as GDPR [Poveda J et al., 2021] [3] in Europe or HIPAA in the United States, adds complexity to the collection and transmission of agricultural data.

4. **Costs and Infrastructure:**

   - **Investment Costs:** Implementing low-latency solutions, including edge computing devices and high-speed networks, can be costly. Smaller or resource-constrained farms may struggle to make these investments.
   - **Infrastructure Development:** Building the necessary infrastructure, such as deploying new communication towers or laying fiber-optic cables, can be time-consuming and expensive in rural areas.

5. **Compatibility and Integration:**

   - **System Compatibility:** Integrating low-latency solutions with existing farm equipment and software systems can be challenging. Compatibility issues may arise, requiring additional effort and resources for seamless integration.
   - **Training and Education:** Farmers and agricultural workers need training to effectively use low-latency technologies and interpret real-time data. Bridging the knowledge gap can be a hurdle.

6. **Power and Energy Supply:**

   - **Edge Device Power:** Edge computing devices require a stable power supply. Ensuring uninterrupted power in the field, especially in remote locations, can be a logistical challenge.
   - **Energy Efficiency:** The continuous operation of edge devices and sensors can lead to increased energy consumption, which may not align with sustainability goals.

7. **Scalability:**

   - **Scaling Solutions:** As farms grow or evolve, scaling low-latency solutions to accommodate additional sensors or data sources can be complex. Ensuring that the system remains responsive and efficient at scale is crucial.

8. **Environmental Factors:**

   - **Harsh Conditions:** Farming environments can be harsh, with exposure to extreme temperatures, humidity, and dust. Ensuring the durability and reliability of low-latency equipment in these conditions is essential.

Addressing these challenges requires a coordinated effort from farmers, technology providers, policymakers, and stakeholders in the agricultural industry. Overcoming these obstacles is essential to unlock the full potential of low-latency solutions in agriculture and realize the benefits of efficient, data-driven farming practices.

5.2 Edge Computing and Low-Latency Solutions

Edge computing and low-latency solutions play a pivotal role in addressing the challenges posed by latency in agriculture while enabling efficient and data-driven farming practices. Here's an exploration of edge computing and its significance in achieving low-latency solutions in agriculture:

1. **Edge Computing Defined:**
   - **Decentralized Data Processing:** Edge computing is a decentralized computing paradigm that brings data processing closer to the data source, often at or near the "edge" of the network, rather than relying solely on centralized cloud-based servers.
   - **On-Field Processing:** In agriculture, edge computing involves deploying computing devices directly in the field or on farming equipment. These devices process data locally, allowing for immediate analysis and decision-making.
2. **Significance of Edge Computing in Agriculture:**
   - **Real-Time Data Analysis:** Edge computing enables real-time analysis of data generated by sensors, drones, and other devices in the field. This immediate analysis provides farmers with timely insights for decision-making.
   - **Reduced Latency:** By processing data locally, edge computing minimizes latency, ensuring that critical decisions, such as irrigation adjustments or pest control measures, can be implemented swiftly.
   - **Bandwidth Optimization:** Edge computing reduces the need to transmit vast amounts of data to centralized servers, optimizing bandwidth usage and lowering data transmission costs for farmers.
   - **Offline Capabilities:** Edge devices in agriculture can operate offline or with intermittent connectivity, ensuring that data collection and analysis continue even in remote areas.
3. **Low-Latency Solutions in Agriculture:**
   - **Sensor Networks** (Chukwuemeka, P.-I.K., et al., 2018) [7]**:** Deploying sensors directly in the field, such as soil moisture sensors or weather stations, provides real-time data on environmental conditions. Edge computing devices process this data locally, reducing the time it takes to make irrigation or fertilization decisions.
   - **Drone Imaging:** Drones equipped with cameras and multispectral sensors capture high-resolution images of crops. Edge computing on the drone processes these images to identify areas requiring attention, such as disease outbreaks or nutrient deficiencies, with minimal delay.
   - **Livestock Monitoring:** Wearable sensors on livestock animals collect health and behavior data. Edge computing processes this data locally, allowing for immediate detection of anomalies and timely intervention.

- **Equipment Automation:** Edge computing is integral to autonomous farming equipment, allowing tractors and harvesters to make real-time decisions about field navigation and precision tasks.

4. **Benefits of Low-Latency Solutions:**

   - **Real-Time Decision-Making:** Low-latency solutions empower farmers with immediate insights, enabling them to make informed decisions on the spot.
   - **Cost Savings:** By reducing the need for constant data transmission to centralized servers, low-latency solutions can lower data transmission costs and reduce the dependence on expensive, always-on Internet connections.
   - **Scalability:** Smart farming systems can scale easily by adding more edge devices as needed, without overloading centralized servers.
   - **Privacy and Security:** Local data processing at the edge enhances data privacy and security since sensitive agricultural data can be kept on-farm rather than transmitted to external servers.

In summary, edge computing and low-latency solutions are integral to modernizing agriculture, reducing latency challenges, and enabling efficient and data-driven farming practices. These technologies empower farmers with immediate insights and the ability to make informed decisions, leading to improved crop yields, resource efficiency, and sustainability in agricultural operations.

6 Optimizing Resource Management

A tabulated summary of optimizing resource management in agriculture (Table 1).

These aspects highlight various strategies and technologies used in agriculture to optimize resource management, improve sustainability, and enhance productivity.

6.1 Precision Agriculture and Resource Allocation

Precision agriculture and resource allocation go hand in hand in modern farming practices. Precision agriculture involves using advanced technologies and data-driven approaches to optimize the allocation of resources such as water, fertilizers, pesticides, and labor. Here's how precision agriculture enhances resource allocation in farming:

1. **Data-Driven Decision-Making:**

   - Precision agriculture relies on data collected from various sources, including sensors, satellite imagery, and weather stations, to make informed decisions.
   - Data analysis helps farmers understand the specific needs of different areas within a field, allowing for precise resource allocation.

Table 1 Summary of optimizing resource management in agriculture

Aspect	Description
Precision agriculture	Using data and technology to precisely allocate resources such as water, fertilizers, and pesticides, minimizing waste and maximizing crop yields
IoT and sensor networks (Chukwuemeka, P.-I.K. et al., 2018) [7]	Deploying sensors and IoT devices to monitor soil conditions, weather, and crop health in real-time, enabling data-driven resource management
AI and machine learning (Tiwari, S et al., 2018) [8]	Utilizing AI algorithms and machine learning models to analyze data and make predictions, optimizing resource allocation and crop planning
Smart irrigation	Implementing intelligent irrigation systems that adjust water delivery based on real-time weather and soil moisture data, conserving water resources
Crop rotation	Employing crop rotation strategies to improve soil health and nutrient management, reducing the need for external inputs
Cover crops	Planting cover crops between main crops to prevent soil erosion, enhance soil fertility, and reduce the need for chemical inputs
Organic farming	Embracing organic farming practices that focus on natural resource management (Nicholson, et al., 2021) [2], sustainable soil health, and reduced chemical use
Sustainable livestock	Implementing sustainable livestock management practices to optimize feed, reduce waste, and minimize environmental impact
Renewable energy	Integrating renewable energy sources, such as solar and wind, to power farm operations and reduce reliance on fossil fuels
Data-driven insights	Leveraging data analytics to gain insights into resource usage patterns and make informed decisions for efficient management

2. **Variable Rate Technology (VRT):**
    - VRT allows farmers to adjust the application of resources based on real-time data and field conditions.
    - For example, VRT systems can vary the rate of fertilizer application, irrigation, or planting density within a single field, optimizing resource use.
3. **Efficient Irrigation:**
    - Soil moisture sensors and weather data are used to determine when and how much irrigation is needed.
    - Precision irrigation ensures that water is allocated precisely where it is needed, reducing water wastage and energy consumption.
4. **Nutrient Management:**
    - Soil testing and analysis help farmers understand nutrient levels in the soil.
    - This information guides the precise application of fertilizers, minimizing over-application and nutrient runoff.
5. **Pest and Disease Control:**
    - Precision agriculture integrates data on pest and disease pressures with GPS technology.

- This enables farmers to target pesticide applications to specific areas of the field, reducing chemical usage while maintaining crop health.

6. **Livestock Management:**
   - Wearable sensors and RFID tags are used for monitoring the health and behavior of livestock.
   - This data helps optimize feeding schedules and resource allocation for animal welfare and productivity.
7. **Predictive Analytics:**
   - Machine learning (Tiwari, S et al., 2018) [8] and AI algorithms analyze historical and real-time data to predict crop yields, weather patterns, and disease outbreaks.
   - Predictive analytics guide resource allocation decisions, allowing farmers to proactively address challenges.
8. **Soil Conservation:**
   - Precision agriculture promotes sustainable soil management practices, such as no-till farming and cover cropping, to improve soil health.
   - These practices optimize resource allocation by reducing erosion and enhancing nutrient retention.
9. **Labor Efficiency:**
   - Precision agriculture automates certain tasks, such as planting and harvesting, using GPS-guided machinery.
   - Labor resources are allocated efficiently, and the risk of human error is reduced.
10. **Monitoring and Feedback:**
    - Continuous monitoring of field conditions and resource usage provides real-time feedback to farmers.
    - Farmers can adjust their resource allocation strategies based on this feedback for ongoing improvement.
11. **Economic Viability:**
    - Precise resource allocation not only conserves resources but also improves crop yields and quality, leading to increased profitability for farmers.

Precision agriculture revolutionizes resource allocation by optimizing the use of inputs while minimizing waste. By harnessing technology and data-driven insights, farmers can achieve sustainable and efficient resource allocation practices, ultimately benefiting both their agricultural operations and the environment.

6.2 Edge Computing'S Role in Resource Optimization (Table 2)

Edge computing's role in agriculture is instrumental in resource optimization, enabling real-time data analysis, precision agriculture, efficient irrigation, and sustainable farming practices, among other benefits. It empowers farmers with timely insights and supports data-driven decision-making, ultimately enhancing the economic and environmental sustainability of farming operations.

Table 2 Overview of how edge computing plays a significant role in resource optimization in agriculture

Aspect of resource optimization	Role of edge computing in agriculture
Data processing	– Edge devices process data locally in real-time, reducing the need to transmit large volumes of data to centralized servers. This minimizes latency and optimizes bandwidth usage – Immediate data analysis enables timely decision-making for resource allocation
Precision agriculture	– Edge computing supports precision agriculture by providing on-field data processing for sensors and devices – It enables variable rate technology (VRT), allowing for precise resource allocation based on specific field conditions – Real-time data analysis informs decisions on irrigation, fertilization, and pest control, optimizing resource use
Efficient irrigation	– Soil moisture sensors, when integrated with edge computing, allow for immediate assessment of moisture levels, optimizing irrigation schedules – Edge devices process weather data and sensor information to determine precise irrigation needs, reducing water wastage – Drip and micro-irrigation systems can be controlled with low-latency edge computing for efficient water distribution
Nutrient management	– Soil testing data is processed locally through edge computing, enabling precise and immediate fertilizer application – Data analytics on nutrient levels guide resource allocation for fertilization, preventing over-application and nutrient runoff
Pest and disease control	– Edge devices analyze data from pest and disease sensors in real-time, allowing for targeted and timely pesticide applications – Integration of GPS technology enables accurate and efficient resource allocation for pest management – Early detection of pests and diseases through edge computing minimizes chemical usage while maintaining crop health
Livestock management	– Wearable sensors and RFID tags on livestock transmit data to edge devices for real-time monitoring – Edge computing processes livestock health and behavior data, optimizing resource allocation for animal welfare and productivity – Efficient feeding schedules are generated based on data analysis

(continued)

Table 2 (continued)

Aspect of resource optimization	Role of edge computing in agriculture
Predictive analytics	– Machine learning and AI algorithm (Tiwari, S et al., 2018) [8] run on edge devices to predict crop yields, weather patterns, and disease outbreaks – Predictive analytics guide resource allocation decisions for proactive problem-solving – Edge computing ensures immediate access to predictive insights
Soil conservation	– Sustainable practices like no-till farming and cover cropping are enhanced through edge computing – Real-time monitoring of soil conditions helps optimize resource allocation for soil health and erosion prevention – Nutrient cycling and retention are improved
Labor efficiency	– Edge computing automates tasks using GPS-guided machinery, optimizing labor allocation – Labor resources are efficiently allocated, and the risk of human error is minimized
Monitoring and feedback	– Continuous monitoring of field conditions and resource usage provides real-time feedback to farmers through edge computing – Farmers can adjust their resource allocation strategies based on immediate feedback for ongoing improvement – Edge devices facilitate timely course correction
Economic viability	– Edge computing supports increased profitability by optimizing resource allocation, conserving resources, and improving crop yields and quality – Efficient resource use contributes to the economic sustainability of farming operations

7 Autonomous Machinery and Edge Intelligence

Autonomous machinery and edge intelligence are two key components of modern agriculture that work together to revolutionize farming practices. Here's a breakdown of their roles and how they intersect:

Autonomous Machinery:

- **Definition:** Autonomous machinery in agriculture refers to vehicles and equipment that can perform tasks without direct human intervention. These machines are equipped with sensors, GPS technology, and onboard computing systems.
- **Roles and Functions:**
  - **Planting and Seeding:** Autonomous tractors can precisely plant seeds at optimal depths and spacing, ensuring efficient use of resources.
  - **Harvesting:** Self-driving combines and harvesters can autonomously collect and process crops, minimizing waste and increasing harvesting efficiency.

- **Weeding and Pest Control:** Autonomous robots (Acosta, J.A, et al., 2011) [15] equipped with cameras and AI algorithms can identify and remove weeds or apply targeted pesticide treatments.
- **Fertilization:** Autonomous machinery can distribute fertilizers with precision, reducing overuse and minimizing environmental impact.

• **Benefits:**

- **Labor Efficiency:** Autonomous machinery reduces the need for manual labor, addressing labor shortages and allowing farmworkers to focus on more skilled tasks.
- **Resource Optimization:** These machines optimize resource use, including fuel, water, fertilizers, and pesticides, contributing to sustainable farming.
- **Data Collection:** Autonomous machinery can collect valuable data on field conditions, crop health, and equipment performance, which can be used for decision-making.

Edge Intelligence:

• **Definition:** Edge intelligence involves the processing and analysis of data at or near the source of data generation, rather than sending it to a centralized cloud server. In agriculture, edge intelligence typically involves onboard computing systems within autonomous machinery.
• **Roles and Functions:**

- **Real-Time Data Analysis:** Edge intelligence allows autonomous machinery to analyze data from sensors and cameras in real-time. For example, a self-driving tractor can adjust its planting depth based on soil conditions detected by onboard sensors.
- **Decision-Making:** Autonomous machinery equipped with edge intelligence can make autonomous decisions on tasks like steering, adjusting planting rates, or applying treatments based on pre-programmed algorithms and real-time data.
- **Communication:** Edge intelligence enables autonomous machinery to communicate with other equipment and share data, facilitating coordinated operations within a field.

• **Benefits:**

- **Reduced Latency:** Edge intelligence minimizes data transmission delays, allowing for immediate decision-making and responses to changing field conditions.
- **Privacy and Security:** Onboard processing enhances data privacy and security, as sensitive information remains within the equipment rather than being transmitted to external servers.
- **Reliability:** Edge intelligence ensures that autonomous machinery can function independently even in areas with limited network connectivity.

Intersection of Autonomous Machinery and Edge Intelligence:

- Autonomous machinery relies on edge intelligence to function effectively. These onboard computing systems process data from various sensors and cameras in real-time, enabling autonomous decision-making.
- Edge intelligence enables autonomous machinery to adapt to changing conditions and make split-second decisions, such as adjusting planting depth, steering to avoid obstacles, or applying treatments only where needed.
- Data collected by autonomous machinery with edge intelligence can be used to optimize resource allocation, such as determining precise planting rates or adjusting irrigation schedules based on real-time soil moisture levels.

In summary, the combination of autonomous machinery and edge intelligence is a powerful force in modern agriculture. These technologies work together to enhance efficiency, reduce resource waste, and improve the overall sustainability and productivity of farming operations.

7.1 Autonomous Farming Equipment

A tabulated summary of autonomous farming equipment (Table 3).

These autonomous farming equipment and technologies are transforming agriculture by increasing efficiency, reducing labor costs, and enabling precise and sustainable farming practices.

Autonomous farming equipment, also known as self-driving or robotic farming machinery, is revolutionizing the agriculture industry. These advanced machines can perform various tasks without direct human intervention, significantly increasing efficiency, precision, and productivity in farming operations. Here are key aspects of autonomous farming equipment:

1. **Types of Autonomous Farming Equipment:**

   - **Autonomous Tractors:** These are self-driving tractors equipped with GPS, sensors, and advanced control systems. They can perform tasks like plowing, planting, and harvesting with high precision.
   - **Harvesting Robots:** Autonomous harvesters and robots (Acosta, J.A, et al., 2011) [15] are designed to pick fruits, vegetables, or crops with minimal human intervention. They use computer vision and robotic arms for precise harvesting.
   - **Weeding and Pest Control Robots:** These machines are equipped with cameras and AI algorithms to identify and remove weeds or apply targeted pesticide treatments, reducing the need for chemical use.
   - **Autonomous Seeders and Planters:** These machines can autonomously plant seeds at optimal depths and spacing, ensuring efficient use of resources.

Table 3 Summary of autonomous farming equipment

Type of equipment	Description
Autonomous tractors	Self-driving tractors equipped with GPS, sensors, and AI algorithms that can perform tasks such as plowing, planting, and harvesting
Autonomous harvesters	Machines capable of autonomously harvesting crops like fruits, vegetables, and grains with precision and efficiency
Autonomous planters	Equipment that automatically plants seeds at precise intervals and depths, optimizing crop establishment
Autonomous sprayers	Self-driving sprayers equipped with sensors and AI to apply fertilizers, pesticides, and herbicides with accuracy and reduced chemical usage
Autonomous weeders	Machines designed to identify and remove weeds with minimal human intervention, reducing the need for herbicides
Autonomous seeders	Equipment capable of autonomously seeding fields with precision, ensuring even crop distribution
Autonomous livestock	Robots (Acosta, J.A, et al., 2011) [15] and systems for managing livestock, including autonomous feeders, milkers, and herders
Autonomous drones	UAVs (unmanned aerial vehicles) equipped with cameras and sensors to monitor crops, livestock, and land, providing valuable data for precision agriculture
Autonomous soil sensors [68]	Sensors buried in the ground to monitor soil conditions and nutrient levels, allowing for precise irrigation and fertilization
Autonomous data analysis	AI-powered software and platforms (Chopra, A.K, et al., 2009) [12] that analyze data from autonomous equipment to optimize farming practices and resource allocation

- **Precision Sprayers:** Autonomous sprayers equipped with sensors and AI can precisely apply fertilizers, pesticides, and herbicides only where needed, reducing overuse.
- **Unmanned Aerial Vehicles (UAVs) or Drones:** Drones equipped with cameras and sensors are used for aerial crop monitoring, mapping, and even crop spraying in some cases.

2. **Key Technologies:**

- **GPS and GNSS (Global Navigation Satellite Systems):** These technologies provide accurate positioning data to guide autonomous equipment within fields.
- **Sensors:** Various sensors, including soil moisture sensors, weather sensors, and cameras, collect data on field conditions, crop health, and environmental factors.
- **Artificial Intelligence (AI):** AI algorithms process sensor data to make real-time decisions, such as adjusting planting rates or applying treatments based on field conditions.
- **Edge Computing:** Onboard computing systems, known as edge computing, enable autonomous equipment to process data locally, reducing latency and improving decision-making.

3. **Benefits of Autonomous Farming Equipment:**
   - **Increased Efficiency:** Autonomous equipment can work around the clock, in varying weather conditions, and with consistent precision, leading to increased operational efficiency.
   - **Resource Optimization:** These machines optimize resource use, including fuel, water, fertilizers, and pesticides, contributing to sustainable and environmentally friendly farming practices.
   - **Labor Savings:** Autonomous equipment reduces the need for manual labor, addressing labor shortages and allowing farmworkers to focus on more skilled tasks.
   - **Data Collection:** Autonomous machinery collects valuable data on field conditions and crop health, which can be used for decision-making, yield predictions, and long-term farm planning.
   - **Precision Farming:** These machines enable precise and targeted applications of inputs, leading to higher crop yields, better quality produce, and reduced waste.
4. **Challenges and Considerations:**
   - **High Initial Costs:** The upfront investment in autonomous farming equipment can be significant, especially for smaller farms.
   - **Maintenance and Repairs:** Autonomous machinery requires regular maintenance and may need specialized technicians for repairs.
   - **Data Privacy and Security:** Collecting and managing data from autonomous equipment requires robust cybersecurity measures to protect sensitive information.
   - **Regulatory and Liability Issues:** There may be regulatory hurdles and liability concerns related to the use of autonomous farming equipment, especially in areas with strict regulations.
   - **Integration with Existing Systems:** Integrating autonomous equipment with existing farm infrastructure and management systems can be complex and require additional investment.

Autonomous farming equipment represents a significant step forward in modern agriculture, offering the potential for increased productivity, sustainability, and profitability. As technology continues to advance and costs decrease, the adoption of autonomous machinery is likely to continue growing, transforming the way farms operate around the world.

7.2 Edge Computing for Safe and Efficient Autonomy

Edge computing plays a crucial role in ensuring safe and efficient autonomy in various domains, including autonomous farming equipment. It enables real-time data processing, decision-making, and control at the edge of the network, closer to the source of data generation. In the context of autonomous farming equipment, here's how edge computing contributes to safe and efficient autonomy:

1. **Real-Time Data Processing:**
   - **Onboard Computing:** Edge computing involves onboard computing systems within autonomous farming equipment. These systems can process data from various sensors, cameras, and other sources in real-time.
   - **Low Latency:** Edge computing minimizes data transmission delays by processing information locally. This low-latency processing is critical for ensuring timely responses and safe operation in dynamic farm environments.
2. **Immediate Decision-Making:**
   - **Autonomous Control:** Edge computing enables autonomous machinery to make real-time decisions based on processed data. For instance, a self-driving tractor can adjust its path or speed autonomously to avoid obstacles detected by onboard sensors.
   - **Safety Algorithms:** Safety-critical algorithms run at the edge to ensure that autonomous equipment operates within predefined safety parameters. These algorithms can override commands if necessary to prevent accidents.
3. **Enhanced Privacy and Security:**
   - **Data Isolation:** Edge computing keeps sensitive data within the equipment, reducing the need to transmit it to external servers. This isolation enhances data privacy and security.
   - **Security Protocols:** Edge devices can employ robust security protocols and encryption to protect against unauthorized access and cyber threats.
4. **Reliable Operation in Low Connectivity Areas:**
   - **Offline Capabilities:** Edge devices can operate with intermittent or no connectivity, which is common in rural farming areas. They continue to process data and make decisions, ensuring uninterrupted operation.
5. **Efficient Resource Utilization:**
   - **Resource Optimization:** Edge computing supports resource-efficient operation by processing data related to resource allocation, such as precise planting or targeted pesticide application.
6. **Scalability:**
   - **Modular Architecture:** Edge computing solutions can be modular, allowing farms to scale their autonomous fleets by adding more equipment with minimal impact on central server infrastructure.

7. **Localized Data Insights:**
   - **Immediate Feedback:** Edge computing provides immediate insights into field conditions, equipment performance, and crop health. This localized feedback helps farmers make informed decisions quickly.
8. **Integration with Cloud Resources:**
   - **Hybrid Approach:** While edge computing handles real-time processing, it can also be integrated with cloud-based resources for more extensive data analysis, long-term planning, and data storage.
9. **Safety Overrides:**
   - **Emergency Protocols:** Edge computing can incorporate emergency protocols that prioritize safety in unforeseen situations. For example, autonomous equipment can stop or slow down if it detects anomalies or unsafe conditions.

In summary, edge computing is a critical enabler of safe and efficient autonomy in autonomous farming equipment. It allows for immediate data processing, real-time decision-making, enhanced privacy and security, and reliable operation in low-connectivity areas. By leveraging edge computing, farmers can harness the benefits of autonomous machinery while ensuring safety, resource optimization, and responsiveness in their agricultural operations.

8 Sustainable Agriculture Through Edge Technologies

Sustainable agriculture is increasingly reliant on cutting-edge technologies, including edge computing, to address environmental, economic, and social challenges in farming practices. Here's how edge technologies contribute to sustainable agriculture:

1. **Precision Resource Management:**
   - **Data-Driven Decisions:** Edge devices process data from various sensors and sources to provide real-time insights into soil conditions, weather patterns, and crop health.
   - **Variable Rate Technology (VRT):** Edge computing supports VRT, enabling precise allocation of resources such as water, fertilizers, and pesticides based on specific field conditions. This minimizes waste and environmental impact.
2. **Efficient Water Usage:**
   - **Precision Irrigation:** Soil moisture sensors and weather data, processed at the edge, allow for precise irrigation scheduling, reducing water wastage and conserving this critical resource.

3. **Reduced Chemical Usage:**
   - **Targeted Application:** Edge devices equipped with cameras and AI algorithms can identify and treat weeds, pests, and diseases with high precision. This reduces the need for broad-spectrum chemicals[73] and minimizes environmental impact.
4. **Soil Health Management:**
   - **Real-Time Monitoring:** Edge computing enables continuous monitoring of soil conditions, facilitating the implementation of sustainable soil management practices such as cover cropping, crop rotation, and reduced tillage.
5. **Renewable Energy Integration:**
   - **Energy Efficiency:** Edge devices can optimize the use of energy resources in farming operations, including the integration of renewable energy sources such as solar panels and wind turbines.
6. **Reduced Emissions:**
   - **Efficient Machinery Operation:** Edge computing supports autonomous and semi-autonomous farm equipment, which can be optimized for fuel efficiency and emissions reduction.
7. **Data-Driven Sustainability:**
   - **Data Analytics:** Edge computing platforms facilitate the analysis of large volumes of data, helping farmers identify trends and make data-driven decisions that enhance sustainability.
8. **Localized Decision-Making:**
   - **Real-Time Feedback:** Edge devices provide immediate feedback on field conditions, equipment performance, and crop health, allowing for rapid adjustments to improve sustainability.
9. **Low Latency for Safety:**
   - **Safety Protocols:** Edge computing allows for immediate safety overrides and emergency responses, crucial for preventing accidents in autonomous farming equipment.
10. **Rural Connectivity: Operational Continuity**—Edge devices can operate in areas with limited connectivity, ensuring that sustainable farming practices can be implemented even in remote locations.
11. **Data Privacy and Security: Protecting Sensitive Data**—Edge computing keeps sensitive farm data within the local network, reducing the risk of data breaches and ensuring data privacy and security.

12. **Regulatory Compliance: Environmental Regulations**—Edge technologies support compliance with environmental regulations by enabling precise resource management and data documentation.
13. **Scalability and Accessibility: Affordable Adoption**—Edge technologies can be adopted incrementally, making them accessible to a wide range of farmers, including smallholders.

In conclusion, edge technologies are pivotal in advancing sustainable agriculture by promoting resource efficiency, reducing environmental impact, enhancing data-driven decision-making, and ensuring the long-term viability of farming practices. As agriculture continues to evolve, the integration of edge computing and related technologies will play a central role in achieving the goals of sustainability, resilience, and productivity in farming.

8.1 Environmental Benefits

Environmental benefits in the context of agriculture refer to the positive impacts and contributions of sustainable farming practices and technologies on the environment. These benefits are crucial for mitigating the environmental challenges associated with traditional farming methods. Here are some key environmental benefits of sustainable agriculture:

1. **Soil Health Improvement:**
   - Sustainable practices like reduced tillage, cover cropping, and organic farming [70] promote soil health and reduce soil erosion, helping to maintain fertile and productive land.
2. **Water Conservation:**
   - Precision irrigation systems, soil moisture monitoring, and responsible water management reduce water wastage and minimize the environmental impact of agricultural runoff, which can contain pollutants.
3. **Reduced Chemical Usage:**
   - Sustainable farming practices, including integrated pest management and precision agriculture, minimize the need for synthetic pesticides and fertilizers, lowering chemical runoff into water bodies.
4. **Biodiversity Preservation:**
   - Sustainable farming methods create habitat diversity on farms, supporting pollinators and beneficial insects. This, in turn, helps maintain biodiversity and ecosystem services.

5. **Carbon Sequestration:**
    - Practices such as agroforestry, afforestation, and cover cropping sequester carbon dioxide in soil and vegetation, mitigating greenhouse gas emissions and combating climate change.
6. **Reduced Greenhouse Gas Emissions:**
    - Sustainable agriculture promotes the use of energy-efficient machinery and practices that reduce emissions, contributing to climate change mitigation.
7. **Enhanced Nutrient Cycling:**
    - Sustainable agriculture encourages nutrient cycling through practices like composting and crop rotation, reducing nutrient runoff and eutrophication of water bodies.
8. **Reduced Deforestation:**
    - Sustainable farming discourages deforestation for agriculture by promoting practices that make better use of existing land.
9. **Reduced Soil and Water Pollution:**
    - Sustainable agriculture minimizes the leaching of chemicals[73] into soil and water, reducing contamination and its adverse effects on ecosystems.
10. **Improved Air Quality:** Reduced use of chemical fertilizers and pesticides in sustainable farming helps maintain better air quality by reducing the release of volatile organic compounds[70].
11. **Resilience to Climate Change:** Sustainable practices enhance the resilience of agricultural systems to extreme weather events, helping farms adapt to the changing climate.
12. **Wildlife Habitat Creation:** Sustainable farms often incorporate wildlife-friendly practices and habitats, providing shelter and food sources for various species.
13. **Preservation of Natural Resources:** Sustainable agriculture ensures the responsible use and preservation of natural resources(Nicholson, et al., 2021) [2] like water, soil, and biodiversity.
14. **Long-Term Sustainability:** By promoting practices that prioritize the environment, sustainable agriculture helps ensure that farming remains viable for future generations.
15. **Compliance with Environmental Regulations:** Sustainable farming practices align with environmental regulations and support farmers in meeting compliance standards.
16. **Economic Benefits for Farmers:** Many sustainable practices lead to cost savings for farmers, making them economically beneficial while also being environmentally responsible.

These environmental benefits demonstrate the potential for agriculture to be a positive force for conservation and environmental protection. By adopting sustainable practices and integrating advanced technologies, agriculture can contribute to a more sustainable and resilient future for both the environment and food production.

8.2 Economic and Social Sustainability

A tabulated summary of economic and social sustainability in agriculture (Table 4).

These aspects highlight the multifaceted nature of sustainability in agriculture, encompassing economic viability and social well-being for farmers and communities.

9 Challenges and Considerations

Challenges and considerations in sustainable agriculture encompass various factors:

1. **Environmental Stewardship:** Balancing increased production with minimizing environmental impacts, such as soil degradation and water pollution.
2. **Economic Viability:** Ensuring that sustainable practices remain financially feasible for farmers, especially in the face of market volatility.

Table 4 Summary of economic and social sustainability in agriculture

Aspect	Economic sustainability	Social sustainability
Profitability	Ensuring farms generate profits and financial stability	Fair wages, access to education, and healthcare for farm labor
Diversification	Expanding income sources through value addition and diversification	Supporting local communities and businesses
Resource efficiency	Optimizing resource use for cost savings	Ensuring safe and fair working conditions
Access to markets	Access to fair and competitive markets	Providing access to safe and affordable food for all
Financial resilience	Preparing for economic risks and shocks	Strategies for community development and engagement
Investment in technology	Adoption of advanced technologies for productivity	Preserving cultural heritage and indigenous knowledge
Farm labor	Ensuring fair wages, safe working conditions, and benefits	Promoting equity, diversity, and inclusion
Community engagement	Contributing positively to the community's social fabric	Land tenure security and stable rural communities
Food security	Ensuring access to nutritious and affordable food	Preserving cultural traditions and heritage
Land tenure and ownership	Secure land tenure and ownership rights	Providing educational and training opportunities

3. **Resource Scarcity:** Addressing challenges related to resource scarcity, such as water scarcity and land degradation.
4. **Climate Change:** Adapting farming practices to changing climate conditions and mitigating greenhouse gas emissions.
5. **Social Equity:** Promoting fair labor practices, gender equality, and equitable access to resources and opportunities in agriculture.
6. **Technological Adoption:** Encouraging the adoption of sustainable technologies while addressing the digital divide in rural areas.

Balancing these challenges and considerations is essential for achieving long-term sustainability in agriculture.

9.1 Data Security and Privacy

A tabulated summary of key aspects of data security and privacy in agriculture (Table 5).

9.2 Infrastructure and Connectivity

Infrastructure and connectivity play pivotal roles in modernizing and advancing agriculture. Here are key considerations related to infrastructure and connectivity in agriculture:

Table 5 Summary of key aspects of data security and privacy in agriculture

Aspect	Data security	Data privacy
Data storage	Secure storage of agricultural data to prevent unauthorized access or breaches	Ensuring that personal and sensitive data is stored safely and access is restricted
Encryption	Encrypting data during transmission and storage to protect against interception	Protecting individual privacy by anonymizing or de-identifying data
Access control	Implementing access controls and user authentication to restrict unauthorized access	Clearly defining data ownership rights and responsibilities
Cybersecurity	Employing cybersecurity measures such as firewalls and intrusion detection systems	Transparent data handling practices, including data-sharing agreements
Software updates (Chopra, A.K, et al., 2009) [12]	Regularly updating software and systems with security patches to prevent vulnerabilities	Obtaining informed consent for data collection and use

Infrastructure:

1. **Transportation:** Well-maintained roads and transportation networks are essential for the efficient movement of agricultural products to markets and distribution centers.
2. **Irrigation Systems:** Infrastructure for irrigation, including canals, pipelines, and reservoirs, is crucial for water management and crop cultivation.
3. **Storage Facilities:** Adequate storage infrastructure, such as silos and cold storage units, is necessary to prevent post-harvest losses and ensure food security.
4. **Energy Supply** (Burhan, M. et al., 2018) [16]**:** Reliable energy infrastructure, including electrification and renewable energy sources, supports modern farming practices and technology adoption.
5. **Research Facilities:** Research institutions and agricultural extension services require infrastructure for conducting experiments, trials, and disseminating knowledge to farmers.
6. **Digital Infrastructure:** Robust digital infrastructure, including broadband Internet access and data centers, is vital for data-driven agriculture and precision farming.

Connectivity:

1. **Broadband Internet:** High-speed Internet connectivity is essential for accessing agricultural information, market prices, and weather forecasts, as well as for remote monitoring and data sharing.
2. **Mobile Networks:** Mobile networks provide real-time communication and access to mobile apps and services, facilitating remote data collection and decision-making.
3. **IoT Connectivity:** The Internet of Things (IoT) relies on connectivity to enable sensors, drones, and smart devices to collect and transmit data for precision agriculture.
4. **Satellite Communication:** Satellite-based communication systems offer connectivity in remote rural areas, ensuring that farmers have access to information and services.
5. **Data Interoperability:** Ensuring that different agricultural technologies and systems can communicate and share data seamlessly enhances efficiency and effectiveness.
6. **Cybersecurity:** Connectivity also brings security challenges, and robust cybersecurity measures are necessary to protect sensitive agricultural data and infrastructure.
7. **Rural Connectivity:** Bridging the digital divide in rural areas is essential to ensure that all farmers, regardless of location, can benefit from modern agricultural practices.

Investment in infrastructure and connectivity is a prerequisite for the adoption of advanced technologies in agriculture. It enhances productivity, enables data-driven decision-making, and supports sustainable farming practices, ultimately contributing to food security and rural development.

10 Future Directions and Innovations

Emerging trends in smart farming (Meena, A.K, et al., 2005) [9] are revolutionizing agriculture with innovative technologies and sustainable practices. Digital twins for crops, powered by advanced modeling and simulation, enable precise crop management and yield predictions. Integrated farm management software provides real-time insights (Chopra, A.K, et al., 2009) [12] for data-driven decision-making, optimizing operations. Artificial intelligence is enhancing pest and disease detection through image analysis, reducing crop losses. Blockchain ensures transparent and traceable supply chains(Burhan. M. et al., 2018) [16], building consumer trust. Vertical and urban farming methods address land scarcity and offer year-round cultivation in controlled environments. Agri-robotics, backed by AI and machine learning (Tiwari, S et al., 2018) [8], automate tasks, cutting labor costs. Climate-resilient crops are developed for resilience in a changing climate, while smart irrigation systems optimize water use.

Evolving edge technologies are at the forefront of this transformation, facilitating real-time data processing and decision-making. Edge computing ensures rapid responses, crucial for autonomous machinery and drones. Edge AI and machine learning (Tiwari, S et al., 2018) [8] enable on-device data analysis, enhancing tasks like crop monitoring and pest detection. Low-power sensors provide continuous monitoring in remote areas. Edge robotics platforms perform tasks autonomously, from weeding to data collection. Wireless connectivity solutions like LoRaWAN extend data collection in rural regions. Edge-to-cloud integration ensures seamless data transfer. Fleet management benefits from real-time insights and optimized routes, while robust edge security safeguards sensitive agricultural data. These evolving edge technologies are reshaping agriculture, enabling efficiency, cost savings, and data-driven innovation.

Conclusion

In conclusion, the world of agriculture is undergoing a profound transformation driven by technology, data, and sustainability imperatives. Emerging trends in smart farming are reshaping traditional practices and revolutionizing the way we produce food. From digital twins for crops to blockchain-enabled supply chains, these trends are enhancing efficiency, reducing waste, and ensuring food safety (Wild, S.R, et al., 1992) [14]. Evolving edge technologies are playing a pivotal role in this transformation, enabling real-time data processing and decision-making at the field level. Edge computing, AI, low-power sensors, and robotics are empowering farmers to make informed choices, optimize resource use, and increase productivity.

The future of agriculture is one of sustainability, where innovative solutions address the challenges of a growing population, climate change, and resource scarcity. By embracing these emerging trends and leveraging evolving edge technologies, agriculture is poised to become more efficient, resilient, and environmentally friendly. As we move forward, the collaboration between technology innovators, farmers, and policymakers will be essential to unlock the full potential of smart farming (Kim, S.W, et al., 2015) and ensure food security for generations to come.

References

1. S.K. Routray, A. Javali, L. Sharma, A.D. Ghosh, A. Sahoo, Internet of Things based precision agriculture for developing countries, in *Proceedings of the 2019 International Conference on Smart Systems and Inventive Technology (ICSSIT), Tirunelveli, India, 27–29 November 2019*, pp. 1064–1068. [Google Scholar]
2. C.F. Nicholson, E.C. Stephens, B. Kopainsky, P.K. Thornton, A.D. Jones, D. Parsons, J. Garrett, Food security outcomes in agricultural systems models: Case examples and priority information needs. Agric. Syst. **188**, 103030 (2021) [Google Scholar] [CrossRef]
3. J. Poveda, Insect Frass in the development of sustainable agriculture. A review. Agron. Sustain. Dev. **41**, 1–10 (2021) [Google Scholar] [CrossRef]
4. M. Arora, B. Kiran, S. Rani, A. Rani, B. Kaur, N. Mittal, Heavy metal accumulation in vegetables irrigated with water from different sources. Food Chem. **111**, 811–815 (2008) [Google Scholar] [CrossRef]
5. P.C. Nagajyoti, K.D. Lee, T.V.M. Sreekanth, Heavy metals, occurrence and toxicity for plants: A review. Environ. Chem. Lett. **8**, 199–216 (2010) [Google Scholar] [CrossRef]
6. P.K. Rai, S.S. Lee, M. Zhang, Y.F. Tsang, K.-H. Kim, Heavy metals in food crops: health risks, fate, mechanisms, and management. Environ. Int. **125**, 365–385 (2019) [Google Scholar] [CrossRef]
7. P.-I.K. Chukwuemeka, N.U. Hephzibah, Potential health risk from heavy metals via consumption of leafy vegetables in the vicinity of Warri refining and petrochemical company, Delta State, Nigeria. Ann. Biol. Sci. **6**, 30–37 (2018) [Google Scholar]
8. S. Tiwari, C. Lata, Heavy metal stress, signaling, and tolerance due to plant-associated microbes: an overview. Front. Plant Sci. **9**, 452 (2018) [Google Scholar] [CrossRef][Green Version]
9. A.K. Meena, G.K. Mishra, P.K. Rai, C. Rajagopal, P.N. Nagar, Removal of heavy metal ions from aqueous solutions using carbon aerogel as an adsorbent. J. Hazard. Mater. **122**, 161–170 (2005) [Google Scholar] [CrossRef]
10. P.K. Rai, Heavy metal phytoremediation from aquatic ecosystems with special reference to Macrophytes. Crit. Rev. Environ. Sci. Technol. **39**, 697–753 (2009) [Google Scholar] [CrossRef]
11. S. Kohzadi, B. Shahmoradi, E. Ghaderi, H. Loqmani, A. Maleki, Concentration, source, and potential human health risk of heavy metals in the commonly consumed medicinal plants. Biol. Trace Elem. Res. **187**, 41–50 (2019) [Google Scholar] [CrossRef]
12. A.K. Chopra, C. Pathak, G. Prasad, Scenario of heavy metal contamination in agricultural soil and its management. J. Appl. Nat. Sci. **1**, 99–108 (2009) [Google Scholar] [CrossRef][Green Version]
13. W. Feng, Z. Guo, X. Xiao, C. Peng, L. Shi, H. Ran, W. Xu, A dynamic model to evaluate the critical loads of heavy metals in agricultural soil. Ecotoxicol. Environ. Saf. **197**, 110607 (2020) [Google Scholar] [CrossRef] [PubMed]
14. S.R. Wild, K.C. Jones, Organic chemicals entering agricultural soils in sewage sludges: screening for their potential to transfer to crop plants and livestock. Sci. Total Environ. **119**, 85–119 (1992) [Google Scholar] [CrossRef]
15. J.A. Acosta, A. Faz, S. Martínez-Martínez, R. Zornoza, D.M. Carmona, S. Kabas, Multivariate statistical and GIS-based approach to evaluate heavy metals behavior in mine sites for future reclamation. J. Geochem. Explor. **109**, 8–17 (2011) [Google Scholar] [CrossRef]
16. M. Burhan, R.A. Rehman, B. Khan, B.-S. Kim, IoT elements, layered architectures and security issues: a comprehensive survey. Sensors **18**, 2796 (2018) [Google Scholar] [CrossRef] [Green Version]

Impact of Cloud Computing on the Future of Smart Farming

J. Immanuel Johnraja, P. Getzi Jeba Leelipushpam, C. P. Shirley, and P. Joyce Beryl Princess

1 Traditional Farming: An Overview

Traditional farming which is a practice that involves the intensive use of inherent knowledge, primitive tools, various natural resources, organic manures, and fertilizers and the cultural beliefs of farmers can be defined as primitive farming. Around 50% of the world's population still uses it and it is shown in Fig. 1 [37].

1.1 Key Characteristics of Traditional Farming

1.1.1. *Subsistence Farming:* Traditional farming is mainly subsistence, where only sufficient quantities of food are produced by the farmer to feed his family and sell surpluses on the local market. The focus shall be on the cultivation of crops required for survival, like rice, wheat, and vegetables [7].

1.1.2. *Use of Natural Resources:* The traditional farming method is based on the exploitation of natural resources, for example, soil, water, and seeds. To enrich the soil and increase the growth of crops, farmers use organic manures and compost. They are also using their animals' strength to plow the land and for carrying goods.

1.1.3. *Crop Rotation and Intercropping:* A significant technique used in traditional farming is crop rotation and intercropping. Crop rotation allows for the sowing of various crops over some time in order to preserve soil fertility and

J. I. Johnraja · P. G. J. Leelipushpam · C. P. Shirley (✉) · P. J. B. Princess
Department of Computer Science and Engineering, Karunya Institute of Technology and Sciences, Coimbatore, Tamilnadu, India
e-mail: shirleycp@karunya.edu

S. Balasubaramanian et al. (eds.), *Intelligent Robots and Drones for Precision Agriculture*, Signals and Communication Technology,
https://doi.org/10.1007/978-3-031-51195-0_18

Fig. 1 Traditional farming

avoid soil depletion. To achieve maximum productivity and minimize the risk of failure of crops, intercropping involves planting various crops in a single field.

1.1.4. *Extensive farming with simple tools and crops:* Extensive farming with simple tools and crops harkens back to a time when farming was more about the breadth of land than technological intensiveness. Instead of focusing on maximizing yield per unit of land, as in intensive farming, extensive farming relies on vast tracts of land and typically yields less produce per hectare. The reliance on basic tools—like the hand plow, scythe, and hoe—means that the relationship between the farmer and the land is direct and unmediated by complex machinery. Such a method, while less efficient in terms of yield, can be more sustainable since it often avoids the exhaustive use of chemicals and heavy machinery that can degrade soil health over time.

1.1.5. *Usage of Simple and Traditional Means:* Simple and traditional means of farming are a testament to time-honored agricultural practices that have sustained communities for centuries. Instead of relying on advanced machinery, chemical fertilizers, and genetically modified organisms, these methods harness the natural synergy between land, water, and seeds. Farmers using such techniques often employ hand tools like hoes, sickles, and wooden plows, following rhythms dictated by nature rather than industrial timetables. Crop rotation, composting, and natural pest management are cornerstones of this approach, ensuring that the land remains fertile and productive over generations.

1.1.6. *Usage of Natural Seeds and Manure:* In traditional farming, the use of natural seeds preserves the genetic diversity of crops, ensuring resilience against pests, diseases, and changing climate conditions. These seeds, often passed down through generations, are adapted to local soils and weather patterns, optimizing their growth potential. Complementing this, the application of organic manure not only enriches the soil with vital nutrients but also promotes its microbial health, leading to more robust and sustainable crop yields.

1.1.7. *Sustainability and Resource Efficiency:* Sustainability and resource efficiency are intrinsic to traditional farming methods, which have evolved over generations to work in harmony with local ecosystems. These practices often prioritize the health of the soil, water conservation, and biodiversity, ensuring the land remains fertile for future crops. By leveraging nature's processes and cycles, traditional farming reduces waste and minimizes the need for external inputs, making it both environmentally and economically sustainable [9].

1.2 Impact and Implications

Soil organic matter is reduced by the primitive method of framing, and in a short period crops will take over the nutrient content of the soil. This is why farmers are forced to go somewhere else for farming. This results in the process of deforestation where the forest lands are converted to farms, ranches, or urban use. Jungle rainforests have the highest concentration of deforestation. The forests, which led to the situation of deforestation, had to be savagely cut and burned to allow for cultivation. Soil erosion is a process that is caused by natural physical forces of water and wind, as well as factors associated with agricultural activity such as plowing, to remove topsoil.

Traditional farming, rooted in age-old practices, plays a pivotal role in preserving biodiversity by maintaining a diverse array of crops adapted to various environments. These methods, often devoid of synthetic chemicals, reduce environmental pollution, safeguarding both the land and water sources. Traditional farming can occasionally produce lower yields than modern agricultural techniques, which could present problems for feeding expanding populations. Finding a balance between traditional practices and contemporary needs is crucial because traditional farmers may experience economic constraints and competitiveness as the global economy grows more intertwined [8].

2 The Transformation of Traditional Farming into Smart Farming

Precision agriculture, which was first introduced in the early 1990s, marked the beginning of the evolution of smart farming. Precision agriculture is a method of farming where technical instruments are used to focus on specific farm areas that require attention, such as the moisture content of the soil or nutrient shortages. Utilizing data analysis tools, smart farming technology combines input optimization and waste reduction techniques to increase yields while lowering expenses [1].

In recent years, a diverse set of cutting-edge technologies have been gathered by the smart farming movement. When used in conjunction with a camera or sensor, a drone can provide invaluable information on the health of a crop and its potential yield. In order to make informed decisions about irrigation and fertilizer application, farmers can utilize soil sensors to measure soil moisture, temperature, and nutrient content [2]. Farmers everywhere may now make use of IoT-enabled sensors and tools to optimize their weather, irrigation, and fertilizer management. For data to facilitate well-informed choices, it must be consolidated in one place [3, 4].

2.1 Traditional Farming

The old farming method is one of the ancient ways of agriculture which has existed for hundreds of years. To grow crops and raise livestock, it is necessary to use indigenous methods of cultivation. The sun, rain, and soil are mainly relied upon by farmers in order to guarantee crop health and productivity. They cultivate and harvest their crops in traditional methods, giving plants and animals what they need to survive. Traditional farming is still an effective means of food production even though it may be technologically less advanced than smart farming. Farmers have an excellent understanding of the land and vegetation that may grow on it, so they are able to take care of their cattle as well as deal with pests, insects, or diseases. This experience will typically be passed down to younger generations, enabling them to adapt to changing conditions and develop their methods over time [12].

2.2 Smart Farming

Smart farming aims at using modern technologies, in order to increase the effectiveness and productivity of agriculture. Everything from using sensors and drones for crop monitoring to the use of artificial intelligence that can predict weather conditions or optimize irrigation schedules could be included. Smart farming can help to increase farmers' yields and reduce their use of resources over time, which may lead to improvements in agricultural sustainability. Nevertheless, some opponents argue that intelligence farming can help to consolidate and exacerbate existing problems such as water scarcity [6, 10].

3 Introduction to Smart Farming: An Overview

Smart farming aims to enhance the quality and quantity of products so that they are useful for humans by using advanced technology, information, and communication in agriculture. The transformation from traditional farming to smart farming involves the following key components (Fig. 2).

3.1 Sensing Technologies

An agricultural sensor is one of the sensors applied in smart farming. The sensors provide information that will help farmers to monitor and optimize the crops in response to changing environmental conditions. Electrochemical sensors facilitate the collection, processing, and mapping of soil's chemical data. Usually, it's fitted onto specially designed sleds. They provide the relevant details needed for agricultural purposes including the level of nutrients in the soil and pH values [11].

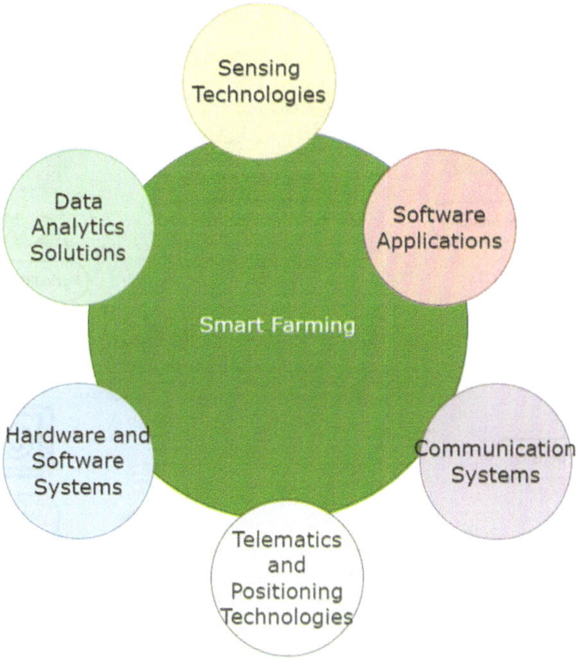

Fig. 2 Smart farming

3.2 Software Tools and Applications

With the use of agriculture software applications specifically tailored to meet agricultural needs and challenges, farmers will be able to enhance their efficiency and control over all aspects of their operations. The aim is to speed up the process of making important managerial decisions through these agricultural software tools. But farming applications are much more than a means of saving time; they also help farmers to make informed and predictive decisions. By referring to the past crop rotation data, obtained from satellite images combined with technical recommendations for growing specific types of crops, software programs and mobile applications can recommend the most profitable planting plan (Fig. 3) [38].

3.3 Communication Systems

Old farming practices and poor resource management are frequent problems, for example, lack of appropriate storage facilities, excessive pesticide usage, or overirrigation. The asymmetry of information can be directly addressed by appropriate

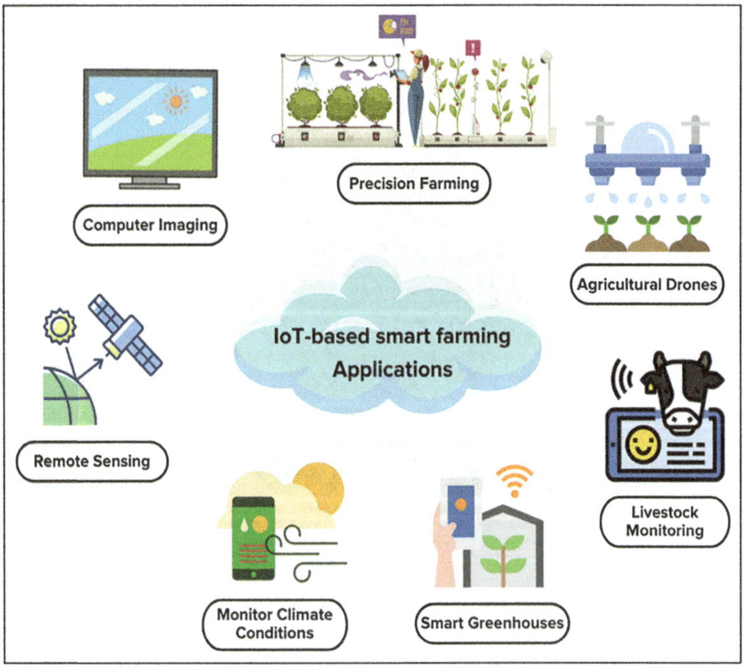

Fig. 3 Web applications for smart farming

communication channels and tools for farmers. There are a lot of current protocols that can be used, such as Wi-Fi 802.11n, LoRaWAN, WiMAX Ethernet, LRWPAN, BLE Bluetooth Low Energy RFID, MQTT, Zigbee, Sigfox, etc. [13].

3.4 Telematics and Positioning Technologies

To achieve minimum effort and optimum results, farmers need to select the best means of tracking and managing their machines. Telematics solutions that monitor the location, speed, and maintenance of these vehicles can be used in order to reduce downtime and increase their efficiency through smart farming. With telematics data, farmers can monitor the performance and usage information, make better-informed decisions for daily usage such as seed and fertilizer distribution and long-term usage such as predictive maintenance, and maximize their equipment's ROI [14].

3.5 Hardware and Software Systems

Smart farming involves the usage of various sensors for soil monitoring, water, light, humidity, and temperature. Technologies in telecommunications such as advanced network technologies and GPS are involved. Various hardware and software for specific applications, enabling the use of IIoT solutions, robotics, and automation are used. Numerous agricultural management software solutions such as Farmbrite, Agrivi, Farmlogs, Granular, Conservis, Agworld, Croptracker, Trimble, Agriculture, Farmlogic, Harvest Profit, etc. are available for modern agriculture. Each of them has its own specific functions.

3.6 Data Analytics Solutions

Analytics along with the Internet of Things represent a number of use cases in agriculture and are, for example, remote monitoring of agricultural machinery and its performance, analytics for the monitoring of farm operations and to improve their efficiency, and predictive analytics for accurate weather forecast. Digital soil and crop mapping can be carried out using data science. Farmers and agronomists may use this information to optimize the use of their land, thereby gaining better understanding of what crops are needed. Moreover, the forecasting of weather patterns and crop yields can also be based on data science.

3.7 Human-Machine Association

A variety of tasks, such as planting, irrigation, pest control, and soil analysis, can be performed by introducing various flexible machines. The benefits of higher productivity, lower labor charges, and a reduction in the usage of harmful chemicals are available to farmers when these processes are automated. The majority of the agricultural work in today's farm is done by tractors and machines that till the ground, plant seeds, and perform other tasks. Tillage harvesting equipment loosens the soil, kills weeds or competing plants, and prepares the area for planting.

3.8 Sustainability and Resource Efficiency

Smart farming emphasizes resource efficiency and sustainability. Resource efficiency is measured in terms of how effectively resources such as land, energy, rainwater, underground water, and labor are used in agricultural production. It is a topic of great importance in sustainable agriculture because it can help mitigate negative environmental effects on agricultural production while maximizing productivity and profitability.

The transformation from traditional farming to smart farming represents a leap forward in the agricultural sector. It involves the integration of various sensors, software programs and applications, communication systems, data analytics, and positioning technologies to create a highly responsive, efficient, and productive agricultural environment. This evolution not only enhances productivity but also shapes agricultural sectors' ability to meet dynamic market demands while promoting sustainability and human-machine collaboration.

4 Cloud Computing: Understanding Its Fundamental Concepts

Cloud indicates a network or the Internet. This mentions a technology used to store, receive, and access data from a network instead of local disk drives by using distant servers on the Internet. All sorts of data such as files, pictures, documents, audio, and video can be included. To understand more about cloud computing, let us explore its fundamental concepts.

4.1 Cloud Computing Architecture: Key Components

Cloud computing is an on-demand service that you can use via the Internet to manage and run your applications. In cloud computing, the "cloud" denotes only a network or servers that are used by your web applications, databases, and so on. Compute, storage, database, networking, and security are some of the fundamental key components of cloud computing [15]. Let's delve into the key components and types of cloud computing architecture.

4.1.1 Compute

The processing capacity required by the system and application to handle the data and perform a variety of calculations may be considered as compute. This computing power required for the server can be obtained from a cluster of virtual machines on the cloud instead of having it installed in an existing local data center. A virtual machine's computing capability depends only on the hardware resources allocated to its host computer. The CPU, storage, memory, and network bandwidth of the machine in which the VM is running are referred to as compute resources. Today, most cloud service providers have done a fine job of preparing an extensive set of pre-configured servers that can be used for all kinds of tasks.

4.1.2 Storage

The main advantage of storing data in a cloud environment is that it allows the user to increase their storage capacity without having to maintain or buy additional local hard drives. If a disk failure occurs, it will be hard to prevent data corruption. Data can be stored on a physical storage server which is hosted by the cloud service provider for an indefinite period in the cloud environment through logical pools. The user can store various types of data, e.g., pictures, files, and backups.

4.1.3 Databases

A database is a system for the storage and management of structured and unorganized information. Cloud service providers typically manage and offer cloud databases as a service. This means that the user no longer has any responsibility for maintaining or updating the components of a database instance, such as updates to operating systems and software patches. In actuality, scalability and availability of cloud databases are also very high.

4.1.4 Networking

The cloud is a vast network of computers that interact, share, and integrate to offer specific services to its clients. Cloud service providers guarantee that their infrastructures can constantly maintain a quick network connection in order to meet the expectations of their end users. The user's application may be disseminated globally thanks to the cloud's ability to provide a worldwide link.

4.1.5 Security

The cloud's remote, secure data centers are where data is kept. This indicates that hazards to users like theft and data breaches are rare. When using a cloud service, the user's obligation is more heavily centered on data management. Users will be able to gain access to a set of tools in the cloud environment that allow them to achieve higher security levels. With regard to encryption and decryption of data, users are responsible for themselves. To access the applications, the users have the option to authenticate or authorize selected users and services.

4.2 Characteristics of Cloud Computing

Cloud computing offers a wide range of characteristics that might be beneficial to customers, along with those specified by the National Institute of Standards Technology (NIST) [16]. Let us have a brief discussion on them:

4.2.1 On-Demand Self Service

Computer services like server time and network storage can be automatically provisioned with cloud computing. There will be no need for interaction with the cloud service provider. To view the cloud services and monitor their usage, provisioning and de-provisioning the services, the customers using the cloud can access their cloud accounts through a web self-service portal. No human administrators are required to access the cloud computing services.

4.2.2 Broad Network Access

Broad access to the Internet is another essential feature associated with cloud computing. Cloud services are made available to the user over the network on various portable devices, such as smartphones, tablets, notebooks, or desktop computers. Internet is used for public clouds; local area networks are used for private clouds. In cloud computing and wide network connectivity, both latency and bandwidth are crucial factors affecting the quality of service.

4.2.3 Resource Pooling

Multiple customers using the multitenant model can share physical resources with one another through resource pooling. These models assign and reassign physical and virtual resources to multiple users on the basis of demand. Multitenancy supports the sharing of application and infrastructure between customers while maintaining privacy and security. Although the customer may not have been able to determine exactly where their resources are located, he or she is free to specify a location at a higher level of abstraction such as country, state, and data center. Some of the resources that customers can pool are memory, processing, and bandwidth.

4.2.4 Rapid Elasticity

In order for customers to scale quickly based on demand, cloud services may be flexibly configured and released, sometimes automatically. There are virtually unlimited capabilities available for provisioning. These capacities can be used by customers in any quantity at any time. Without additional fees or charges, customers also have the option of scaling up their cloud use, capacity, and costs. With rapid elasticity, the user need not buy computer hardware. Rather, the cloud service provider's computing resources can be used.

4.2.5 Measured Service

A metering capability that is appropriate for the type of service will optimize resource utilization at an abstraction level in cloud systems. For example, a measured service may be used for monitoring the storage used, processing capacity utilized, bandwidth consumed, and users utilizing them. The payment shall be made on the basis of the actual consumption of the customer by means of a payment for what, what you use, model. For both consumers and the service providers, a transparent experience will be achieved with monitoring, control, and report on use of resources.

4.2.6 Resiliency and Availability

Cloud computing resilience refers to the fact that, in case of an error, a service can be quickly repaired. The speed of its servers, databases, and networks to get back up and running after a disruption is an indication of cloud resilience. A copy of stored data shall be created by cloud services in order to avoid loss of data. Copy version from the other server shall be restored when one server loses data for any reason.

In cloud computing, availability is an adherent concept. Cloud services offer the advantage of being able to view them remotely, which means that you do not have restrictions on geographical location when using their resources.

4.2.7 Flexibility

As their businesses grow, companies will have to expand. The cloud service provides clients with more flexibility of movement, and they don't have to restart the server anymore. In order to avoid overspending on the resources they do not need, they can also choose from a range of payment options.

4.2.8 Remote Work

Users benefit from cloud computing by being able to work in a remote location. Remote workers can safely and rapidly obtain corporate data through the use of their equipment, e.g., laptops or mobile phones. Remote workers can also communicate with each other and use the cloud effectively to carry out their tasks.

4.3 Cloud-Based Services

Cloud computing refers to the practice of storing, managing, and processing data rather than local or personal computers on an international network of remote servers that are maintained in a web site. Cloud providers are companies that offer these types of computing services and typically provide them for free or at a price determined by usage. The foundations of cloud computing are grids and clusters. Five broad categories account for the majority of cloud computing services (Fig. 4).

4.3.1 Software as a Service (SaaS)

An Internet-based method of offering services and applications is software as a service. Users are only given access to the Internet so they can avoid having to install or update software or deal with other laborious hardware and software administration tasks. Users can deploy and run programs on their computer or data center with its assistance, saving money on hardware expenditures and software upkeep. SaaS originates from the cloud services provider, which offers users entire software solutions for pay-as-you-go purchases. As long as they don't need to be downloaded or installed, the majority of Internet browsers can be used to access software as a service applications. The application as a service might be referred to as web-based software, on-demand software, or hosting software [17].

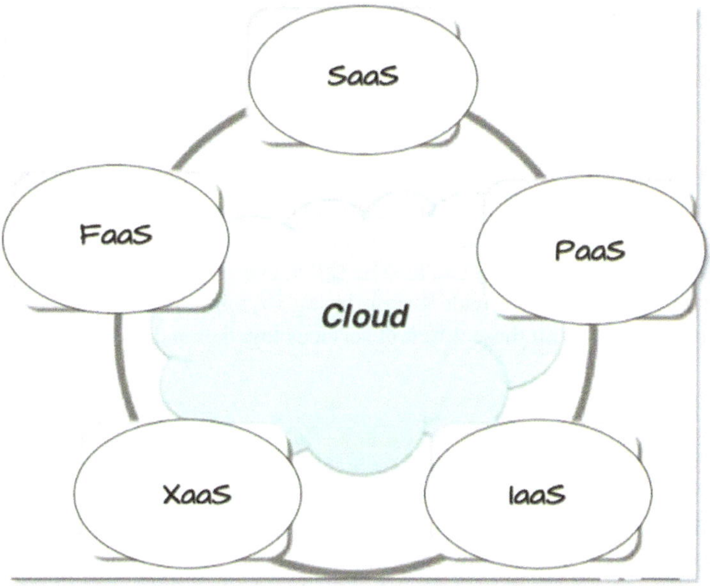

Fig. 4 Cloud-based services

4.3.2 Platform as a Service (PaaS)

Platform as a service is the area of cloud computing that offers a setting for programmers to create Internet-based apps and services. Users can access the PaaS service by using just a web browser because it is located in the cloud. The PaaS provider's own infrastructure must host all hardware and software. As a result, PaaS enables customers to avoid installing their own proprietary hardware or software for the creation and management of a new application. Consumers are in charge of deploying and configuring apps in an application hosting environment but not the cloud infrastructure itself, including the network, servers, operating system, and storage environment [18].

4.3.3 Information as a Service (IaaS)

Services for infrastructure IaaS refers to the method by which IT services are outsourced to support various processes. The term "infrastructure as a service," or "IaaS," refers to a service that enables the outsourcing of infrastructure, including web servers, various devices, networking hardware, and other items. Hardware as a service (HaaS) is another name for IaaS. IaaS users are charged according on the number of users, typically by the hour, week, or month. Customers may also be charged based on how much virtual machine space they utilize by some providers. IaaS only offers the underlying operating systems, security, networking, and servers needed to construct applications and services and to deploy development tools, databases, etc. [19].

4.3.4 Anything as a Service (XaaS)

"Anything as a service" (XaaS) also known as "everything as a service" denotes a general category of services related to both cloud computing and remote access. It also recognizes the enormous number of products, tools, and technologies now being delivered as a service to users via the Internet. Essentially, it is possible to turn any IT function into a service for enterprise consumption. Instead of paying for the service through an upfront purchase or license, it is made available as a discretionary consumption model. XaaS can also be called everything as a service. Today, the majority of cloud service providers are offering anything that can be classified as a service by combining all these different services together and providing some other services [20].

4.3.5 Function as a Service (FaaS)

Function as a service (FaaS), is a cloud-computing service that enables the customers to execute code as a response to events, without having the need for managing the complex infrastructure associated with building and launching the related microservice applications. When hosting software applications on the Internet, it is normally necessary to provide and run a virtual or physically based server, as well as operating system and web server services. With FaaS, a cloud service provider provides full automatic management of physical equipment, virtual operating system, and web server software. This ensures that the developers are focused exclusively on specific functions in their application code [21].

5 Integration of Cloud Computing in Smart Farming

The integration of cloud computing in smart farming leads to a transformative approach to optimizing production processes, enhancing quality control, and enabling predictive maintenance. Cloud computing techniques play a vital role in analyzing large volumes of data generated by interconnected sensors, devices, and machines in a smart farming environment. By extracting valuable insights from the reference data obtained from the cloud, other smart farming facilitates real-time decision-making, production optimization, and adaptive farming can be carried out [22–24]. Here's a broad overview of how cloud computing can be integrated into smart farming.

5.1 Data Management

A number of sensors are being used in smart farming using the Internet of Things (IoT) to monitor environmental conditions. The sensor's primary task is to gather data across the field and send it into the cloud. A number of basic measurements are set up in the cloud, and these will be compared to sensed data [5].

5.2 Data Collection and Retrieval

Data collection and retrieval are the most common uses of cloud-based software in agriculture. In order to provide accurate and rapid information, it stores a large amount of data on weather, crop patterns, soil quality, harvesting, and satellite imagery. The cloud stores all the data relating to the farm, making it easy to access. In this way, the data can be used more rapidly to identify a remedy in order to avoid large losses when crops are afflicted with symptoms similar to those seen 10 years earlier.

5.3 Data Processing and Analysis

Database management in the cloud environment allows decision-makers to make precise decisions and connects all data sources that are available for farms such as weather data, market data, agricultural data, GIS, and water availability. Weather data agricultural data, geographic information system, and water availability, all forms of old and existing data and current information must be analyzed in detail before valuable information is made available on optimal requirements to plant seeds, water, or pesticides. Whenever there is a discrepancy in the growth of crops, these systems have an alert system to communicate with the farmers. Therefore, in the event of a pest attack, these systems work effectively to inform farmers of relevant information.

5.4 Data Storage and Dissemination

Data storage is the core of predictive analysis. Storage of data has always been a hardware-based system; thus, the infrastructure requires constant monitoring and updates. The data was permanently lost if the hardware had been damaged. Today, agricultural technology has evolved to cloud-based systems that mean there is no need for investment in hardware purchases or maintenance. All information is available 24 hours a day and can be accessed from your computer or mobile device.

Other devices and solutions can also benefit from this approach of data storage for further analysis. With more data on farms available, it's easier for farmers to gain an accurate picture of agriculture output and management, the detection of pests and diseases, as well as yield estimates.

5.5 Real-Time Data Analysis

The deployment of scalable cloud computing with powerful analytic tools to identify patterns in data and obtain new information is called cloud analytics. Businesses increasingly rely on data analysis to achieve competitive advantage, advance scientific knowledge, or improve human life in a variety of ways. With cloud-based analytics tools, data collected from various sensors can be processed and analyzed in real time. This offers farmers immediate insights into soil health, weather conditions, and crop status, facilitating timely interventions and decision-making.

5.6 Remote Monitoring and Control

Through cloud integration, farmers can remotely monitor and control various farming operations via smartphones, tablets, or computers. This involves using cloud-based software to modify irrigation, track equipment, and even deploy drones.

5.7 Enhanced Decision Support Systems

When machine learning and artificial intelligence are linked into cloud platforms, a powerful decision-support system for farmers can be created. These systems can forecast agricultural yields, evaluate disease risks, and optimize irrigation schedules thanks to the power of cloud computing.

5.8 Continuous Improvement

The ability of cloud computing to continuously learn and adapt is one of the key advantages of smart farming. Models can be updated with fresh data, allowing them to develop over time as they take into account differences in production processes and real-world experiences.

5.9 *Cost Efficiency*

Farms can save the expenses of maintaining and upgrading their on-site IT infrastructure by implementing cloud computing. Additionally, cloud service models often operate on a pay-as-you-go basis, which means farms only pay for the computing resources they actually use.

The integration of cloud computing into smart farming offers immense potential to transform the way products are produced, quality is ensured, and resources are managed. By harnessing the power of the cloud and AI, farmers can achieve higher efficiency, improved product quality, reduced costs, and enhanced sustainability, thereby shaping the future of the agricultural industry.

6 Implementation of Cloud Computing in Various Stages of Smart Farming

Cloud computing is revolutionizing smart farming by offering data-driven insights, predictive capabilities, and process optimization across various stages of agriculture [25–27]. Let's explore how cloud computing is implemented in each of these stages with reference to Fig. 5 [36].

6.1 *Data Collection and Sensors*

Cloud computing enables smart farming by collecting data from various sensors placed in the field. In the foundation of smart farming lies the ability to collect data. Sensors placed in various parts of the farm gather crucial information such as soil moisture, temperature, humidity, and plant health. Farmers can make data-driven decisions by transferring this data to cloud platforms, which enable them to gain a real-time overview of the field conditions.

6.2 *Remote Monitoring*

Through cloud-connected cameras and sensors, farmers can remotely watch their crops in real time, enabling them to make educated decisions regarding irrigation, pest management, and harvesting. Regardless of physical distance, cloud computing enables a continuous link between the farm and the farmer. Farmers can monitor crop development, keep a watch on their fields, and even see the first signs of pest infestations with cloud-connected cameras and sensors without having to physically be there.

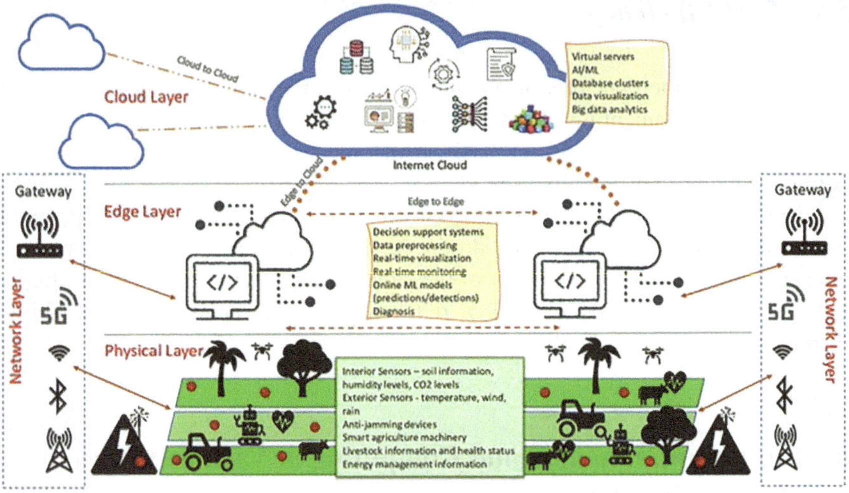

Fig. 5 Multilayer smart farming architecture

6.3 Data Storage

Huge amounts of agricultural data, such as historical crop yield data and meteorological data, may be stored securely and flexibly on cloud systems. The capacity to securely store a massive amount of data is one of the key benefits of cloud computing. A trustworthy reference point for upcoming farming operations can be found in the cloud storage of historical information on crop yields, pest activity, and weather trends.

6.4 Data Analysis

Cloud-based analytics solutions can process the collected data to provide insights into crop health, yield projections, and ideal planting times. Using sophisticated cloud-based analytics technologies, farmers can glean insights from their data that is beneficial to them. These systems may predict crop yields, identify trends, and suggest the most effective agricultural practices based on past results.

6.5 Weather Forecasting

Farmers may organize their activities and make weather-informed decisions with the use of cloud computing, which has access to real-time meteorological data. For farming, timely and precise weather information is essential. Farmers may schedule their activities properly when real-time weather forecasts are integrated into cloud platforms. This integration lowers the dangers brought on by erratic weather fluctuations.

6.6 Precision Agriculture

By maximizing the use of resources, such as water and fertilizers, as well as a reduction in waste and an increase in crop yields, cloud-based platforms enable precision agriculture. Precision agriculture, which involves applying farming resources in precisely the right amounts where they are needed, is a practice that heavily relies on cloud-based technologies. Cloud systems may advise farmers on when and where to irrigate, fertilize, or apply pesticides by analyzing data from the field, maximizing resource use.

6.7 Remote Control

In addition to monitoring, farmers can act remotely thanks to the cloud. Cloud-enabled gadgets provide farmers unparalleled control over their operations from any location, whether it is to change irrigation settings, turn on greenhouse lights, or use drones for monitoring. Through cloud-connected devices, farmers may remotely control irrigation systems, drones, and other equipment, improving efficiency and lowering the requirement for on-site presence.

6.8 Mobile Applications

Mobile apps enabled by the cloud make farming more practical and effective by enabling farmers to access data and manage farm operations from their smartphones or tablets. The widespread use of smartphones and tablets has made cloud-powered mobile apps essential tools for farmers. These apps ensure that farmers have access to their operations at their fingertips by offering real-time updates, alarms, and even remote control.

6.9 Machine Learning and AI

Artificial intelligence (AI) and machine learning algorithms are supported by cloud computing and can be used to recognize crop illnesses, forecast insect infestations, and suggest the best farming techniques. The true power of cloud computing is unleashed when combined with machine learning and AI. These technologies can automatically detect crop diseases from images, predict pest activities based on past patterns, and even recommend crop rotation schedules for optimal soil health.

6.10 Inventory Management

Cloud solutions also extend to inventory management. Keeping track of seeds, equipment, fertilizers, and other resources becomes straightforward with cloud-based systems. Automated alerts can notify farmers when stocks are low, ensuring seamless operations.

6.11 Supply Chain Optimization

From the farm to the marketplace, cloud platforms can monitor and manage the entire supply chain. Real-time updates about crop availability, demand patterns, and distribution logistics help in reducing wastage and improving market responsiveness. Cloud computing aids in supply chain management by providing real-time information on crop availability and demand, improving distribution, and reducing waste.

6.12 Farm-to-Table Traceability

Consumers today are more conscious about the origin of their food. Cloud solutions can track produce from its source farm to the dining table, ensuring transparency, and safety, and building trust among consumers.

6.13 Energy Efficiency

Smart farming systems in the cloud can optimize energy consumption, such as controlling irrigation pumps and ventilation systems based on real-time data. With cloud-based energy management systems, farms can reduce their carbon footprint. Whether it's optimizing the operation hours of irrigation pumps based on soil moisture levels or managing greenhouse temperatures, the cloud ensures energy is used judiciously.

Cloud computing has ushered in a new era in agriculture. With its myriad applications in data collection, analysis, remote operations, and supply chain management, it has transformed the way farming is perceived and conducted. As technology continues to evolve, the synergy between cloud computing and farming promises even more groundbreaking solutions in the future.

7 Real-Time Case Studies for the Application of Cloud Computing in Smart Farming

The fusion of cloud computing with smart farming has given birth to innovative agricultural solutions around the world. Numerous real-world case studies attest to the transformative power of this integration, illustrating how farms—both big and small—are harnessing the potential of the cloud for increased productivity and sustainability.

7.1 Cropin's Smart Farming with Cloud Computing

CropIn, an agtech company based in India, has been at the forefront of integrating cloud computing with smart farming practices. CropIn offers SaaS solutions that help farmers, agribusinesses, and governments optimize agricultural operations by utilizing the power of the cloud. Using sensors, satellite imaging, or manual input, their platform gathers enormous volumes of field data and processes it in real time on a cloud-based platform. Predictive analytics, crop monitoring, and risk assessment are made possible by this. The platform also provides farmers with practical advice on resource management, insect control, and the best times to harvest. CropIn's integration of cloud computing with conventional farming has therefore improved farmers' ability to make decisions, dramatically increased crop yields, decreased losses, and paved the road for sustainable and scalable agricultural operations [29].

7.2 Precision Irrigation in California

Cloud-based solutions were used by a vineyard in California to combat the area's regular drought circumstances. The vineyard was equipped with sensors to track the soil moisture in real time. The precise amount of water that each section of the vineyard needed was then determined by using this data, which was then uploaded to the cloud for analysis. As a consequence, the vineyard preserved its ideal grape yield and quality while cutting its water use by up to 25% [30].

7.3 Livestock Disease Prediction in Kenya

In Kenya, smallholder farmers faced challenges with crop diseases, particularly maize lethal necrosis disease. A cloud-based system was deployed that utilized AI to analyze data from satellite imagery, weather forecasts, and local sensors. The system would then predict potential disease outbreaks, allowing farmers to take preventive measures in time. As a result, agricultural yields were raised, and disease outbreak-related losses were reduced [31].

7.4 Livestock Monitoring in Australia

Keeping track of the whereabouts and health of the cattle is a difficult undertaking in Australia's huge ranches. One ranch combined a cloud system with livestock wearing sensors. These sensors sent data to the cloud while continuously tracking the animals' position and health indicators. Farmers could then access this information from their smartphones, ensuring that sick animals receive prompt medical attention and identifying any stray animals [32, 33].

7.5 Supply Chain Management in Brazil

Brazil's coffee industry utilized cloud computing to streamline its complex supply chain. From the time beans were harvested to when they reached the consumer, cloud-based platforms tracked and stored data on bean quality, storage conditions, transportation, and more. This ensured that consumers received the best quality coffee and could even trace their coffee's journey from farm to cup [34].

7.6 Drone Surveillance in India

In India, a vast rice farm employed drones equipped with cameras and sensors to monitor crop health. These drones sent images and data to a cloud platform that used AI to detect pest infestations, diseases, and other issues. The system would then notify farmers in real time, allowing them to address problems promptly and minimize crop damage [35].

These case studies underscore the transformative potential of integrating cloud computing into smart farming practices. From water conservation and disease prevention to livestock management and supply chain optimization, the cloud is reshaping agriculture. As technology continues to evolve, we can expect even more innovative solutions that bolster food security, sustainability, and economic viability in the agricultural sector.

8 Applications of Cloud Computing in Smart Farming

Smart farming is characterized by the integration of various digital technologies, automation of systems, data exchange mechanisms, and advanced analytics in various agricultural sectors. Cloud computing plays a vital role in enabling and enhancing various applications within smart farming:

- *Data readiness anytime and anywhere:* In today's hyper-connected world, the demand for real-time access to information has never been greater. Cloud computing has emerged as a transformative solution to this challenge, ensuring data readiness anytime and anywhere. With the cloud, data storage is not confined to a single physical location or device; instead, it's hosted on virtual servers, accessible from any Internet-connected device. This ubiquitous accessibility empowers professionals across various sectors, from business executives to scientists, to retrieve, analyze, and act upon their data irrespective of their geographical location. Furthermore, cloud platforms often come with built-in tools for data analysis, collaboration, and backup, ensuring that the data is not only accessible but also meaningful and secure. The era of waiting to get back to the office to access critical files or databases is long gone; with the cloud, the world truly has its data at its fingertips.
- *Local and global communication:* In the realm of smart farming, the blend of local and global communication has revolutionized agricultural practices. At the local level, on-farm communication between sensors, equipment, and centralized systems ensures real-time monitoring and immediate response to changing conditions. For instance, a soil moisture sensor might relay data to an irrigation system to optimize water distribution, ensuring plants receive the right amount of hydration. On a global scale, smart farming benefits from a vast interconnected network of data sources. Farmers can access global weather forecasts, satellite imagery, or market trends to make informed decisions. They can also collaborate with experts from around the world, sharing insights, research, and innovations. This fusion of local and global communication not only enhances efficiency and productivity but also fosters a collaborative global community dedicated to sustainable agricultural practices and food security.
- *Improve the economic condition of the nation:* The integration of cloud computing into smart farming holds transformative potential for a nation's economy. By optimizing agricultural processes, cloud platforms enhance productivity, leading to higher yields with lower resource input. This efficiency translates to increased profits for farmers, fostering rural development and strengthening the backbone of many national economies. Furthermore, real-time data access ensures rapid response to market demands, reducing wastage and ensuring maximum returns on produce. On a larger scale, the data-driven insights from cloud platforms can guide national agricultural policies and strategies, promoting sustainable farming practices that ensure long-term food security. Moreover, as nations embrace these advanced technologies, they invariably position themselves at the forefront of agricultural innovation, attracting investments, and partnerships, and creating

new job opportunities in the tech-agri domain. In essence, by intertwining cloud computing with agriculture, nations can catalyze a ripple effect of economic growth, sustainability, and global competitiveness.

- *Enhance the GDP of the nation:* Harnessing the transformative potential of cloud computing in smart farming can significantly enhance a nation's GDP. As farmers tap into the vast reservoirs of data-driven insights made available through the cloud, they can optimize irrigation, improve pest control, and predict harvest yields with greater accuracy. This amplification of productivity and reduction in resource waste can lead to a more prosperous agricultural sector. Furthermore, by integrating cloud-based solutions, the entire supply chain, from farm to table, can be streamlined, reducing overhead costs and driving up profit margins. The resultant upswing in the agricultural output not only ensures food security but also positions the sector as a robust contributor to the nation's economic growth.
- *Ensure food security level:* Ensuring food security is a critical global challenge, and cloud computing in smart farming is emerging as a vital tool in this endeavor. By leveraging the advanced analytics and storage capabilities of the cloud, farmers can access real-time data on soil health, weather patterns, and pest activities. This facilitates the timely and precise deployment of resources, leading to increased crop yields and reduced losses. Predictive analytics, powered by the cloud, allow farmers to anticipate potential threats or changes in demand, thereby ensuring a consistent and sufficient food supply. Furthermore, with cloud-based platforms, knowledge sharing becomes seamless, enabling farmers across regions to adopt best practices, utilize innovative solutions, and collaboratively tackle challenges. By integrating cloud technology into the agricultural fabric, we can establish a resilient framework that elevates food security levels globally.
- *Motivation of farmers and researchers:* The infusion of cloud computing into smart farming has created an inspiring nexus for both farmers and researchers, energizing their collaborative efforts. For farmers, the cloud offers not just a tool, but a treasure trove of insights, demystifying age-old agricultural challenges and revealing optimized paths to yield enhancement. Every piece of data becomes a source of empowerment, transforming traditional farming into a precision-focused endeavor. For researchers, this vast and ever-growing database of agricultural metrics opens avenues for innovation, enabling them to devise advanced algorithms, forecast models, and tailor solutions for varied farming scenarios. The direct and tangible benefits witnessed by farmers from these innovations further motivate researchers to push the boundaries of technological application. As both parties witness the transformative effects of their combined efforts, their motivation soars, setting the stage for a renaissance in agriculture, driven by the harmonious blend of tradition and technology.
- *Reduction of technical issues:* Cloud computing in smart farming has been instrumental in drastically reducing technical issues that once plagued the agricultural sector. Traditional farming practices, reliant on standalone systems and disjointed technologies, often encountered inefficiencies and technical breakdowns that impeded optimal farming operations. With the advent of cloud-based solutions, farmers now benefit from centralized data storage, regular system

updates, and real-time troubleshooting. Learning can analyze customer feedback, market trends, and historical data to inform product design and development. This can lead to products that better meet customer needs and preferences.

- *Rural-urban movement:* The advent of cloud computing in smart farming has the potential to reshape the dynamics of rural-urban movement. Historically, limited opportunities and technological stagnation in rural areas have driven populations toward urban centers in search of better prospects. However, with the integration of cloud-based solutions into agriculture, rural zones are now becoming hubs of technological innovation. Smart farming, powered by the cloud, provides farmers with access to real-time data, sophisticated analytics, and global marketplaces, thereby enhancing the profitability and appeal of agricultural endeavors. As a result, the younger generation may find incentives to remain in or return to their rural roots, armed with the tools to revolutionize traditional farming practices.
- *Improved market price of food, seeds, and other products:* Cloud computing's impact on smart farming transcends the fields and directly influences the market dynamics of food, seeds, and other agricultural products. By leveraging the vast analytical capabilities of the cloud, farmers can gain real-time insights into market demands, crop yields, and potential disruptions. This facilitates a more predictive and dynamic pricing model. For instance, if data indicates a potential shortfall in a particular crop yield, farmers can adjust prices accordingly to reflect the anticipated scarcity. Similarly, cloud platforms enable seamless global communication, allowing farmers to tap into international markets, compare prices, and ensure they're getting the best value for their products. For seed producers, cloud-based databases can store genetic information and track seed performance across different geographies and climates, thereby assisting in setting a value based on its proven productivity. Furthermore, by reducing inefficiencies and wastage in the supply chain through cloud-enhanced logistics, the overall cost of bringing products to market can be reduced, leading to fair and competitive pricing. In essence, cloud computing in smart farming provides a transparent, data-driven foundation for stabilizing and improving the market prices of agricultural outputs.
- *Smart analytics and insights:* Cloud computing in smart farming acts as a catalyst for unlocking the true potential of smart analytics and insights. By pooling together vast amounts of data from diverse sources, such as soil sensors, weather forecasts, and satellite imagery, cloud platforms can process and analyze this data in real-time. This results in actionable insights that would be inconceivable through traditional means. Farmers can gain a comprehensive understanding of their fields down to the minutest detail, be it the moisture content of a specific patch or the predicted pest activity based on environmental conditions.
- Smart analytics powered by the cloud enables precise decision-making, from determining the optimal time for sowing seeds to identifying the best harvest window. Predictive analytics can also foresee potential challenges, allowing farmers to mitigate risks before they manifest. Moreover, these insights are not confined to individual farms. The collaborative nature of cloud platforms means

that insights from one region can benefit another, fostering a global community of informed and empowered farmers.

- *Supply chain optimization:* Cloud computing's intersection with smart farming has revolutionized supply chain optimization in agriculture. Traditionally, the agricultural supplychain faced inefficiencies stemming from information silos, unpredictable weather changes, and transportation hitches. The cloud offers a unified platform where real-time data from various sources converges, enabling stakeholders at every point in the chain to make informed decisions. For instance, predictive analytics can forecast harvest yields, allowing distributors to align transportation and storage resources ahead of time. Similarly, retailers can access this data to manage inventory and reduce food wastage. By having a holistic view of the supply chain, from sowing to sales, potential bottlenecks can be identified and addressed proactively. Furthermore, cloud-based platforms facilitate seamless communication among farmers, distributors, and retailers, ensuring synchronized operations. In essence, cloud computing in smart farming transforms the agricultural supply chain from a traditionally fragmented and reactive system into a cohesive, transparent, and agile entity, leading to reduced costs, minimized wastage, and maximized value delivery to the end consumer.

Overall, cloud computing plays a crucial role in driving innovation and efficiency across various aspects of smart farming, helping farmers adapt to the rapidly changing technological landscape and stay competitive in the global market.

9 Challenges in Integrating Cloud Computing with Smart Farming

Integrating cloud computing with smart farming presents several challenges that need to be addressed to ensure successful implementation and maximize the benefits of both technologies [28]. Here are some of the key challenges:

- *Infrastructure Deficiency:* Many rural areas lack the necessary infrastructure, such as high-speed Internet and data centers, to fully harness cloud computing capabilities.
- *High Initial Costs:* Setting up sensors, and IoT devices, and integrating them with cloud platforms require a significant initial investment.
- *Digital Literacy:* Farmers, especially in developing nations, might not be well-versed in advanced technologies, making adoption challenging.
- *Data Security Concerns:* Storing sensitive agricultural data on the cloud raises concerns about potential breaches and misuse.
- *Reliability Issues:* Internet connectivity, especially in remote farming areas, can be intermittent, affecting real-time data access and decision-making.
- *Integration Hurdles:* Integrating various existing farming technologies with new cloud-based platforms can be complex.

- *Regulatory Challenges:* In many countries, regulations related to data storage, transfer, and cloud computing are still evolving, posing legal challenges.
- *Data Overload:* While cloud computing can handle vast amounts of data, farmers might get overwhelmed by the sheer volume of information and struggle to make sense of it.
- *Cultural Resistance:* Traditional farmers might resist transitioning to tech-based practices due to deep-rooted beliefs or fear of the unknown.
- *Vendor Lock-In:* Committing to one cloud service provider might hinder flexibility and make it challenging to switch or integrate other services in the future.
- *Scalability Issues:* As farms grow and diversify, ensuring that cloud solutions scale efficiently can become a challenge.
- *Limited Customization:* Off-the-shelf cloud solutions might not cater to the unique requirements of every farm, necessitating customization which can be costly and time-consuming.
- *Maintenance and Upgrades:* Regular system maintenance and software upgrades might require technical expertise which not all farmers possess.
- *Privacy Concerns:* Sharing data about crop yields, farming techniques, and other proprietary information on the cloud might raise concerns about intellectual property rights.
- *Environmental Concerns:* The energy consumption of vast data centers powering cloud services has environmental implications, which could be a point of contention for eco-conscious farmers.

Overcoming these challenges requires a collaborative effort among cloud architects, domain experts, farmers, and cloud service providers. Successful integration of cloud computing with smart farming involves careful planning, continuous monitoring, and adaptation to ensure that the technology delivers the desired benefits while mitigating potential risks.

10 Conclusion

In conclusion, cloud computing emerges as a transformative force in the realm of smart farming, acting as the digital backbone that supports a new age of agricultural efficiency and innovation. Its capability to store, process, and provide real-time data analytics brings unparalleled advantages to farmers, from optimizing resource use to predicting market trends. While challenges do exist in its integration, the overarching benefits, such as increased yields, reduction in wastage, and supply chain optimization, position cloud computing as an indispensable tool for the future of farming. As the world grapples with increasing food demand and climate uncertainties, leveraging the cloud's prowess in smart farming becomes not just an opportunity but a necessity for sustainable agricultural growth.

References

1. J.C.K. Wells, J.T. Stock, Life history transitions at the origins of farming: "A model for understanding how niche construction impacts human growth, demography and health". Front. Endocrinol. **11**, Article number: 325 (2020)
2. A.D. Boursianis, M.S. Papadopoulou, P. Diamantoulakis, A. Liopa-Tsakalidi, P. Barouchas, G. Salahas, et al., Internet of things (IoT) and agricultural unmanned aerial vehicles (UAVs) in smart farming: A comprehensive review. Internet of Things **18**, 100187 (2022)
3. R. Nolz, G. Kammerer, P. Cepuder, Calibrating soil water potential sensors integrated into a wireless monitoring network. Agric. Water Manag. **116**, 12–20 (2013)
4. M.S. Rao, S. Modi, R. Singh, K.L. Prasanna, S. Khan, C. Ushapriya, Integration of cloud computing, IoT, and Big Data for the development of a novel smart agriculture model. 2023 3rd international conference on advance computing and innovative technologies in engineering (ICACITE), Greater Noida, India, pp. 2779–2783, 2023. https://doi.org/10.1109/ICACITE57410.2023.10182502
5. S.A. Ali, G. Keerthana, T. Niranjanaa, K. Pavithra, C. Rekha, Cloud computing in smart farming. Int. Res. J. Eng. Technol **07**(08), 200 (2020)
6. M. Saban, M. Bekkour, I. Amdaouch, J. El Gueri, B. Ait Ahmed, M.Z. Chaari, J. Ruiz-Alzola, A. Rosado-Muñoz, O. Aghzout, A smart agricultural system based on PLC and a cloud computing web application using LoRa and LoRaWan. Sensors **23**, 2725 (2023). https://doi.org/10.3390/s23052725
7. B. Achmad, Sanudin, M. Siarudin, A. Widiyanto, D. Diniyati, A. Sudomo, A. Hani, E. Fauziyah, E. Suhaendah, T.S. Widyaningsih, et al., Traditional subsistence farming of smallholder agroforestry systems in Indonesia: A review. Sustainability **14**, 8631 (2022). https://doi.org/10.3390/su14148631
8. I. Kaushik, N. Prakash, Applicability of IoT for smart agriculture: Challenges & future research direction, 2021 IEEE World AI IoT Congress (AIIoT), Seattle, WA, USA, pp. 0462–0467, 2021. https://doi.org/10.1109/AIIoT52608.2021.9454209
9. S. Sushitha, H. Hegde, "Smart farming-a key to sustainable agriculture development in India" - a study. 2022 4th international conference on smart systems and inventive technology (ICSSIT), Tirunelveli, India, pp. 63–71, 2022. https://doi.org/10.1109/ICSSIT53264.2022.9716494
10. R. Vijaya Saraswathi, J. Sridharani, P. Saranya Chowdary, K. Nikhil, M. Sri Harshitha, K. Mahanth Sai, Smart farming: The IoT based future agriculture. 2022 4th international conference on smart systems and inventive technology (ICSSIT), Tirunelveli, India, pp. 150–155, 2022. https://doi.org/10.1109/ICSSIT53264.2022.9716331
11. E.S. Mohamed, A.A. Belal, S.K. Abd-Elmabod, M.A. El-Shirbeny, A. Gad, M.B. Zahran, Smart farming for improving agricultural management. Egypt. J. Remote Sens. Space Sci. **24**(3, Part 2), 971–981., ISSN 1110-9823 (2021). https://doi.org/10.1016/j.ejrs.2021.08.007
12. R. Singh, G.S. Singh, Traditional agriculture: A climate-smart approach for sustainable food production. Energ. Ecol. Environ **2**, 296–316 (2017). https://doi.org/10.1007/s40974-017-0074-7
13. A. Atmaja, A. el Hakim, A. Wibowo, L. Pratama, Communication Systems of smart agriculture based on wireless sensor networks in IoT. J. Robot. Control **2** (2021). https://doi.org/10.18196/jrc.2495
14. M.P. Nootjaroen, Adoption of tractor technology by Thai rice farmers: The case of Kubota tractors with telematics systems. Master's Thesis, Thammasat University, Bangkok, Thailand (2020). [Google Scholar]
15. N. Banger, K. Pallavi, S.J. Shetty, A review paper on cloud computing architecture, types, advantages and disadvantages. Int. J. Adv. Res. Sci. Commun. Technol **2**(2), 14 (2022)
16. M. Hamdaqa, L. Tahvildari, Cloud computing uncovered: A research landscape, in *Advances in Computers*, ed. by A. Hurson, A. Memon, vol. 86, (Elsevier, 2012), pp. 41–85. ISSN 0065-2458, ISBN 9780123965356. https://doi.org/10.1016/B978-0-12-396535-6.00002-8

17. A.S. Rumale, D.N. Chaudhari, Cloud computing: Software as a service. 2017 second international conference on electrical, computer and communication technologies (ICECCT), Coimbatore, India, pp. 1–6, 2017. https://doi.org/10.1109/ICECCT.2017.8117817
18. C. Lv, Q. Li, Z. Lei, J. Peng, Z. Wu, T. Wang, PaaS: A revolution for information technology platforms. 2010 International Conference on Educational and Network Technology, Qinhuangdao, pp. 346–349, 2010. https://doi.org/10.1109/ICENT.2010.5532150
19. S.M.A. Amri, L. Guan, Infrastructure as a service: Exploring network access control challenges. 2016 SAI computing conference (SAI), London, UK, pp. 596–603, 2016. https://doi.org/10.1109/SAI.2016.7556042
20. Y. Duan, G. Fu, N. Zhou, X. Sun, N.C. Narendra, B. Hu, Everything as a service (XaaS) on the cloud: Origins, current and future trends. 2015 IEEE 8th international conference on cloud computing, New York, NY, USA, pp. 621–628, 2015, https://doi.org/10.1109/CLOUD.2015.88
21. M. Szalay, P. Mátray, L. Toka, Real-time FaaS: Towards a latency bounded serverless cloud. IEEE Trans. Cloud Comput **11**(2), 1636–1650 (2023). https://doi.org/10.1109/TCC.2022.3151469
22. M.L. Rathod, A. Shivaputra, H. Umadevi, K. Nagamani, S. Periyasamy, Cloud computing and networking for SmartFarm AgriTech. J. Nanomater. **2022**, Article ID 6491747, 7 pages (2022). https://doi.org/10.1155/2022/6491747
23. A.V. Turukmane, M. Pradeepa, K. Shyam Sunder Reddy, R. Suganthi, Y. Md Riyazuddin, V.V. Satyanarayana Tallapragada, Smart farming using cloud-based Iot data analytics, 100806, ISSN 2665-9174. Measure. Sensors **27** (2023). https://doi.org/10.1016/j.measen.2023.100806
24. C. Ramakrishna, B. Venkateshwarlu, J. Srinivas, S. Srinivas, IoT based smart farming using cloud computing and machine learning. Int. J. Innov. Technol. Explor. Eng. ISSN: 2278-3075 (Online) **9**(1) (2019)
25. M. Junaid, A. Shaikh, M.U. Hassan, A. Alghamdi, K. Rajab, M.S. Al Reshan, M. Alkinani, Smart agriculture cloud using AI based techniques. Energies **14**, 5129 (2021). https://doi.org/10.3390/en14165129
26. K.S.N. Prasad, C.N. Sirisha, C.N. Kumar, A. Deekshitha, D. Gunninka, Cloud computing in agriculture-an affordable way to achieve smart farming. Int. J. Comput. Sci. Commun. (ISSN: 0973-7391) **12**(2), 52–61 (2021)
27. Y. Kalyani, N.V. Bermeo, R. Collier, Digital twin deployment for smart agriculture in Cloud-Fog-Edge infrastructure. Int. J. Parallel Emerg. Distrib. Syst (2023). https://doi.org/10.1080/17445760.2023.2235653
28. S. Choudhary, R. Jadoun, H. Mandoria, Role of cloud computing technology in agriculture fields. Comput. Eng. Intell. Syst **7**, 2222–2863 (2016)
29. https://www.cropin.com/smart-farming
30. G. Jha, F. Nicolas, R. Schmidt, K. Suvočarev, D. Diaz, I. Kisekka, K. Scow, M.A. Nocco, Irrigation Decision Support Systems (IDSS) for California's Water–Nutrient–Energy Nexus. Agronomy **2022**, 12 (1962). https://doi.org/10.3390/agronomy12081962
31. P.N. Nyangau, J.M. Nzuma, P. Irungu, M. Kassie, Evaluating livestock farmers knowledge, beliefs, and management of arboviral diseases in Kenya: A multivariate fractional probit approach. PLoS Negl. Trop. Dis. **15**(9), e0009786 (2021). https://doi.org/10.1371/journal.pntd.0009786
32. https://www.zdnet.com/article/australias-ceres-tag-converting-livestock-into-iot-devices/
33. W.P. Andrew, Progress with livestock welfare in extensive production systems: Lessons from Australia. Front. Veterin. Sci **8** (2021). https://doi.org/10.3389/fvets.2021.674482. ISSN: 2297-1769, https://www.frontiersin.org/articles/10.3389/fvets.2021.674482
34. W. Silva, P. Guarnieri, J.M. Carvalho, Sustainable practices in the coffee supply chain in the Cerrado Mineiro Region, Brazil. J. Oper. Supply Chain Manag **11**, 90–100 (2018). https://doi.org/10.12660/joscmv11n2p90-100
35. S. Vijayakumar, H. Madireddy, S. Bhusarapu, M. Kumar, Rapolu, R. Sundaram, Drone application in rice cultivation: Experiences from ICAR-IIRR trails. Indian Farm **72**, 3–6 (2022)

36. M. Gupta, M. Abdelsalam, S. Khorsandroo, S. Mittal, Security and privacy in smart farming: Challenges and opportunities. IEEE Access., 1–1 (2020). https://doi.org/10.1109/ACCESS.2020.2975142
37. https://www.jagranjosh.com/general-knowledge/traditional-agriculture-and-its-impact-on-the-environment-1518096259-1
38. A. Ali, T. Hussain, N. Tantashutikun, N. Hussain, G. Cocetta, Application of smart techniques, internet of things and data mining for resource use efficient and sustainable crop production. Agriculture 13(2), 397 (2023). https://doi.org/10.3390/agriculture13020397

AI Green Revolution: Reshaping Agriculture's Future

R. Thangamani, D. Sathya, G. K. Kamalam, and Ganesh Neelakanta Lyer

1 Introduction

In the midst of a world grappling with intricate challenges like burgeoning populations, climate uncertainty, and the imperative for sustainable resource management, a new era of agricultural transformation is dawning. This era, known as the "AI Green Revolution," is defined by the convergence of cutting-edge technologies, artificial intelligence (AI), and ecological consciousness. It holds the promise of revolutionizing the way we produce food, ensuring that agriculture not only sustains us but also safeguards our planet. Building on the legacy of the Green Revolution of the mid-twentieth century which saw remarkable advancements in crop yields, the AI Green Revolution introduces a paradigm shift that transcends mere productivity. It integrates the power of AI, data-driven insights, and precision techniques to cultivate a future where agriculture is both bountiful and environmentally sustainable.

In this narrative, fields transform into smart landscapes teeming with interconnected devices, drones glide above analyzing crops with pinpoint accuracy, and algorithms process vast amounts of data to optimize every planting, irrigation, and harvesting decision. But beyond the technological marvels, the AI Green Revolution is a call to reimagine agriculture's role in our world—a catalyst for embracing innovative practices that restore ecosystems, preserve biodiversity, and mitigate climate change.

This expedition is one of immense potential, not just for large-scale agribusinesses, but also for smallholders striving to secure their livelihoods and

R. Thangamani (✉) · D. Sathya · G. K. Kamalam
Kongu Engineering College, Perundurai, India
e-mail: thangamani.ai@kongu.edu; Sathya.ai@kongu.edu

G. N. Lyer
National University, Singapore, Singapore

S. Balasubaramanian et al. (eds.), *Intelligent Robots and Drones for Precision Agriculture*, Signals and Communication Technology,
https://doi.org/10.1007/978-3-031-51195-0_19

communities working toward food security. It envisions equitable access to AI-driven knowledge, bridging the digital divide and fostering inclusive growth. As we embark on this transformative expedition, it's essential to unravel the multifaceted layers of the AI Green Revolution. From the ethical considerations of data privacy to the socioeconomic impacts on rural communities, the coming chapters will explore how this revolution is shaping the future of agriculture, redefining its relationship with nature, and inspiring a collective commitment to a greener, more prosperous world.

1.1 The Changing Landscape of Agriculture: Challenges and Opportunities

Agriculture, the foundation of human sustenance, is undergoing a profound transformation driven by a complex interplay of challenges and opportunities [1]. As we navigate this shifting landscape, it becomes evident that the path forward demands a holistic understanding of the multifaceted dynamics that shape the future of food production. This exploration delves into the challenges that agriculture faces and the opportunities that emerge in response.

Challenges

Population Growth and Food Security: The global population is projected to reach 9.7 billion by 2050. Meeting the increased demand for food while ensuring its equitable distribution poses a significant challenge.

- Climate Change and Resource Constraints [2]: Unpredictable weather patterns, rising temperatures, and water scarcity disrupt traditional farming practices. Climate-resilient strategies are vital to safeguarding yields and livelihoods.
- Soil Degradation and Biodiversity Loss: Intensive farming practices have led to soil erosion, nutrient depletion, and loss of biodiversity. Sustainable agricultural approaches are necessary to restore soil health and ecosystems.
- Rural-Urban Migration and Labor Shortages [3]: As populations shift from rural to urban areas, agriculture faces labor shortages. Mechanization and technological solutions are needed to bridge this gap.
- Market Access and Value Chains: Limited access to markets, post-harvest losses, and inefficient supply chains impede profitability for farmers.
- Environmental Impact and Sustainable Practices: Conventional agricultural practices contribute to greenhouse gas emissions, pollution, and deforestation. Sustainable practices are crucial to minimize the sector's environmental footprint.

Opportunities

- Technological Advancements: Emerging technologies like AI, machine learning, IoT [4], and remote sensing offer data-driven insights that optimize resource use, improve productivity, and enhance decision-making.

- Precision Agriculture and Data-Driven Insights: Precision techniques enable site-specific resource allocation, reducing waste and maximizing yield. Real-time data collection empowers informed decision-making.
- Climate-Smart Agriculture [2]: Implementing practices that mitigate climate change impacts, conserve water, and enhance soil health fosters resilience in agriculture.
- Innovative Farming Practices: Agroecology, vertical farming, hydroponics, and aquaponics offer sustainable alternatives that minimize environmental impact.
- Digital Connectivity and Access to Information: Digital platforms provide smallholders with market information, weather forecasts, and best practices, enabling informed choices.
- Economic Diversification: Expanding value-added activities, such as agro-processing and niche market cultivation, can enhance rural incomes [5].
- Policy and Institutional Support: Government policies that prioritize sustainable agriculture, provide incentives, and enhance market access can create an enabling environment.

Navigating this evolving landscape requires a comprehensive approach that leverages innovative technologies, promotes sustainable practices, and ensures inclusivity. As we confront the challenges and seize the opportunities, the transformation of agriculture holds the potential not only to feed a growing population but also to safeguard the planet and foster prosperity for generations to come.

1.2 Defining the AI Green Revolution: AI's Transformative Role in Farming

In the wake of pressing global challenges and rapid technological advancements, a new agricultural era is unfolding—the AI Green Revolution. This transformative movement leverages the power of artificial intelligence to redefine the way farming operates, ushering in a sustainable, efficient, and environmentally conscious approach to food production. At its core, the AI Green Revolution harnesses AI's capabilities to address the intricate complexities of modern agriculture [6].

- Data-Driven Precision: AI empowers farmers with data-driven precision. Sensors [7], drones, and IoT [4] devices gather real-time information about soil health, weather patterns, and crop conditions. AI algorithms process this data to optimize planting, irrigation, and pest control, ensuring resources are used efficiently.
- Predictive Analytics: The AI Green Revolution thrives on predictive analytics. Machine learning algorithms analyze historical and real-time data to anticipate disease outbreaks, optimize planting schedules, and predict crop yields. This foresight enhances decision-making and minimizes risks.
- Resource Efficiency: AI-driven insights enable resource-efficient farming. Precision irrigation systems deliver water precisely where needed, reducing waste. Similarly, the precise application of fertilizers and pesticides minimizes environmental impact.

- Climate Resilience [8]: The AI Green Revolution contributes to climate resilience. AI models analyze weather patterns to predict climate shifts, enabling farmers to adapt their practices accordingly and mitigate the impact of extreme weather events.
- Empowering Smallholders: This revolution is inclusive, extending benefits to smallholders. AI-driven platforms provide small-scale farmers with access to information, market insights, and best practices, leveling the playing field and enhancing their livelihoods.
- Circular Economy Principles: AI encourages circular economy principles in agriculture. By optimizing nutrient cycling, reducing waste, and promoting sustainable practices, the AI Green Revolution contributes to ecosystem restoration and long-term sustainability.
- Ethical Considerations: As AI becomes integral to agriculture, ethical considerations arise. Ensuring data privacy, transparency in algorithms, and equitable access to technology are paramount in this revolution.
- Collaborative Networks: The AI Green Revolution thrives on collaboration. Partnerships between technology companies, researchers, farmers, and governments foster the exchange of knowledge, best practices, and resources.
- Beyond Productivity: Unlike the historical Green Revolution, the AI Green Revolution isn't solely focused on yield increase. It emphasizes sustainable practices, ecosystem health, and equitable distribution of benefits.

As AI's transformative potential converges with the imperatives of sustainable agriculture, the AI Green Revolution emerges as a beacon of hope for a future where food security, environmental stewardship, and technological innovation converge harmoniously. It's a call to reimagine farming as a dynamic, data-driven endeavor—one that ensures prosperity not just for farmers but for the planet as well.

2 Literature Survey

The emergence of the "AI Green Revolution" represents a significant shift in the agricultural landscape, poised to profoundly reshape the future of farming. A comprehensive literature survey reveals that this revolution hinges on the integration of artificial intelligence (AI) and data-driven technologies into agriculture. These innovations hold the potential to revolutionize the industry by optimizing crop management, resource utilization, and sustainability practices. As documented in various studies, AI is playing a pivotal role in transforming agriculture into a more efficient, environmentally friendly, and economically viable sector.

One key aspect highlighted in the literature is AI's capacity to enhance agricultural productivity while minimizing resource inputs. AI-driven solutions, such as predictive analytics and remote sensing technologies, enable farmers to make data-informed decisions regarding crop planting, irrigation, and pest control. This precision agriculture approach results in increased yields, reduced waste, and lower operational costs. Furthermore, AI's contribution to sustainability cannot be overstated. It helps mitigate the environmental footprint of agriculture by reducing the use of water, fertilizers, and pesticides, thereby preserving valuable resources and promoting ecological balance.

2.1 Addressing Agricultural Challenges with AI

Agriculture, the backbone of our global food system, faces an array of challenges, from population growth and climate change to resource scarcity and market inefficiencies [9]. To tackle these complex issues, the integration of artificial intelligence (AI) emerges as a powerful tool. Here's how AI is being harnessed to address these agricultural challenges [10]:

- Precision Farming for Resource Efficiency: AI-powered sensors, drones, and satellites gather real-time data on soil conditions, weather patterns, and crop health. Machine learning algorithms process this information to provide actionable insights, enabling precise resource allocation. This enhances water management, reduces excessive fertilizer use, and optimizes pest control, all contributing to increased resource efficiency.
- Climate Resilience and Adaptation [8]: AI models analyze historical and current climate data to predict weather patterns and shifts. Farmers can make informed decisions on planting, irrigation, and harvesting based on these predictions, enhancing climate resilience and reducing losses due to extreme weather events.
- Disease and Pest Management: AI-driven disease and pest prediction models analyze data from various sources to anticipate outbreaks. This early warning system allows farmers to take preventive measures, reducing crop losses and minimizing the need for chemical interventions.
- Data-Driven Decision-Making: AI processes vast amounts of data to provide real-time insights. Farmers can make well-informed decisions on crop selection, planting times, and market strategies, leading to higher yields and profitability.
- Access to Information for Smallholders: AI-powered mobile apps and platforms provide small-scale farmers with access to weather forecasts, market prices, and best practices. This democratizes information and empowers smallholders to make informed choices.
- Market Access and Value Chain Optimization: AI helps streamline agricultural supply chains by predicting demand, optimizing logistics, and reducing post-harvest losses. This ensures fresher produce reaches consumers and enhances the incomes of farmers.
- Sustainable Practices and Biodiversity Conservation: AI assists in designing sustainable cropping systems by analyzing soil health, biodiversity data, and ecosystem interactions. This promotes regenerative agriculture that enhances biodiversity, soil fertility, and carbon sequestration.

- Monitoring and Management of Large-Scale Farms: For large farms, AI-powered autonomous machinery, remote sensing, and predictive maintenance reduce labor costs and improve farm management efficiency.
- Early Warning Systems and Disaster Management: AI can predict natural disasters like droughts and floods, enabling governments and organizations to implement disaster management strategies to mitigate their impact on agriculture [11].
- Agroecology and Sustainable Practices: AI supports agroecological approaches by analyzing local ecosystems and suggesting suitable crop rotations, cover cropping, and intercropping strategies [12].

Table 1 shows the overview of existing technologies, challenges, and applications in the AI Green Revolution reshaping agriculture's future.

These technologies are collectively contributing to the AI Green Revolution in agriculture, but they also come with various challenges, including issues related to data, adoption, regulation, and ethics. Successful implementation and widespread adoption of these technologies require careful consideration and solutions to these challenges.

Table 1 Overview of existing technologies, challenges, and applications

Technology/ category	Description	Challenges	Applications
Artificial intelligence	Utilizes machine learning and deep learning algorithms	Data privacy and security, access to data, model accuracy	Crop yield prediction, disease detection, pest management
IoT (Internet of Things)	Sensors, devices, and connectivity for data collection	Scalability, data management, power consumption	Soil monitoring, climate control, irrigation automation
Remote sensing	Satellite and drone imagery for real-time data capture	Data processing, accuracy, cost, and accessibility	Crop health assessment, yield estimation, drought monitoring
Precision farming	GPS technology for precise crop management	Initial investment, education, and training	Variable rate application of resources, precision planting
Blockchain	Distributed ledger for transparent and secure data	Adoption and integration challenges, regulatory compliance	Supply chain traceability, food safety, and authenticity
Robotics and automation	Autonomous machines for tasks like planting and harvesting	Cost, technical complexity, and maintenance	Labor-saving, precision tasks, weed control
Data analytics	Advanced data analysis for informed decision-making	Data quality, interpretation, and integration	Market analysis, predictive modeling, resource optimization
Biotechnology	Genetic modification for crop improvement	Ethical concerns, public perception, regulatory approval	Disease resistance, higher yields, nutrient efficiency

2.2 Enhancing Precision Agriculture: Predictive Insights and Decision Support

Precision agriculture is undergoing a revolutionary transformation through the integration of advanced technologies and data-driven approaches [13]. Central to this transformation is the utilization of predictive insights and decision support systems powered by artificial intelligence (AI) and data analytics [13]. These technologies empower farmers with the tools to make informed decisions, optimize resource management, and maximize productivity. Here's how enhancing precision agriculture with predictive insights and decision support is shaping the future of farming shown in Fig. 1.

- Predictive Analytics for Yield Optimization: AI algorithms analyze historical and real-time data, including [14] weather patterns, soil conditions, and crop health, to forecast yield outcomes. This allows farmers to adjust planting strategies, irrigation schedules [15], and other practices to optimize yields.
- Disease and Pest Management: Predictive models use AI to identify early signs of disease or pest outbreaks. Farmers receive timely alerts and recommendations, enabling them to take preemptive actions and minimize crop losses.
- Weather-Responsive Practices: Real-time weather data integration into decision support systems allows farmers to adjust their practices according to changing weather conditions, reducing risks associated with extreme events.

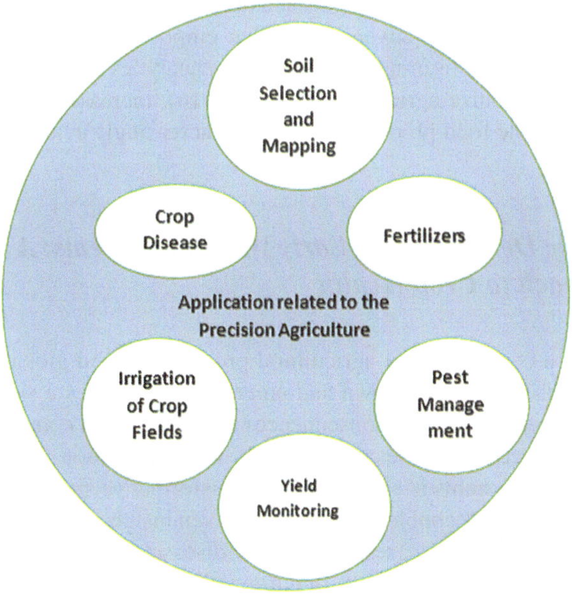

Fig. 1 Applications of precision agriculture

- Efficient Resource Allocation: Decision support systems consider factors such as soil moisture levels, nutrient content, and growth stages to recommend precise irrigation, fertilization, and pesticide application strategies. This minimizes resource waste and environmental impact.
- Risk Mitigation and Crop Insurance: Predictive insights help farmers assess potential risks to their crops, enabling them to make informed decisions about implementing crop insurance or taking measures to mitigate those risks.
- Market-Driven Strategies: AI-powered platforms integrate market data and consumer trends, enabling farmers to align their production with market demands. This reduces food waste and improves profitability.
- Data-Driven Farm Management: AI consolidates data from various sources, such as satellite imagery and sensor networks, providing a comprehensive view of the farm. This holistic perspective informs decisions related to crop rotation, planting density, and overall farm management.
- Remote Monitoring and Precision Interventions: AI-enhanced sensors and drones provide real-time information about crop conditions. Farmers can remotely monitor fields and implement timely interventions without being physically present.
- Long-Term Sustainability: By predicting the impact of different practices on soil health and ecosystem balance, decision support systems guide farmers toward sustainable farming methods that enhance long-term productivity.
- Inclusivity and Accessibility: Decision support systems, often accessible through mobile apps, bridge the gap between technology and farmers. Smallholders and rural communities can leverage these tools to enhance their practices.

The integration of predictive insights and decision support systems into precision agriculture marks a pivotal advancement that empowers farmers to navigate the complexities of modern farming. As AI and data analytics continue to evolve, their potential to revolutionize agricultural practices [16], increase efficiency, and contribute to sustainable food production becomes increasingly evident.

2.3 Disease Detection and Early Warning Systems: A Resilient Approach to Crop Health

Crop health is a cornerstone of agricultural productivity and global food security. The threat of diseases, both known and emerging [17], poses a significant risk to crops and can have devastating consequences for farmers and communities. To bolster resilience and combat these challenges, the implementation of disease detection and early warning systems is emerging as a transformative strategy. This approach leverages cutting-edge technologies and data-driven insights to proactively monitor, detect, and manage crop diseases. Here's how disease detection and early warning systems are fostering resilience in agriculture [12]:

- Real-Time Monitoring with Remote Sensing: Satellite imagery and drones equipped with advanced sensors provide real-time data on crop health, identifying changes in vegetation, moisture levels, and other indicators that signal potential disease outbreaks.
- Data-Driven Disease Identification: Machine learning algorithms analyze large datasets of crop images, symptoms, and environmental conditions to accurately identify diseases. This speeds up diagnosis and enables timely intervention.
- Early Detection and Rapid Response: By detecting diseases in their early stages, farmers can take swift action to mitigate their spread. Early warning systems help prevent outbreaks from reaching epidemic proportions.
- Predictive Modeling: AI-powered predictive models analyze historical data, weather patterns, and disease prevalence to forecast disease risk. Farmers can adjust planting and management strategies accordingly.
- Smart Sensor Networks [7]: IoT-based sensor networks collect real-time data on environmental conditions such as humidity, temperature, and soil moisture. Deviations from normal patterns can indicate disease presence [18].
- Localized Disease Mapping: Geographic information systems (GIS) create localized disease maps, enabling farmers to identify disease hotspots and plan targeted interventions.
- Reducing Chemical Dependency: Timely disease detection reduces the reliance on broad-spectrum pesticides. Targeted interventions minimize chemical usage and environmental impact.
- Cross-Border Disease Management: Early warning systems transcend borders, allowing for timely information sharing and coordinated responses to transboundary diseases.
- Resilient Agricultural Systems: Disease detection and early warning systems contribute to the overall resilience of agricultural systems by minimizing crop losses and preserving farmer livelihoods.

As climate change and globalization reshape disease dynamics, disease detection and early warning systems offer a proactive and technology-driven approach to safeguarding crop health. By equipping farmers with tools to identify and manage diseases promptly, these systems not only protect yields but also promote sustainable and resilient agricultural practices. Embracing this approach empowers farmers to stay ahead of threats, reduce the environmental impact of disease management, and contribute to food security in an increasingly uncertain world.

3 AI's Role in Smart Farming

Smart farming, also known as precision agriculture, represents a paradigm shift in the way we cultivate crops and raise livestock. At the heart of this transformation lies Artificial Intelligence (AI), a technology that harnesses data-driven insights to optimize every aspect of agricultural operations. AI's integration into smart farming [19] is revolutionizing how we produce food, making it more efficient, sustainable, and responsive to the challenges of a rapidly changing world. Here's how AI is shaping the future of smart farming [20]:

- Data Collection and Analysis: AI-driven sensors, drones, and satellite imagery collect vast amounts of data about soil conditions, weather patterns, crop health, and livestock behavior. AI algorithms process this data to provide real-time insights, enabling farmers to make informed decisions.
- Predictive Analytics: AI employs historical and current data to forecast future trends and outcomes. Farmers can predict crop yields, disease outbreaks, and market demands, allowing them to plan accordingly and optimize resource allocation.
- Precision Farming: AI's ability to process data at a granular level enables precision agriculture. Farmers can apply inputs such as water, fertilizers, and pesticides with pinpoint accuracy, reducing waste and increasing efficiency.
- Disease Detection and Management: AI-powered image recognition and data analysis can identify early signs of disease in crops and livestock. Early detection allows for timely interventions, minimizing losses and reducing the need for chemical treatments.
- Autonomous Machinery: AI-driven autonomous vehicles and robots can perform tasks like planting, weeding, and harvesting with precision. This reduces labor costs and enhances efficiency.
- Resource Optimization: AI optimizes resource use by analyzing data to determine the optimal amount of water, nutrients, and other inputs required for healthy crop growth.
- Livestock Monitoring: AI-powered sensors track livestock behavior, health, and feeding patterns. This helps farmers detect health issues early, ensure animal welfare, and improve productivity.
- Market Insights: AI processes market data to provide insights into consumer trends and demands. Farmers can align their production with market needs, reducing food waste and increasing profitability.
- Climate Resilience: AI analyzes weather patterns and climate data to help farmers adapt to changing conditions. It provides recommendations for planting times, irrigation schedules, and crop choices.
- Data-Driven Decision-Making: AI generates actionable insights from complex datasets, empowering farmers to make informed decisions about planting strategies, resource management, and risk mitigation.

In essence, AI's integration into smart farming is driving a shift from traditional, experience-based farming to a data-driven, responsive, and efficient approach. As technological advancements continue, AI's role in agriculture will evolve, further optimizing food production, conserving resources, and ensuring a sustainable and resilient agricultural future.

3.1 AI as an Enabler of Data-Driven Decision-Making

In an era marked by information abundance, the ability to transform data into actionable insights has become a cornerstone of success across various domains. Agriculture, too, is undergoing a data revolution, where the integration of Artificial Intelligence (AI) as an enabler of data-driven decision-making is reshaping the landscape. This convergence is empowering farmers, researchers, and stakeholders with the tools to optimize resource use, enhance productivity, and address challenges with unprecedented precision. Here's how AI is driving data-driven decision-making in agriculture:

- Data Collection and Integration: AI systems gather data from diverse sources such as sensors, satellites, drones, and historical records. This data fusion creates a comprehensive view of the farm, encompassing weather conditions, soil health, crop growth, and more.
- Real-time Insights: AI processes data in real-time, providing farmers with up-to-the-minute insights. This enables swift reactions to changing conditions and minimizes risks associated with crop health, weather patterns, and pest outbreaks.
- Predictive Analytics: AI algorithms analyze historical and current data to predict future trends. Farmers can anticipate disease outbreaks, optimize planting schedules, and plan resource allocation based on AI-generated forecasts.
- Precision Agriculture: AI's granular analysis guides precision agriculture practices. It determines optimal seeding rates, irrigation amounts, and nutrient application, reducing waste and maximizing yields.
- Customization and Personalization: AI tailors recommendations to the specific needs of each field or crop. This individualized approach ensures that interventions are precise and effective.
- Risk Management: AI assesses risk factors like weather events, market fluctuations, and disease prevalence. Farmers can then adopt strategies to mitigate these risks and protect their investments.
- Resource Optimization [21]: AI-driven insights enable efficient resource allocation. By understanding each field's unique conditions, farmers can optimize water usage, reduce chemical inputs, and conserve energy.
- Market Insights: AI analyzes market data, consumer trends, and demand patterns. Farmers can align their production with market needs, reducing waste and enhancing profitability.
- Accessibility and Inclusivity: AI platforms, often accessible through mobile apps, bridge the technology gap for smallholders and remote farmers, ensuring they benefit from data-driven insights.
- Sustainability and Environmental Stewardship: AI supports environmentally conscious decisions. It identifies opportunities for sustainable practices like cover cropping, crop rotation, and integrated pest management.

As AI technologies continue to evolve, their potential to optimize resource use, increase yields, and promote sustainable practices grows, positioning agriculture to meet the challenges of a rapidly changing world. In this data-rich era, AI catalyzes farmers, researchers, and stakeholders to make smarter, more informed choices that enhance productivity while preserving the environment for future generations.

3.2 From Machine Learning to Robotics: Diverse Applications of AI in Agriculture

Artificial Intelligence (AI) is driving a transformative wave across the agricultural sector, offering a wide array of applications that range from data analysis to robotics. This convergence of AI technologies is reshaping traditional farming [22] practices and revolutionizing how we approach agricultural challenges. From harnessing machine learning for predictive insights to deploying robots for labor-intensive tasks, here's a glimpse into the diverse applications of AI in agriculture:

- Precision Agriculture and Data Analytics: Machine learning processes data from various sources—sensors, satellites, and drones—to provide insights on soil health, weather patterns, and crop conditions [23]. These insights inform decisions about resource allocation, planting strategies, and disease management.
- Disease Detection and Management: AI-powered image recognition and data analysis identify early signs of diseases in crops. Farmers receive timely alerts and recommendations for interventions, reducing crop losses and the need for chemical treatments.
- Weather and Climate Analysis: AI models analyze historical climate data and predict weather patterns. This enables farmers to adapt practices and mitigate risks associated with extreme weather events.
- Yield Prediction and Forecasting: Machine learning algorithms analyze historical yield data, weather conditions, and agronomic practices to forecast crop yields. This helps farmers plan for market demands and allocate resources effectively.
- Livestock Monitoring and Management: AI-driven sensors and data analytics monitor livestock behavior, health, [14] and feeding patterns. This early detection of issues improves animal welfare and productivity.
- Autonomous Machinery and Robotics: Robots equipped with AI navigate fields to perform tasks such as planting, weeding, and harvesting. This reduces labor costs, increases efficiency, and mitigates the challenges of labor shortages.
- Supply Chain Optimization: AI optimizes the supply chain by predicting demand, improving logistics, and reducing post-harvest losses. This ensures fresher produce reaches consumers and reduces waste.
- Crop Rotation and Nutrient Management: AI recommends optimal crop rotation and nutrient management strategies based on soil data and historical performance. This enhances soil health and minimizes nutrient runoff.

- Pest Control and Integrated Pest Management: AI analyzes pest data and recommends integrated pest management strategies, reducing reliance on chemical pesticides and promoting sustainable practices.
- Food Traceability and Quality Assurance: Blockchain and AI are combined to track the journey of food products from farm to table. This enhances transparency, food safety, and consumer trust.
- Soil Health and Erosion Prevention: AI assesses soil erosion risk using satellite data and recommends erosion-prevention strategies, preserving soil fertility and preventing environmental degradation.

The applications of AI in agriculture are as diverse as the challenges faced by the industry. From enabling precision farming through data analytics to revolutionizing labor-intensive tasks with robotics, AI's impact spans the entire agricultural value chain. As technology continues to advance, the potential for AI to enhance productivity, sustainability, and food security in agriculture is vast, presenting an exciting future for the field.

4 The Power of Big Data: Collecting, Analyzing, and Interpreting Agricultural Data

In the age of technology, data has emerged as a pivotal force driving innovation and transformation across various industries [17], including agriculture. Big Data, characterized by its volume, velocity, and variety, is reshaping the way we approach farming practices, resource management, and decision-making. In agriculture, [24] harnessing the power of Big Data involves a comprehensive process of collecting, analyzing, and interpreting vast amounts of information to drive informed and sustainable agricultural practices. Here's how the process unfolds [25]:

- Data Integration: The diverse datasets collected from different sources are integrated into a unified platform. This consolidation enables a comprehensive view of the agricultural landscape and facilitates cross-domain analysis.
- Data Storage and Management: Storing and managing Big Data requires robust infrastructure and techniques. Cloud computing and data warehouses provide the capacity to store and access vast datasets efficiently.
- Data Cleaning and Preparation: Raw data often contain errors, missing values, and inconsistencies. Data cleaning and preprocessing ensure that the information used for analysis is accurate and reliable.
- Data Analysis: Advanced analytics, including machine learning and AI, process the data to identify patterns, correlations, and trends. These algorithms unveil insights that might not be apparent through traditional methods.
- Predictive Modeling: Big Data analytics enable the creation of predictive models. These models forecast future scenarios, such as crop yields, disease outbreaks, and market demand, helping farmers plan accordingly.

- Decision Support Systems: Interpretation of analyzed data results in actionable insights. Decision support systems provide farmers with real-time recommendations, aiding them in making informed choices about resource allocation, planting strategies, and more.
- Resource Optimization [21]: Big Data insights guide the optimization of resources such as water, fertilizers, and pesticides. Farmers can apply these resources precisely, reducing waste and environmental impact.
- Risk Mitigation: Analyzing historical data helps farmers identify and manage risks associated with weather events, pests, and market fluctuations. This allows for proactive risk mitigation strategies.
- Continuous Improvement: The cyclical process of collecting, analyzing, and interpreting data allows for continuous improvement. Farmers can refine their practices based on real-time feedback and evolving insights.

The power of Big Data in agriculture lies not just in its volume, but in the actionable insights it provides. By collecting, analyzing, and interpreting agricultural data, farmers and stakeholders can make more informed decisions, optimize resource use, increase productivity, and contribute to sustainable and resilient farming practices. In a rapidly evolving world, Big Data is transforming agriculture into a dynamic, data-driven industry, shaping the future of food production.

4.1 AI-Driven Insights: Optimizing Planting, Irrigation, and Fertilization

In the quest for sustainable and efficient agriculture, artificial intelligence (AI) is emerging as a game-changing tool, particularly in optimizing critical processes such as planting, irrigation, and fertilization. By harnessing the power of AI-driven insights, farmers can make informed decisions that maximize resource efficiency, minimize environmental impact, and enhance overall crop yield. Here's how AI is revolutionizing these fundamental agricultural practices:

- Precision Planting: AI analyzes historical data, soil conditions, weather patterns, and crop performance to determine optimal planting strategies. These insights guide farmers in choosing the right crop varieties, planting densities, and timing for each field, maximizing germination rates and overall productivity.
- Smart Irrigation [26]: AI integrates real-time data from sensors and weather forecasts to precisely manage irrigation. By calculating soil moisture levels, crop water requirements, and rainfall predictions, AI ensures water is applied where and when it's needed, reducing water waste and conserving resources.
- Nutrient Management: AI-driven insights help farmers optimize fertilization practices. By considering factors such as soil nutrient content, crop growth stage, and environmental conditions, AI recommends precise fertilizer application rates and timings, minimizing nutrient runoff and pollution.

- Data-Driven Decision Support: Farmers receive actionable recommendations from AI-powered decision support systems. These systems consolidate data, process it using machine learning algorithms, and provide real-time guidance on planting, irrigation, and fertilization strategies.
- Variable Rate Application: AI enables variable rate application, adjusting planting density, irrigation levels [7], and fertilizer amounts across different zones within a field. This approach optimizes resource allocation based on specific conditions, leading to uniform crop growth.
- Risk Mitigation: AI considers risk factors such as climate variability and soil characteristics when providing recommendations. This helps farmers mitigate the impact of weather events and ensure crop health.
- Continuous Learning and Improvement: As AI algorithms learn from each season's data, insights become more accurate and refined over time. This iterative process supports ongoing improvement in planting, irrigation, and fertilization practices.
- Environmental Stewardship: AI's precise resource allocation reduces the environmental footprint of agriculture. Water, fertilizers, and pesticides are used more efficiently, minimizing runoff and pollution.
- Increased Profitability: AI's optimization strategies lead to higher crop yields, reduced input costs, and improved overall profitability for farmers.
- Adaptive Management: AI's real-time insights allow farmers to adapt their practices to changing conditions. Whether it's adjusting irrigation schedules due to unexpected rain or altering planting strategies based on evolving weather patterns, AI-driven insights enhance adaptability.

In essence, AI-driven insights provide farmers with a data-driven blueprint for optimizing planting, irrigation, and fertilization practices. By leveraging AI's capabilities, agriculture is moving toward a more precise, resource-efficient, and sustainable future, where technology helps strike the balance between productivity and environmental conservation.

5 Automation and Robotics in Agriculture

Automation and robotics have ushered in a new era of efficiency and innovation in the field of agriculture. From planting and harvesting to monitoring and data collection, these technologies are transforming traditional farming practices and revolutionizing the way we produce food. Here's a comprehensive overview of how automation and robotics are reshaping agriculture [27]:

- Precision Planting and Seeding: Automated planting machines use advanced algorithms to precisely space and place seeds in the soil, optimizing germination rates and crop uniformity.

- Autonomous Tractors and Machinery: Self-driving tractors and machinery equipped with GPS and AI navigate fields, performing tasks such as plowing, cultivating, and spraying with precision and minimal human intervention.
- Robotic Harvesting: Robots equipped with sensors and cameras identify ripe fruits and vegetables, ensuring gentle and efficient harvesting. This reduces labor costs and increases efficiency.
- Weed and Pest Control: Robots equipped with AI-driven cameras and robotic arms can target and remove weeds with high precision, reducing the need for chemical herbicides.
- Monitoring and Data Collection: Drones and ground-based robots collect real-time data on soil moisture, crop health, and weather conditions. This data aids in making informed decisions about irrigation, fertilization, and disease management.
- Autonomous Greenhouses: Automated greenhouses regulate temperature, humidity, and lighting conditions to optimize plant growth. Robots can also perform tasks like planting, pruning, and harvesting in controlled environments [28].
- Dairy and Livestock Management: Robotic systems monitor and manage livestock, including milking, feeding, and health monitoring, leading to improved animal welfare and productivity.
- Autonomous Sorting and Packing: Robots equipped with computer vision systems sort and pack harvested produce based on quality and size, reducing post-harvest losses and ensuring consistency.
- Soil Sampling and Analysis: Robotic systems collect soil samples from various parts of a field, providing data for accurate nutrient management and improving soil health.
- Labor Shortage Mitigation: Automation and robotics address labor shortages in agriculture by performing tasks that are labor-intensive, time-consuming, or physically demanding.
- Remote Monitoring and Management: Farmers can remotely monitor and control agricultural operations through mobile apps and connected devices, increasing operational efficiency.
- Reduction of Chemical Usage: Precision application of pesticides and fertilizers by robots minimizes chemical usage, reducing environmental impact.
- 24/7 Operations: Robots and automated systems can work around the clock, optimizing productivity and reducing the constraints of daylight hours.
- Data-Driven Decision-Making: Automated systems provide real-time data, enabling farmers to make informed decisions for resource allocation, planting strategies, and more.
- Sustainable Practices: Automation and robotics enable more precise resource management, reducing waste and supporting sustainable farming practices.

As automation and robotics continue to evolve, they offer the potential to increase agricultural productivity, reduce environmental impact, and enhance the resilience of farming systems. The integration of these technologies into agriculture is paving the way for a more efficient, sustainable, and technologically advanced future in food production.

5.1 Robotic Farming Systems: From Planting to Harvesting

The integration of robotics into agriculture is revolutionizing traditional farming practices, from the initial stages of planting to the final steps of harvesting [5]. These robotic farming systems are changing the way we cultivate crops, offering increased efficiency, precision, and sustainability. Here's an exploration of how robotics is transforming the entire farming process [29]:

- Precision Planting: Robotic planting systems use advanced algorithms and sensors to precisely plant seeds at optimal depths and intervals. This ensures uniform germination and strong crop establishment.
- Automated Irrigation and Fertilization: Robots equipped with sensors [30] can monitor soil moisture levels and nutrient content. They autonomously apply irrigation [26] and fertilizers in precise amounts, reducing waste and enhancing resource efficiency.
- Weed and Pest Management: Robotic systems equipped with cameras and AI can identify and target weeds, removing them with precision. This reduces the need for chemical herbicides and promotes sustainable practices.
- Crop Monitoring and Disease Detection: Drones and ground-based robots equipped with sensors and cameras collect data on crop health, spotting disease outbreaks and enabling timely interventions.
- Precision Spraying: Robotic sprayers precisely apply pesticides or other treatments to specific areas of crops, reducing chemical usage and minimizing environmental impact.
- Autonomous Tractors and Machinery: Self-driving tractors equipped with GPS and AI can perform various tasks, such as plowing, cultivating, and seeding, with minimal human intervention.
- Robotic Harvesting: Robots equipped with sensors and AI can identify ripe fruits and vegetables and harvest them gently and efficiently. This reduces labor costs and post-harvest losses.
- Controlled Environment Farming: Automated greenhouses and vertical farms use robotics to manage lighting, temperature, and humidity conditions, creating optimal growing environments.
- Data-Driven Insights: Robotic systems collect and analyze data to provide real-time insights on crop health, soil conditions, and other factors. These insights guide decision-making for better farming practices.
- Labor Efficiency: Robotic farming systems address labor shortages by performing labor-intensive tasks, ensuring operations continue seamlessly.
- Energy Efficiency: Robots can optimize energy use by precisely managing lighting, irrigation, and climate control systems in indoor farming environments.
- Soil Health and Nutrient Management: Robotic soil sampling and nutrient application systems help maintain soil health and ensure optimal nutrient levels for plant growth.

- Data Integration and IoT: Robotic systems [27] can be part of the Internet of Things (IoT) ecosystem, sharing data and insights with other connected devices for comprehensive farm management.
- Reduced Environmental Impact: Robotic systems enable targeted interventions [27], reducing the need for chemical inputs and minimizing the environmental footprint of farming.
- Continuous Improvement: As robotics and AI technologies advance, these systems continuously learn and improve, adapting to changing conditions and optimizing performance.

6 The Smart Agriculture Process

Smart agriculture encompasses a range of processes and technologies aimed at improving the efficiency, sustainability, and productivity of farming practices [31]. An overview of the smart agriculture process is shown in Fig. 2:

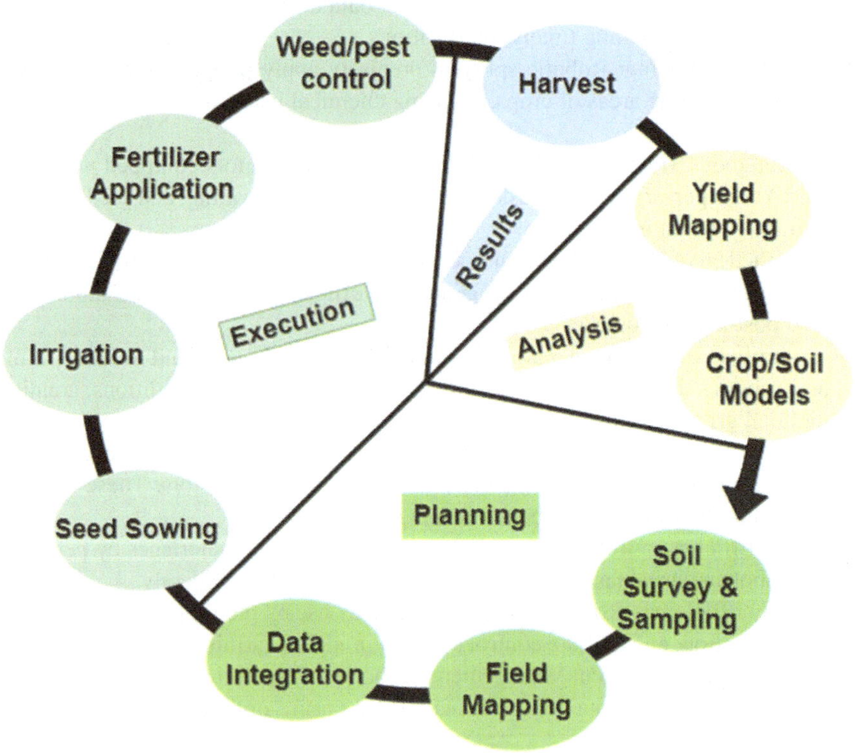

Fig. 2 Smart agriculture process

- Data Collection: The process begins with the collection of data from various sources using IoT sensors, cameras, drones, and other monitoring devices. These sensors can gather data on soil conditions [32], weather, crop health, livestock behavior, and more. This real-time data forms the foundation for smart decision-making.
- Data Transmission: Collected data is transmitted to a central system or cloud platform through wireless or wired networks. This allows for remote access and real-time monitoring, which is crucial for making timely decisions and adjustments.
- Data Analysis: Once the data is collected, it is analyzed using data analytics and machine learning algorithms. This analysis helps in identifying trends [33], patterns, and anomalies, providing valuable insights into the state of the farm and its various components.
- Decision Support: Farmers receive actionable insights and recommendations based on data analysis. These recommendations can include optimal irrigation [34] schedules, precise fertilization plans, disease detection in crops, or health monitoring for livestock. Farmers can use this information to make informed decisions.
- Automation: Smart agriculture often involves automation through the use of autonomous machinery and equipment. This can include automated tractors, drones for precision spraying, and robotic milking machines. Automation reduces the need for manual labor and ensures tasks are carried out with precision and efficiency.
- Resource Management: Smart agriculture optimizes the use of resources such as water, energy, and fertilizers. IoT-enabled systems can adjust irrigation [26] based on soil moisture levels, reduce energy consumption through efficient equipment usage, and minimize fertilizer runoff.
- Monitoring and Control: Through IoT-connected devices, farmers can remotely monitor and control various aspects of their operations. This includes adjusting irrigation systems, managing livestock feeding, and even controlling greenhouse environments for specialized crops.
- Predictive Maintenance: IoT technology can predict when machinery or equipment might require maintenance, reducing downtime and preventing costly breakdowns. This proactive approach ensures that farming operations remain efficient.
- Market Analysis: Smart agriculture systems can also provide insights into market conditions, helping farmers make decisions about crop selection and timing for optimal prices in the market.
- Environmental Sustainability: By optimizing resource use and reducing waste, smart agriculture contributes to environmental sustainability. It minimizes the environmental impact of farming practices, which is crucial for long-term agricultural viability.

Overall, smart agriculture is a data-driven and technology-enabled approach that aims to make farming more efficient, profitable, and sustainable while ensuring food security for a growing global population. It integrates data, connectivity, and automation to transform traditional farming into a high-tech industry.

6.1 Application of IoT on Smart Agricultural

The application of IoT (Internet of Things) in smart agriculture has revolutionized the way farmers manage their crops and livestock. IoT technology involves the use of sensors, data analytics, and connectivity to gather real-time information from various agricultural assets, such as fields, machinery, and animals. These innovations have significantly improved agricultural efficiency and sustainability. For instance, IoT-enabled sensors placed in fields can monitor soil moisture levels, temperature, and nutrient content, allowing farmers to make data-driven decisions about irrigation and fertilization. This precision agriculture approach minimizes resource wastage and maximizes crop yields. Furthermore, IoT-connected livestock monitoring systems can provide farmers with valuable insights into the health and behavior of their animals, helping to prevent disease outbreaks and optimize feeding schedules. Overall, IoT in smart agriculture has the potential to increase food production, reduce environmental impacts, and enhance the overall sustainability of farming practices shown in Fig. 3.

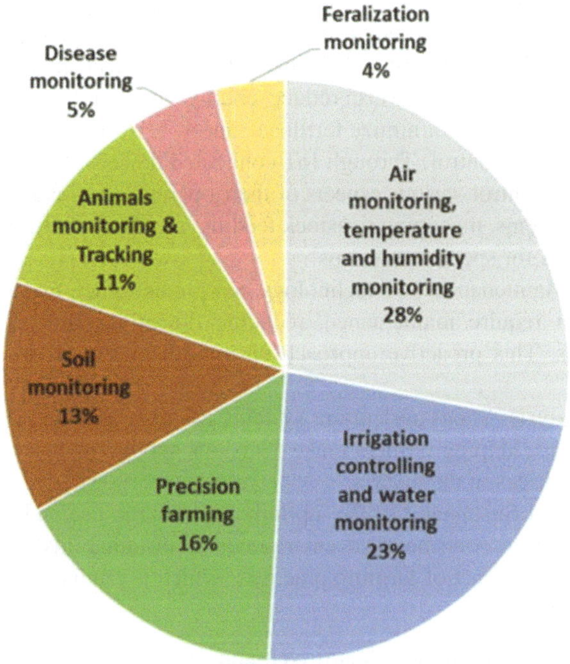

Fig. 3 Application of IoT on smart agriculture

Another critical aspect of IoT in smart agriculture is the automation of various tasks through smart machinery and autonomous vehicles. IoT-connected tractors and drones equipped with sensors and cameras can perform tasks like planting, harvesting, and pest control with high precision and efficiency. This not only reduces labor costs but also minimizes the need for chemical inputs, promoting sustainable farming practices. Additionally, IoT-based farm management platforms enable remote monitoring and control of farm operations, allowing farmers to make timely adjustments and respond to changing environmental conditions. With the help of predictive analytics and machine learning algorithms, IoT can also provide early warnings about weather events or crop diseases, helping farmers mitigate risks and optimize their production strategies. In summary, the application of IoT in smart agriculture is transforming traditional farming [8] into a data-driven, efficient, and sustainable industry that can better meet the growing global demand for food while minimizing its impact on the environment.

The expedition from planting to harvesting is being reimagined by the capabilities of robotic farming systems. These systems offer farmers the potential to increase productivity, reduce resource waste, and promote sustainable practices while also addressing the challenges of labor shortages and changing climate conditions. As technology continues to evolve, the impact of robotics on agriculture promises a more efficient, resilient, and technologically advanced future.

6.2 AI-Powered Drones and Sensors: The Revolutionizing Field Monitoring

Combination is empowering farmers with real-time insights, precise data collection, and targeted interventions that enhance productivity, optimize resource use, and promote sustainable practices. Here's how AI-powered drones and sensors are transforming field monitoring [35]:

- Aerial Surveillance and Imaging: Drones equipped with high-resolution cameras capture detailed images of fields. AI algorithms analyze these images to assess crop health, detect disease, and identify areas that require attention.
- Crop Health Assessment: AI algorithms process drone-captured images to identify early signs of stress, disease, or nutrient deficiencies in crops. This enables timely interventions, reducing losses and improving yields.
- Precision Mapping: Drones use GPS technology to create detailed field maps. AI interprets these maps to provide insights into soil variations, drainage patterns, and other factors that impact crop growth.
- Soil Moisture and Nutrient Analysis: Ground-based sensors and drones equipped with sensors measure soil moisture levels and nutrient content. AI analyzes this data to optimize irrigation and fertilization strategies.

- Pest and Weed Detection: AI-powered sensors and cameras on drones identify pests and weeds in real-time. This enables targeted interventions, reducing the need for broad-spectrum pesticides.
- Weather Monitoring: Drones equipped with weather sensors collect real-time data on temperature, humidity, wind, and other environmental factors. AI interprets this data to provide insights for decision-making.
- Irrigation Management: Sensors detect soil moisture levels, and AI recommends precise irrigation schedules based on crop needs and weather forecasts, conserving water and ensuring optimal growth.
- Disease Outbreak Prediction: AI analyzes historical data, current conditions, and disease patterns to predict disease outbreaks. Farmers receive alerts and recommendations to mitigate risks.
- Yield Estimation: Drones capture data on plant height, density, and other parameters. AI uses this data to estimate crop yields, aiding in harvest planning and market projections.
- Data-Driven Insights: AI processes the vast amount of data collected by drones and sensors, providing actionable insights for decision-making in real-time.
- Reduced Resource Usage [36]: AI-powered field monitoring enables precise resource allocation, reducing water waste, chemical usage, and other inputs.
- Environmental Sustainability: By targeting interventions and optimizing practices, AI-powered field monitoring supports sustainable farming practices and reduces the environmental impact of agriculture.
- Increased Efficiency: AI-powered drones and sensors cover large areas quickly and accurately, enhancing monitoring efficiency and reducing labor requirements.
- Rapid Response to Emergencies: In the event of weather-related disasters or disease outbreaks, AI-powered field monitoring provides rapid, data-driven insights for effective response and recovery.

The synergy of AI-powered drones and sensors is transforming field monitoring into a dynamic, data-driven process. By providing farmers with real-time, accurate insights, these technologies are helping to maximize yields, conserve resources, and promote environmentally responsible farming practices. As technology continues to advance, the potential for AI-powered field monitoring to drive agricultural innovation and sustainability is limitless.

7 Sustainability and Resource Optimization

The pursuit of sustainability has become a cornerstone of modern agriculture. As the global population grows and environmental challenges escalate, optimizing resources while minimizing negative impacts is essential. Through innovative technologies, data-driven insights, and sustainable practices, agriculture [37] is striving to balance productivity with environmental stewardship. Here's how the principles of sustainability and resource optimization are being integrated into the agricultural landscape:

- Precision Agriculture: Precision agriculture leverages technology such as GPS, sensors, and data analytics to optimize resource use. Farmers apply inputs like water, fertilizers, and pesticides with precision, reducing waste and minimizing their impact on ecosystems.
- Efficient Water Management [36]: Water scarcity is a pressing concern. Techniques like drip irrigation, soil moisture sensors, and AI-driven irrigation management ensure water is used efficiently, reducing overuse and runoff.
- Nutrient Management: Balancing nutrient inputs with crop needs is crucial. Soil testing, nutrient analysis, and AI-driven recommendations help farmers apply fertilizers judiciously, minimizing nutrient runoff into water bodies.
- Organic Farming: Organic farming practices avoid synthetic pesticides and fertilizers, focusing on natural alternatives. This approach promotes soil health, reduces chemical pollution, and enhances biodiversity.
- Cover Cropping and Crop Rotation: Cover crops prevent soil erosion, enhance soil structure, and capture carbon. Crop rotation disrupts pest and disease cycles and helps maintain soil fertility.
- Integrated Pest Management (IPM): IPM combines biological, cultural, and chemical control methods to manage pests sustainably. This reduces reliance on chemical pesticides and preserves beneficial insects.
- Renewable Energy Adoption: Solar panels, wind turbines, and other renewable energy sources power farms, reducing reliance on fossil fuels and decreasing greenhouse gas emissions.
- Data-Driven Decision-Making: Data analytics and AI-driven insights guide farmers to make informed decisions about resource allocation, planting strategies, and risk management.
- Erosion Prevention and Soil Conservation: Practices like contour plowing, terracing, and maintaining grassed waterways prevent soil erosion and protect valuable topsoil.
- Sustainable Livestock Management: Practices that prioritize animal welfare, rotational grazing, and efficient feed conversion help reduce the environmental impact of livestock farming.
- Agroforestry and Biodiversity Enhancement: Planting trees and incorporating diverse plant species on farms enhance biodiversity, provide habitat for beneficial organisms, and sequester carbon.
- Sustainable Supply Chains: Integrating sustainability throughout the supply chain ensures that farming practices are aligned with environmental and social considerations.
- Circular Economy Practices: Recycling agricultural waste, adopting regenerative practices, and reducing food waste contribute to a circular economy that maximizes resource use efficiency.
- Carbon Sequestration: Techniques like agroforestry, cover cropping, and no-till farming help capture carbon dioxide from the atmosphere and store it in soils and vegetation.

Sustainability and resource optimization are not only crucial for the well-being of our planet but also for ensuring food security in the face of changing climate patterns and population growth. As technology and agricultural practices continue to evolve, the integration of these principles will drive the future of farming, fostering resilience, productivity, and a healthier environment.

7.1 Precision Water Management: Efficient Irrigation with AI

Water is a precious resource, and in agriculture, its efficient use is vital for sustainability and productivity. Precision water management [13], empowered by artificial intelligence (AI), is transforming irrigation practices by providing farmers with data-driven insights and tools to optimize water usage [38]. Here's how AI is revolutionizing irrigation for more efficient and sustainable farming [39]:

- Real-Time Monitoring: AI-powered sensors and IoT devices continuously monitor soil moisture levels, weather forecasts, and crop needs in real-time. This data is used to make informed irrigation decisions.
- Data Analysis and Predictive Insights: AI algorithms analyze historical and real-time data to predict future irrigation requirements based on crop growth stages, weather conditions, and soil characteristics.
- Variable Rate Irrigation: AI determines different irrigation rates for various parts of a field, accounting for variations in soil composition and water needs. This ensures uniform plant growth and efficient water distribution.
- Automated Irrigation Scheduling [15, 26, 38]: AI-driven systems automatically adjust irrigation schedules based on the specific needs of each field, reducing water wastage and increasing efficiency.
- Soil Moisture-Based Irrigation: AI interprets data from soil moisture sensors to determine optimal irrigation timings and amounts, preventing over-irrigation and waterlogging.
- Weather Integration: AI considers weather forecasts, evaporation rates, and rainfall predictions to fine-tune irrigation schedules, avoiding unnecessary watering during rainy periods.
- Mobile Apps and Notifications: AI-powered apps provide farmers with real-time data and notifications, allowing them to remotely control and monitor irrigation systems.
- Drip and Micro-irrigation Optimization: AI optimizes drip and micro-irrigation systems by calculating precise water amounts needed for each plant, minimizing water waste.
- Water Use Efficiency: AI-powered irrigation reduces water wastage, conserving this valuable resource and promoting sustainable water use.
- Water Stress Avoidance: AI detects signs of water stress in plants through data analysis, prompting timely irrigation interventions to prevent crop losses.

- Crop-Specific Recommendations: AI considers the water needs of different crops, ensuring tailored irrigation strategies that promote healthy growth and yield.
- Environmental Conservation: By preventing over-irrigation, AI-driven systems reduce the risk of nutrient leaching and runoff, preserving water quality in surrounding ecosystems.
- Reduced Energy Costs: Efficient irrigation practices powered by AI also lead to energy savings, as less energy is required for water pumping and distribution.
- Increased Crop Resilience: Proper irrigation through AI-guided practices helps crops better withstand drought conditions and other environmental stresses.
- Food Security Enhancement: Precision water management ensures consistent and optimized irrigation, contributing to a stable food supply.

AI's integration into irrigation practices is a key step toward sustainable agriculture in a world challenged by water scarcity and climate change. By maximizing water efficiency, minimizing waste, and promoting optimal plant growth, AI-powered precision water management is playing a pivotal role in shaping a resilient and productive agricultural future.

7.2 Reducing Chemical Usage: Precision Pest and Weed Management

The sustainable management of pests and weeds is a critical aspect of modern agriculture. Precision pest and weed management, driven by advanced technologies and data-driven approaches, aims to minimize the use of chemical pesticides and herbicides while effectively safeguarding crops. Here's how precision management techniques are revolutionizing pest and weed control [40]:

- Integrated Pest Management (IPM): IPM combines various strategies, including biological control, cultural practices, and chemical interventions as a last resort. This holistic approach reduces chemical reliance and promotes natural predators and beneficial organisms.
- Remote Sensing and Imaging: Drones equipped with cameras and sensors provide real-time imaging of fields, enabling early detection of pest outbreaks and allowing for targeted interventions.
- AI-Powered Pest Detection: AI algorithms process images and data to identify pests and diseases. This enables timely and accurate pest identification for prompt action.
- Trapping and Monitoring Systems: AI-enhanced traps and monitoring systems detect pest activity. This data guides decisions on when and where to apply interventions.
- Beneficial Insect Promotion: Precision management practices foster populations of beneficial insects that prey on pests, reducing the need for chemical treatments.

- Resistant Crop Varieties: Plant breeding techniques are used to develop crop varieties that are naturally resistant to specific pests or diseases.
- Trap Crops and Border Planting: Specially planted trap crops or border plants attract pests away from main crops, reducing the need for chemical treatments.
- Herbicide-Resistant Weed Control: Precision genetic modification has led to the development of crops resistant to specific herbicides. This enables targeted weed control without affecting the main crop.
- Site-Specific Application: AI-driven systems analyze data to determine where and when chemical treatments are needed, minimizing overuse and reducing environmental impact.
- Autonomous Weed Detection and Removal: Robots equipped with cameras and AI algorithms can identify and remove weeds with precision, minimizing the need for herbicides.
- Biological Control Agents: The introduction of natural predators or pathogens that target specific pests can control their populations without resorting to chemicals.
- Cultural Practices: Practices like crop rotation, intercropping, and adjusting planting dates disrupt pest life cycles and reduce the need for chemical treatments.
- Disease Forecasting Models: AI-powered models analyze weather and other data to predict disease outbreaks. Early warnings enable timely interventions.
- Reduced Environmental Impact: Precision management techniques reduce chemical runoff, minimize soil and water contamination, and protect nontarget organisms.
- Improved Soil Health: Reduced chemical usage enhances soil microbial diversity and overall soil health, supporting plant growth.

Precision pest and weed management not only addresses environmental concerns but also contributes to healthier crops and safer food products. By embracing data-driven strategies, innovative technologies, and holistic approaches, modern agriculture is finding ways to strike a balance between protecting crops and preserving ecosystems.

8 Economic and Environmental Benefits

The AI Green Revolution is ushering in a transformative era for agriculture, with profound economic and environmental benefits that hold the promise of reshaping the future of farming. On the economic front, this revolution is boosting productivity by enabling farmers to make data-driven decisions. Through advanced technologies like machine learning and precision farming, it empowers farmers to optimize resource allocation, reduce input costs, and enhance labor efficiency. The result is higher crop yields, improved crop quality, and greater economic returns for farmers. In an era marked by increasing global demand for food and volatile market conditions, these economic benefits are vital for the sustainability of the agricultural sector [28].

In parallel, the AI Green Revolution is delivering substantial environmental advantages. Minimizing resource wastage through precise management of water, fertilizers, and pesticides contributes to resource efficiency and conservation. This translates into reduced water consumption, lower energy usage, and minimal chemical runoff, leading to a substantial decrease in the environmental footprint of agriculture. Moreover, the adoption of sustainable farming practices encouraged by AI technologies helps preserve soil health, conserve biodiversity, and mitigate the impact of climate change. These environmental benefits align with global efforts to combat environmental degradation and climate-related challenges, making AI-driven agriculture an essential component of environmental stewardship.

The AI Green Revolution also fosters resilience to climate change by offering adaptive solutions for farmers. With the capacity to monitor changing weather patterns and adapt crop management accordingly, it enhances farmers' ability to mitigate risks associated with climate volatility. Furthermore, it promotes climate-smart farming practices [41], including soil carbon sequestration, which plays a pivotal role in mitigating greenhouse gas emissions. Ultimately, the economic and environmental benefits of the AI Green Revolution are tightly intertwined, underpinning its pivotal role in reshaping agriculture's future by ensuring food security, economic stability, and environmental sustainability on a global scale.

The AI Green Revolution brings together economic and environmental benefits by promoting sustainable and efficient agricultural practices that not only increase farmers' profitability but also reduce the environmental footprint of agriculture. These combined benefits make AI-driven agriculture a crucial element in addressing global challenges related to food security, climate change, and resource scarcity shown in Table 2.

8.1 Increasing Productivity and Yield: AI's Contribution to Food Security

The global challenge of providing food security for a growing population has prompted the agricultural sector to harness the capabilities of artificial intelligence (AI) to enhance productivity and yield. AI's integration into various aspects of farming has the potential to revolutionize food production, ensuring a stable and sufficient food supply. Here's how AI is contributing to food security by increasing productivity and yield [42]:

- Precision Resource Management: AI analyzes data from sensors, satellites, and drones to optimize the application of water, fertilizers, and pesticides. This targeted approach maximizes resource efficiency and crop yields.
- Early Disease Detection: AI-powered systems process vast amounts of data to identify signs of disease or pest infestations early. This timely intervention prevents crop losses and ensures higher yields.

Table 2 Economic and environmental benefits

Benefits	Economic benefits	Environmental benefits
Increased productivity	Higher crop yields and quality	Reduced resource wastage (water, fertilizer, pesticides)
	Improved labor efficiency	Preservation of natural habitats
	Greater economic returns for farmers	Lower greenhouse gas emissions
Cost savings	Reduced input costs (water, fertilizers, pesticides)	Lower energy consumption
	Lower operational and labor costs	Conservation of biodiversity
Resource efficiency	Precise resource allocation (water, nutrients)	Preservation of soil health
	Reduced water consumption	Enhanced water quality and conservation [41, 43, 44]
	Minimal waste due to precision farming practices	Reduced land degradation
Environmental stewardship	Sustainable farming practices	Mitigation of climate change
	Reduced chemical runoff and pollution	Reduced use of synthetic chemicals
Resilience to climate change	Improved adaptability to changing weather patterns	Climate-smart farming practices
	Better risk management in volatile markets	Enhanced soil carbon sequestration
Economic diversification	Opportunities for ag-tech industry growth	Reduced dependency on traditional farming methods

- Climate-Resilient Crops: AI-assisted breeding techniques enable the development of crop varieties that are more resilient to changing climate conditions, enhancing yield stability.
- Crop Monitoring and Analysis: AI-driven drones and sensors monitor crops for stress, nutrient deficiencies, and other factors. This real-time data guides interventions that boost productivity.
- Predictive Analytics: AI algorithms use historical and current data to predict future crop yields, enabling better planning and decision-making.
- Automated Planting and Harvesting: Robotics and AI streamline planting and harvesting processes, reducing labor requirements and increasing efficiency.
- Soil Health Management: AI interprets soil data to recommend optimal planting strategies, contributing to healthy soil and improved crop yields.
- Data-Driven Decision-Making: AI processes vast datasets to provide farmers with actionable insights, guiding them in making informed choices for improved productivity.
- Efficient Pest Management: AI helps in identifying and targeting specific pests, minimizing damage and losses to crops.
- Sustainable Intensification: AI optimizes land use, enabling higher yields without expanding agricultural land, thus reducing pressure on natural ecosystems.

8.2 *Minimizing Environmental Impact: AI's Role in Sustainable Farming*

Sustainable farming practices are vital for preserving the environment and ensuring the long-term viability of agriculture. AI plays a crucial role in promoting such practices by enabling data-driven decision-making that minimizes environmental impact. Here's how AI contributes to sustainable farming by reducing its environmental footprint:

- Precision Application of Inputs: AI-driven systems apply inputs such as water, fertilizers, and pesticides precisely where and when needed, reducing waste and pollution.
- Reduced Chemical Usage: AI identifies optimal times for chemical applications, minimizing the amount needed and decreasing chemical runoff.
- Soil Health Monitoring: AI analyzes soil data to determine soil health, enabling practices that enhance soil fertility and minimize erosion.
- Biodiversity Enhancement: AI-driven practices promote biodiversity by supporting integrated pest management, pollinator habitats, and sustainable crop rotation.
- Efficient Water Management: AI optimizes irrigation practices, minimizing water usage and conserving this valuable resource.
- Carbon Sequestration: AI assists in implementing practices like cover cropping and agroforestry, which sequester carbon from the atmosphere.
- Waste Reduction: AI helps prevent overproduction and food waste by aligning production with demand, reducing the environmental impact of surplus food.
- Energy Efficiency: AI-driven automation and optimization reduce energy consumption in various farming processes.
- Adaptive Planning: AI adjusts farming strategies based on real-time environmental conditions, ensuring minimal impact on ecosystems.
- Conservation of Ecosystems: AI enables targeted interventions that protect surrounding ecosystems from contamination and disturbance.

The combination of AI's capabilities with sustainable farming practices holds the promise of not only increasing food security but also safeguarding the environment for future generations. By embracing AI-driven innovation, agriculture [45] can strike a harmonious balance between productivity and environmental preservation.

8.3 *Challenges and Ethical Considerations*

While the integration of artificial intelligence (AI) in agriculture holds tremendous promise for improving productivity and sustainability, it also presents a range of challenges and ethical considerations. Addressing these issues is crucial to ensure that AI-driven advancements benefit both farmers and the environment while upholding ethical standards. Here are some key challenges and ethical considerations in agricultural AI:

Challenges:

- Data Privacy and Ownership: The collection of extensive data raises concerns about who owns and controls the data generated by AI systems, as well as how it is used and shared.
- Data Quality and Bias: AI models rely on accurate and unbiased data. If the data used for training is incomplete or biased, it can lead to inaccurate predictions and decisions.
- Access to Technology: Access to AI-driven technology may be limited for small-scale and resource-constrained farmers, exacerbating inequalities in the agricultural sector.
- Technical Infrastructure: Adequate infrastructure, such as high-speed Internet and reliable electricity, is necessary for the effective deployment of AI in remote agricultural areas.
- Skilled Workforce: The adoption of AI requires a workforce skilled in data analysis, programming, and AI implementation, which may be a challenge in some regions.
- Cost of Implementation: The initial investment required for AI technology and infrastructure can be a barrier, particularly for small-scale farmers.
- Algorithm Complexity: Understanding and interpreting AI algorithms can be challenging, leading to potential mistrust or lack of transparency.
- Risk of Overreliance: Overreliance on AI systems without farmer expertise could lead to disengagement from the decision-making process.
- Environmental Impact: The energy consumption associated with AI infrastructure could counteract the sustainability goals of precision agriculture.

Ethical Considerations:

- Data Privacy and Security: The collection and storage of sensitive agricultural data raise concerns about privacy and the potential for data breaches.
- Transparency and Accountability: AI algorithms should be transparent, explainable, and accountable, enabling farmers to understand decisions made by AI systems.
- Equitable Access: Ethical considerations require that AI-driven benefits are accessible to all farmers, irrespective of their size or location.
- Fair Compensation: Farmers providing data for training AI models should be fairly compensated for their contributions.
- Environmental Impact Assessment: The environmental impact of AI deployment should be carefully evaluated to ensure that sustainability goals are not compromised.
- Avoiding Monopolies: Ensuring competition and preventing monopolies in the AI agricultural technology sector is crucial for a diverse and accessible market.
- Informed Consent: Farmers should be adequately informed about the implications of using AI technology and provide informed consent.

- Ethical Treatment of Animals: AI systems used in livestock management should prioritize the welfare and ethical treatment of animals.
- Long-Term Consequences: The long-term effects of AI on agricultural practices, ecosystems, and rural communities should be considered.

Balancing the potential benefits of AI in agriculture with these challenges and ethical considerations is essential to ensure that technological advancements are deployed responsibly and ethically. A collaborative effort involving governments, researchers, farmers, and industry stakeholders can guide the development and adoption of AI solutions that contribute positively to agriculture while upholding ethical standards.

9 Data Privacy and Security in Agricultural Systems

As agriculture becomes increasingly data-driven, concerns about data privacy and security are paramount. The use of technology and AI in farming generates substantial amounts of sensitive data, including crop information, location data, and farmer practices. Safeguarding this data is essential to maintain trust among farmers and ensure the integrity of agricultural systems. Here's how data privacy and security are managed in agricultural AI:

- Secure Data Storage and Transmission: Data should be stored in encrypted formats and transmitted securely to prevent unauthorized access.
- Access Control and Authentication: Implement strict access controls, requiring proper authentication for anyone accessing the data.
- Anonymization and Aggregation: Personal and sensitive data can be anonymized or aggregated to protect individual privacy while still providing valuable insights.
- Consent and Ownership: Farmers should have clear knowledge of how their data is used, and they should provide informed consent for data collection and utilization.
- Data Sharing Agreements: When sharing data with third parties, clear agreements should be established to define how the data will be used and protected.
- Regular Auditing: Regular audits of data storage and processing systems can identify vulnerabilities and ensure compliance with data protection regulations.
- Compliance with Regulations: Adherence to data protection regulations, such as GDPR, CCPA, and regional agricultural data laws, is crucial.
- Educating Farmers: Farmers should be educated about data privacy risks and best practices to ensure responsible data management.

Balancing AI with Human Expertise: Ensuring Equitable Access to Technology

While AI holds great potential for transforming agriculture, ensuring equitable access to these technologies is essential to avoid exacerbating existing disparities. Balancing AI with human expertise acknowledges the importance of local knowledge and fosters inclusivity in technological adoption. Here's how to ensure equitable access to AI technology in agriculture:

- Training and Capacity Building: Provide training and support for farmers to understand and effectively use AI tools, regardless of their scale of operation.
- Local Context Integration: Develop AI solutions that consider local practices, languages, and cultural factors to ensure relevance and adoption.
- Accessible User Interfaces: AI systems should have intuitive user interfaces that accommodate users with varying levels of technological literacy.
- Collaboration and Partnerships: Collaborate with local organizations, NGOs, and governments to facilitate technology adoption and provide resources.
- Affordable Solutions: Develop cost-effective AI solutions that cater to the financial constraints of small-scale farmers.
- Open Data and Knowledge Sharing: Foster a culture of open data and knowledge sharing, promoting equitable access to insights and solutions.
- Rural Connectivity: Improve Internet connectivity in rural areas to ensure that farmers can access and benefit from AI technologies.
- Local Innovation Hubs: Establish local innovation centers that provide training, support, and resources for adopting AI in agriculture.
- Policy Considerations: Governments can play a role in ensuring that AI technology is accessible to all by implementing policies that encourage inclusivity and prevent monopolies.

By prioritizing data privacy, security, and equitable access, the agricultural sector can harness the benefits of AI while respecting individual rights and promoting fair and inclusive technological advancement. This approach ensures that AI enhances farming practices and supports the livelihoods of all farmers, regardless of their scale or location.

9.1 Empowering Farmers and Communities

The integration of artificial intelligence (AI) in agriculture holds the potential to empower farmers and rural communities by providing them with access to valuable insights, tools, and resources. This technological transformation can enhance agricultural practices, improve livelihoods, and contribute to sustainable rural development. Here's how AI is empowering farmers and communities:

- Access to Information: AI-powered platforms offer real-time information on weather forecasts, market prices, and best farming practices, enabling informed decision-making.
- Knowledge Sharing: AI facilitates the exchange of knowledge and expertise among farmers, both locally and globally, fostering a community of learning.
- Crop Management Guidance: AI-driven advice on planting, irrigation, and disease management helps farmers optimize their crop yields and minimize losses.

- Financial Management: AI tools assist in financial planning, budgeting, and optimizing input costs, contributing to better financial outcomes for farmers.
- Market Access: AI-based market insights and price predictions help farmers make strategic decisions on when and where to sell their produce.
- Smallholder Empowerment: AI levels the playing field for small-scale farmers by providing them with access to advanced tools and resources typically available to larger operations.
- Digital Payments and Transactions: AI-powered payment platforms enable secure and convenient transactions, reducing the risks associated with handling cash.
- Resource Optimization: AI-driven recommendations for water, fertilizer, and pesticide use enhance resource efficiency and reduce costs.
- Skill Development: Farmers gain digital literacy skills as they interact with AI tools, contributing to their personal growth and adaptability.
- Climate Resilience: AI-generated climate information helps farmers adapt to changing weather patterns and make decisions that enhance resilience.
- Inclusive Innovation: Customized AI solutions can address specific challenges faced by marginalized and underserved farming communities.
- Women's Empowerment: AI-enabled tools can empower women in agriculture by providing them with access to information and decision-making tools.
- Conservation Agriculture: AI supports sustainable farming practices that protect soil health, water resources, and biodiversity.
- Entrepreneurship Opportunities: AI-powered platforms can help farmers explore value-added opportunities, such as processing and niche markets.
- Community Development: AI-driven improvements in agriculture can lead to enhanced food security, increased income, and overall improved quality of life in rural communities.

The empowerment of farmers and rural communities through AI requires a collaborative effort from governments, organizations, researchers, and technology providers. By tailoring AI solutions to the needs and contexts of farmers, the potential benefits of this technology can be harnessed to drive positive change and sustainable development in agricultural communities around the world.

9.2 Digital Divide: Bridging the Gap for Small-Scale Farmers

The digital divide, characterized by unequal access to technology and information, poses significant challenges for small-scale farmers, often limiting their ability to benefit from technological advancements like artificial intelligence (AI). Bridging this gap is essential to ensure that small-scale farmers can also harness the potential of AI for improved agricultural practices and livelihoods. Here's how the digital divide can be addressed for the benefit of small-scale farmers:

- Accessible Technology: Develop user-friendly AI tools and applications that are tailored to the needs and capabilities of small-scale farmers.
- Affordable Solutions: Create cost-effective AI solutions that cater to the financial constraints of small-scale farming operations.
- Local Language Support: Offer AI platforms in local languages to enhance usability for farmers with varying levels of literacy.
- Training and Capacity Building: Provide training and support to small-scale farmers, helping them understand and effectively utilize AI tools.
- Mobile-Friendly Interfaces: Develop AI applications that can be accessed and used on basic smartphones, which are more prevalent in rural areas.
- Extension Services: Leverage AI to provide virtual extension services that offer expert advice and guidance to small-scale farmers.
- Community Hubs: Establish community centers equipped with AI-enabled tools where farmers can access information and receive training.
- Public-Private Partnerships: Collaborations between governments, NGOs, and private sector entities can provide resources and support to bridge the digital gap.
- Internet Connectivity: Invest in improving Internet connectivity in rural areas, enabling small-scale farmers to access online resources.
- Local Innovation Ecosystems: Foster innovation hubs that offer training, support, and resources for adopting AI in agriculture.
- Tailored Solutions: Develop AI applications that address specific challenges faced by small-scale farmers, such as pest control, irrigation, and market access.
- Awareness Campaigns: Raise awareness among small-scale farmers about the benefits of AI and how it can enhance their productivity and income.
- Financial Inclusion: Integrate digital payment solutions that enable small-scale farmers to conduct transactions securely and conveniently.
- Farmer-Centric Design: Involve small-scale farmers in the design and development of AI solutions to ensure their relevance and usability.
- Government Support: Governments can play a crucial role by implementing policies that promote technology access and adoption in rural areas.

Addressing the digital divide for small-scale farmers is not only essential for their prosperity but also for ensuring global food security and sustainable agriculture. By providing these farmers with the tools and knowledge to embrace AI, we can empower them to overcome challenges, make informed decisions, and contribute to a more equitable and resilient agricultural sector.

9.3 Rural Revitalization: AI's Potential to Strengthen Agricultural Communities

The integration of artificial intelligence (AI) in agriculture has the potential to play a pivotal role in revitalizing rural communities and driving sustainable development. By enhancing agricultural practices, improving livelihoods, and fostering innovation, AI can contribute to the growth and resilience of agricultural communities. Here's how AI can strengthen rural areas and promote rural revitalization:

- Economic Diversification: AI-driven innovations can open doors to new revenue streams and value-added activities, reducing dependence solely on traditional farming.
- Enhanced Agricultural Productivity: AI-powered precision agriculture increases crop yields and efficiency, leading to improved income for farmers.
- Access to Information: AI provides farmers with real-time data on weather, market prices, and best practices, empowering them to make informed decisions.
- Skill Development: The adoption of AI encourages skill development among rural residents, fostering a tech-savvy workforce.
- Entrepreneurship Opportunities: AI can help identify niche markets and value-added opportunities, encouraging the development of agri-entrepreneurs.
- Youth Engagement: AI-driven technologies can attract the younger generation to agriculture by offering innovative and tech-driven approaches.
- Sustainable Practices: AI promotes sustainable agriculture, preserving natural resources and ensuring long-term economic viability.
- Digital Infrastructure Development: The deployment of AI necessitates improved Internet connectivity, benefiting rural areas with enhanced communication and access to services.
- Smart Infrastructure: AI can facilitate the development of smart rural infrastructure, including smart irrigation and energy-efficient systems.
- Market Access: AI-powered market insights enable farmers to connect with larger markets, improving their bargaining power.
- Community Engagement: AI platforms can facilitate communication and collaboration among farmers, fostering a sense of community and shared knowledge.
- Agricultural Innovation Hubs: Establishing AI-driven innovation centers in rural areas provides resources, training, and support for tech adoption.
- Climate Resilience: AI helps farmers adapt to changing climate conditions, ensuring the sustainability of agricultural practices.
- Food Security: AI's contributions to improved productivity and efficient resource management enhance local food security.
- Sustainable Tourism: AI-powered agritourism initiatives can attract visitors interested in experiencing advanced and sustainable agricultural practices.
- Government Support: Government policies and incentives can promote the adoption of AI in agriculture and support rural development initiatives.

The revitalization of rural areas through AI integration requires a holistic approach that involves collaboration between governments, private sectors, local communities, and technology providers. By leveraging the transformative power of AI, rural communities can not only enhance their agricultural productivity but also foster innovation, entrepreneurship, and long-term sustainability, ultimately contributing to the holistic development of the entire region.

10 A Vision for the Future: AI's Long-Term Impact on Agriculture

The path ahead in agriculture is illuminated by the promising potential of artificial intelligence (AI) to revolutionize farming practices, enhance sustainability, and ensure food security. AI-enabled agriculture represents a journey toward more efficient [44], resilient, and responsive farming systems. By embracing AI, we pave the way for a future where data-driven insights, precision resource management, and innovative technologies harmonize to create a new era of agricultural advancement [41].

- Data-Driven Decision-Making: AI empowers farmers with real-time data, enabling them to make informed decisions that optimize crop yields, reduce waste, and increase profitability.
- Precision Resource Management: Through AI-driven insights, the allocation of water, fertilizers, and pesticides becomes precise, conserving resources and minimizing environmental impact.
- Climate Resilience: AI equips farmers to adapt to changing climate conditions, making agriculture more resilient against unpredictable weather patterns.
- Sustainable Practices: AI fosters the adoption of sustainable practices that safeguard soil health, water resources, and biodiversity.
- Customized Solutions: Tailored AI applications address specific challenges faced by diverse farming communities, fostering inclusivity and equitable growth.
- Collaboration and Knowledge Sharing: AI platforms facilitate collaboration, enabling farmers to exchange experiences, insights, and best practices on a global scale.
- Empowered Rural Communities: AI revitalizes rural areas by offering new economic opportunities, skill development, and access to modern technologies.
- Ethical and Responsible Deployment: The ethical integration of AI prioritizes data privacy, transparency, and accountability to ensure a responsible agricultural transformation.

Looking ahead, the long-term impact of AI on agriculture holds the promise of a more resilient, productive, and sustainable global food system. This vision encompasses not only increased yields and optimized resource use but also a harmonious relationship between human ingenuity and technological innovation. As AI continues to evolve, agriculture will embrace a future where the well-being of farmers, communities, and the planet is paramount.

- Food Security: AI's potential to enhance productivity and improve resource management contributes to global food security by ensuring a consistent and abundant supply of quality produce.
- Ecosystem Preservation: AI-driven sustainable practices protect ecosystems, conserve biodiversity, and mitigate the impact of agriculture on the environment.
- Empowered Farmers: AI empowers farmers of all scales with tools and knowledge to navigate challenges, make informed decisions, and succeed in a rapidly changing agricultural landscape.

- Technological Synergy: AI synergizes with other cutting-edge technologies such as robotics, nanotechnology, and biotechnology to create holistic solutions for complex agricultural problems.
- Data-Driven Innovation: AI's ability to analyze vast datasets accelerates the discovery of innovative solutions, unlocking new pathways for agricultural advancement.
- Inclusive Growth: AI ensures that the benefits of agricultural progress reach all stakeholders, from smallholders to large agribusinesses, creating a more equitable industry.

10.1 The Synergy of Human Innovation and AI Advancements

The convergence of human innovation and AI advancements marks a pivotal moment in agricultural history. It's a synergistic relationship where human creativity, experience, and intuition combine with AI's computational power and predictive capabilities. This synergy elevates agriculture to new heights of efficiency, sustainability, and prosperity, ensuring that the future of farming is not only shaped by technology but guided by human values and aspirations. As we navigate this evolving landscape, the vision of AI-enabled agriculture promises a future where humanity and technology work hand in hand to nourish the world sustainably.

AI's Transformation of Agriculture

Artificial intelligence (AI) and agriculture are ushering in a profound transformation that redefines how we produce food, manage resources, and ensure sustainability. From precision resource management to data-driven decision-making, AI is revolutionizing the agricultural landscape, offering solutions to some of the most pressing challenges facing the industry. This transformation is not just about technology; it's about enhancing the well-being of farmers, fostering rural development, and safeguarding our environment for future generations.

AI's integration in agriculture empowers farmers with actionable insights, optimizes resource allocation, and enhances productivity. It bridges gaps, breaks down barriers, and democratizes access to information and innovation. Moreover, AI's positive impact extends beyond fields and farms, touching every facet of the global food supply chain, from production to distribution and consumption. As we embrace this transformation, we must also recognize the importance of ethical considerations, data privacy, and the equitable distribution of benefits. Ensuring that AI-enabled agriculture is responsible, inclusive, and sustainable will be essential in reaping the full benefits of technological advancement.

Collaboration between farmers, researchers, policymakers, and technology developers will play a pivotal role. By harnessing the power of AI and combining it with human expertise, we can propel agriculture into an era of unprecedented efficiency, resilience, and innovation. This fusion of human ingenuity and AI-driven capabilities will shape the future of agriculture, creating a world where food security, environmental sustainability, and economic prosperity coexist harmoniously.

10.2 A Greener and More Efficient Agricultural Future with AI

The vision of a greener and more efficient agricultural future powered by artificial intelligence (AI) is not just a distant aspiration; it's a tangible reality that is reshaping the way we cultivate, manage resources, and nourish the planet. The harmonious integration of AI and agriculture holds the promise of sustainable food production, environmental conservation, and enhanced livelihoods. As we draw this exploration to a close, it's evident that AI's transformative influence on agriculture is steering us toward a future that is both prosperous and ecologically responsible.

Through AI-enabled precision and data-driven insights, agriculture is becoming a greener endeavor, minimizing waste, conserving resources, and reducing the environmental footprint. AI's capabilities are unlocking new levels of efficiency in irrigation, pest control, and energy usage, contributing to the overarching goal of sustainable farming practices. This transformation is not just limited to individual farms; it extends to entire ecosystems, preserving biodiversity, mitigating climate change, and promoting responsible land stewardship. The efficiency gains enabled by AI extend beyond environmental considerations. They embrace economic sustainability by improving crop yields, optimizing resource allocation, and ensuring resilient farming systems. These advancements translate to improved livelihoods for farmers, strengthened rural communities, and enhanced global food security.

As we move forward, collaboration between agricultural stakeholders, technology developers, policymakers, and local communities will be essential. By fostering an ecosystem of innovation, knowledge-sharing, and responsible adoption, we can collectively steer the trajectory of AI's impact on agriculture toward a future that is ecologically harmonious and economically vibrant. In embracing the potential of AI to cultivate a greener and more efficient agricultural future, we are not just redefining the way we grow food; we are reimagining our relationship with the planet and the generations that will inherit it. As we harness the power of AI, we commit to leaving a legacy of sustainable practices, resilient ecosystems, and nourished communities. This is not merely a technological advancement; it's a shared commitment to shaping a future where agriculture and nature thrive in unison.

11 Conclusion

The AI Green Revolution is undeniably reshaping the future of agriculture, offering a promising vision of sustainable, efficient, and productive farming practices. By harnessing the power of artificial intelligence, farmers are gaining unprecedented insights into their operations, enabling data-driven decisions that optimize resource use, reduce waste, and increase crop yields. AI-driven tools, such as predictive analytics and autonomous machinery, are enhancing productivity and mitigating the impact of climate change by allowing farmers to adapt to shifting weather patterns and environmental conditions. Additionally, AI-powered precision agriculture is contributing to greater food security by ensuring that resources are used more efficiently and effectively, ultimately allowing for the production of more food with fewer inputs. As the world faces the challenges of feeding a growing global population while confronting climate-related disruptions, the AI Green Revolution represents a beacon of hope in addressing these complex agricultural and environmental issues.

However, it's important to acknowledge that the AI Green Revolution is not without its challenges and ethical considerations. Issues related to data privacy, accessibility, and the digital divide must be addressed to ensure that the benefits of AI in agriculture are accessible to all farmers, regardless of their location or resources. Additionally, responsible AI development and deployment are essential to prevent unintended consequences and ensure the long-term sustainability of agriculture. In conclusion, the AI Green Revolution holds immense promise for the future of agriculture, offering the potential to revolutionize how we produce food while addressing critical global challenges. To fully realize this potential, it is crucial to strike a balance between technological advancement, ethical considerations, and inclusive access to AI-driven solutions.

References

1. F.J.J. Joseph, A review of IoT implementations in environment and agriculture. JARES **6**(1&2), 1–5 (2019)
2. T. Wheeler, J. von Braun, Climate change impacts on global food security. Science **341**(80), 508–513 (2013). https://doi.org/10.1126/science.1239402
3. C. Prabakar, K.S. Devi, S. Selvam, Labour scarcity—Its immensity and impact on agriculture. Agric. Econ. Res. **24**, 373–380 (2011)
4. T. Rajesh, K.T. Kumar, M.M. Balan, M.S. Sidhartha, E-agriculture based on AI and IOT. IJRASET **7**(II), 343–347 (2019)
5. S. David, R.S. Anand, M. Sagayam, Enhancing AI based evaluation for smart cultivation and crop testing using agro-datasets. JAIS **2**(1), 149–167 (2020)
6. M.S. Patange, P.G. Student, M.Z. Farooqui, IOT assisted farming. RJSE **3**(7), 736–739 (2019)
7. R. Divya, R. Chinnaiyan, Reliable AI-based smart sensors for managing irrigation resources in agriculture—A review, in *ICCNCT*, (Springer, Singapore, 2019), pp. 263–274
8. G. Adamides, N. Kalatzis, A. Stylianou, N. Marianos, F. Chatzipapadopoulos, M. Giannakopoulou, G. Papadavid, V. Vassiliou, D. Neocleous, Smart farming techniques for climate change adaptation in Cyprus. Atmos. **11**(6), 557 (2020). https://doi.org/10.3390/atmos11060557
9. S.S. Mane, M.S. Mane, U.S. Kadam, S.T. Patil, Design and development of cost effective real time soil moisture based automatic irrigation system with GSM. IRJET **6**(9), 1744–1751 (2019)
10. O. Elijah, T.A. Rahman, I. Orikumhi, C.Y. Leow, M.N. Hindia, An overview of internet of things (IoT) and data analytics in agriculture: Benefits and challenges. IEEE Internet Things **J5**, 3758–3773 (2018). https://doi.org/10.1109/JIOT.2018.2844296
11. A. Agrawal, M.K. Arya, Optimize and customize irrigation mechanism and crops prevention from pests. JPP **5**, 17–19 (2018)
12. S. Kumar, G. Chowdhary, V. Udutalapally, D. Das, S.P. Mohanty, *gCrop: Internet-of-Leaf-Things (IoLT) for Monitoring of the Growth of Crops in Smart Agriculture* (IEEE, 2019)

13. H. Bag, D.K. Chaurasiya, Precision farming: The bright future for Indian agriculture. BRT **2**(5), 362–364 (2020)
14. A.B. McBratney, B.M. Whelan, *The Null Hypothesis of Precision Agriculture* (SheffieldAcademic Press, Sheffield, 1999)
15. S. Kumar, Artificial intelligence in Indian irrigation. IJSRCSEIT **5**(5), 149–167 (2019)
16. S. Adebayo, E.O. Ogunti, F.K. Akingbade, O. Oladimeji, A review of decision support system using mobile applications in the provision of day to day information about farm status for improved crop yield. Per. Eng. Nat. Sci **6**, 89–99 (2018)
17. S. Singh et al., A systematic study on Big Data in IOT and agriculture. JTGRS **21**(6), 566–572 (2019)
18. P.M. Gupta, M. Salpekar, P.K. Tejan, *Agricultural Practices Improvement Using IoT Enabled SMART Sensors* (IEEE, 2018)
19. D. Sivaganesan, Design and development AI-enabled edge computing for intelligent-IOT applications. JTCSST **1**(2), 84–94 (2019)
20. V. Dharmaraj, C. Vijayanand, Artificial Intelligence (AI) in Agriculture. IJCMAS **7**(12), 2122–2128 (2018)
21. M.A. Supreetha, M.R. Mundada, J.N. Pooja, Design of a smart water-saving irrigation system for agriculture based on a wireless sensor network for better crop yield, in *ICCCE 2018. ICCCE 2018. Lecture Notes in Electrical Engineering*, ed. by A. Kumar, S. Mozar, (2019), p. 93–104. https://doi.org/10.1007/978-981-13-0212-1_11
22. T. Duckett, S. Pearson, S. Blackmore, B. Grieve, W.-H. Chen, G. Cielniak, J. Cleaversmith, J. Dai, S. Davis, C. Fox, P. From, I. Georgilas, R. Gill, I. Gould, M. Hanheide, A. Hunter, F. Iida, L. Mihalyova, S. Nefti-Meziani, G. Neumann, P. Paoletti, T. Pridmore, D. Ross, M. Smith, M. Stoelen, M. Swainson, S. Wane, P. Wilson, I. Wright, G.-Z. Yang, Agricultural robotics: The future of robotic agriculture. arXiv:1806.06762v2 (2018)
23. A. Goap, D. Sharma, A.K. Shukla, C. Rama Krishna, An IoT based smart irrigation management system using machine learning and open source technologies. Comput. Electron. Agric **155**, 41–49 (2018). https://doi.org/10.1016/j.compag.2018.09.040
24. V. Puranik, A. Ranjan, A. Kumari, *Automation in Agriculture and IoT* (IEEE, 2019)
25. B.V. Raju, An IOT based low cost agriculture field monitoring system. JASC **VI**(IV), 128–136 (2019)
26. S. Kumar, P. Raja, G. Bhargavi, *A Comparative Study on Modern Smart Irrigation System and Monitoring the Field by Using IOT* (IEEE, 2018)
27. S.M. Pedersen, S. Fountas, S. Blackmore, Agricultural robots-applications and economic perspectives: Chapter 21, in *Service Robot Applications*, (IntechOpen, London, 2008), pp. 369–382
28. K.R. Gowri, Greenhouse monitoring and scheming based IoT technology. EPRA IJRD **4**(4), 316–321 (2019)
29. A.V. Sujith, K.C. Sekhar, Automated agriculture as a service using IoT. IJARCSSE **7**(5), 925–930 (2017)
30. V. Lohchab, M. Kumar, G. Suryan, V. Gautam, R.K. Das, *A Review of IoT Based Smart Farm Monitoring* (IEEE, 2018)
31. D. Kumari, R. Pandita, M. Mittal, An agricultural perspective in the Internet of Things. IJCSE **6**(5), 107–110 (2018)
32. N.P. Bhatta, N. Thangadurai, Utilization of IOT and AI for agriculture applications. IJEAT **8**(5), 2731–2735 (2019)
33. K. Nalinaksh, V. Rishiwal, M. Ganzha, M. Paprzycki, *An Internet of Things (IoT) Solution Framework for Agriculture in India and Other Third World Countries* (IEEE, 2018)
34. N.K. Verma, A. Usman, *Internet of Things (IoT): A Relief for Indian Farmers* (IEEE, 2016)
35. D.R. Vincent, N. Deepa, D. Elavarasan, K. Srinivasan, S.H. Chauhdary, C. Iwendi, Sensors driven AI-based agriculture recommendation model for assessing land suitability. Sensors **19**(17), 3667 (2019)

36. S.K. Abd-Elmabod, H. Mansour, A.A.E.F. Hussein, E.S. Mohamed, Z. Zhang, M. AnayaRomero, A. Jordán, Influence of irrigation water quantity on the land capability classification. Plant Arch **2**, 2253–2561 (2019)
37. P.S. Kambar, A study on the role of E-technology to take over agriculture distress in India. IJCAS **8**(12(A)), 335–341 (2018)
38. L.S. Pereira, T. Oweis, A. Zairi, Irrigation management under water scarcity. Agric. Water Manag **57**, 175–206 (2002). https://doi.org/10.1016/S0378-3774(02)00075-6
39. A.T. Ahmed, F. El Gohary, V.A. Tzanakakis, A.N. Angelakis, Egyptian and Greek water cultures and hydro-technologies in ancient times. Sustainability **12**(22), 9760 (2020)
40. P. Silakari, P. Silakari, L. Bopche, A. Gupta, *Smart Ultrasonic Insects & Pets Repeller for Farms & Inventories Purpose* (IEEE, 2018)
41. M.K. Abdel-Fattah, S.K. Abd-Elmabod, A.A. Aldosari, A.S. Elrys, E.S. Mohamed, Multivariate analysis for assessing irrigation water quality: A case study of the Bahr Mouise Canal. Eastern Nile Delta. Water **12**(9), 2537 (2020)
42. D. Sinwar, V.S. Dhaka, M.K. Sharma, G. Rani, AI-based yield prediction and smart irrigation, in *Internet of Things and Analytics for Agriculture*, vol. 2, (Springer, Singapore, 2020), pp. 155–180
43. M. Abu-hashim, E. Mohamed, A.-E. Belal, Identification of potential soil water retention using hydric numerical model at arid regions by land-use changes. Int. Soil Water Conserv. Res **3**(4), 305–315 (2015)
44. M.K. Abdel-Fattah, E.S. Mohamed, E.M. Wagdi, S.A. Shahin, A.A. Aldosari, R. Lasaponara, M.A. Alnaimy, Quantitative evaluation of soil quality using Principal Component Analysis: The case study of El-Fayoum depression Egypt. Sustainability **13**(4), 1824 (2021)
45. B. Ragavi, L. Pavithra, P. Sandhiyadevi, G.K. Mohanapriya, S. Harikirubha, *Smart Agriculture with AI Sensor by Using Agrobot* (IEEE, 2020)

Cloud Computing for Smart Farming: Applications, Challenges, and Solutions

Justin Rajasekaran, Saleem Raja Abdul Samad, and Pradeepa Ganesan

1 Introduction

Farming has been vital to humans for millennia. Recent years have witnessed the implementation of numerous advanced technologies because of rapid population growth. The term "smart farming" refers to the application of cutting-edge communication and technology in the agricultural sector [1]. Modern-day smart farming drives worldwide demand. It involves using contemporary technology and sensors in agriculture. Agricultural land is typically monitored by sensors and managed by software. Using cutting-edge technology, yield of the agriculture land increases with less labor intervention. It also agricultural practices by minimizing the consumption of resources such as fertilizers and water. Using varied farming methods can make farming more environmentally friendly and create higher-quality products [2]. In smart farming, cloud computing helps farmers in numerous ways. Globally, there is a great deal of data pertaining to weather, crops, and soil that needs to be gathered and analyzed. The information regarding farming must be gathered, after which it must be archived and analyzed. Thus, cloud computing is an excellent technology that can be utilized to gather, analyze, and store agricultural data. The data kept in the cloud comprises soil, crop, and ecommerce data that have been gathered from diverse sources. This data can be used to track the meteorological conditions on the farm. Only authorized users have access to this information [3] as shown in Fig. 1. Cloud computing offers a diverse range of services to support the implementation of smart farming practices. Table 1 shows the growth of IoT-based adoption in agriculture sector from year 2000 to 2016 and forecasts of year

J. Rajasekaran (✉) · S. R. A. Samad · P. Ganesan
Information Technology Department, College of Computing and Information Sciences, University of Technology and Applied Sciences-Shinas, Shinas, Oman
e-mail: Justin.rajasekaran@utas.edu.om; https://doi.org/10.1007/979-0-12345-123-113

S. Balasubaramanian et al. (eds.), *Intelligent Robots and Drones for Precision Agriculture*, Signals and Communication Technology,
https://doi.org/10.1007/978-3-031-51195-0_20

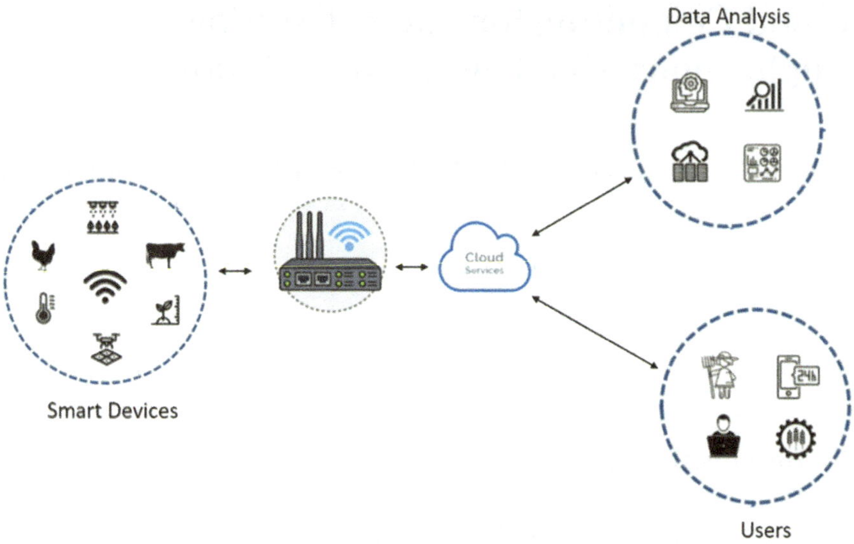

Fig. 1 End-to-end interaction between various stakeholders

Table 1 The growth of IoT-based adoption in agriculture sector [5]

Year	Data analysis
2000	525 million farms connected to IoT
2016	540 million farms till date are connected to IoT
2035	780 million farms would be connected to IoT
2050	2billion farms are likely to be connected to IoT

2035–2050 [4]. This chapter discusses the various roles of cloud computing in smart farming, as well as the necessary infrastructure and challenges.

2 Smart Farming and Precision Agriculture (PA)

Smart farming is a broader concept that incorporates the improvement of various aspects of farming through the use of technology and data-driven solutions [6]. The primary objective of smart farming is to improve the efficiency, sustainability, and data-driven nature of agricultural operations [7]. It entails incorporating Big Data analytics, cloud computing, IoT sensors, artificial intelligence, and cloud computing into a range of farming operations. It encompasses sustainable practices, livestock management, supply chain optimization, and precision agriculture. Precision agriculture is a subset of smart farming [8]. It is primarily concerned with increasing crop yields and optimizing resource utilization (including water, fertilizers, and pesticides) through the application of precision techniques driven by data and technology. The purpose of precision agriculture is to maximize crop yield and minimize waste [9]. It incorporates soil type, geography, weather, plant growth, and yield data while managing crops. In order to gather real-time data, sensors are installed in the field to detect the soil's moisture content, temperature, and surrounding air. Upon the completion of data collection, the utilization of analytical software enables farmers to access vital insights pertaining to crop rotation, as well as determine the most advantageous periods for planting and harvesting.

3 Cloud Computing

Cloud computing makes huge data storage affordable and scalable. In accordance with the National Institute for Standards and Technology [10], "Cloud Computing is based on pay-per-use services for enabling convenient, on-demand network access to a shared pool of configurable computing resources such as servers, networks, and services that can be rapidly provisioned and released with minimal management effort or service provider interaction." Cloud computing services can be categorized into the three categories as shown in Fig. 2.

Cloud users can select the best appropriate model depending on their individual requirements from among several models that provide varying degrees of abstraction and administration.

3.1 Infrastructure as a Service (IaaS)

Virtualized computing resources are made available to users via the Internet by IaaS. It provides standard IT services like virtual machines (VMs), storage, networking, and even additional features like firewalls and load balancers [11].

IaaS gives users greater control over their infrastructure by letting them handle data, operating systems, and apps. Hosting web applications, operating virtual servers, and cloud data storage are the common use cases.

Fig. 2 Cloud services

3.2 Platform as a Service (PaaS)

PaaS abstracts infrastructure management further by providing a development and deployment platform in addition to underlying infrastructure components. It offers tools, frameworks, and runtime environments for building, testing, and deploying applications [12].

PaaS is great for developers and businesses that focus on making and deploying apps. It makes development easier by getting rid of the need to manage infrastructure details. DevOps, web, and mobile app development are common use cases.

3.3 Software as a Service (SaaS)

SaaS delivers fully functional software applications over the Internet. Users access and use software applications hosted in the cloud without needing to worry about underlying infrastructure, maintenance, or updates [13].

SaaS is beneficial for end-users and businesses looking to access software solutions without the complexities of software installation and management. Common SaaS applications include email services, office productivity suites (Microsoft 365, Google Workspace), customer relationship management (CRM) tools, and collaboration platforms [14].

Cloud computing offers unparalleled scalability, allowing businesses to easily scale up or down their resources based on demand. Cloud computing enables users to access their applications and data from anywhere with an Internet connection and on various devices. By migrating to the cloud, organizations can significantly reduce their capital expenses and eliminate the need for costly infrastructure maintenance [15].

4 Integration of Smart Farming and Cloud Computing

The implementation of cloud computing in smart farming encompasses the amalgamation of diverse technologies and components with the aim of enhancing agricultural methodologies, augmenting productivity, and mitigating resource utilization. The key components for cloud-enabled smart farming [16, 17] are listed in Table 2.

4.1 Sensors

To transform conventional agricultural practices into data-driven, highly efficient operations, the use of sensors and Internet of Things (IoT) devices is crucial in smart farming.

Table 2 Key components of smart farming with cloud integration

Sensors	*Communication*
Soil moisture sensors Weather stations GPS-enabled tractors and machinery Livestock monitoring devices Crop health sensors (camera) Water quality sensors Smart irrigation systems	Wireless networks (Wi-Fi, LoRaWAN, NB-IoT) Edge computing devices for local data processing Data aggregators and gateways Cloud-connected IoT platforms for data transmission
Cloud infrastructure	*Data analytics*
Public or private cloud platforms (AWS, Azure, Google Cloud) Cloud storage for data storage and retrieval Scalable computing resources for data analysis Data security and encryption measures	Big Data analytics tools for processing large datasets Machine learning models for predictive analytics AI algorithms for crop disease detection and yield prediction [18] Data visualization tools for farmers and stakeholders
Decision support systems	*Mobile application*
Crop management software Weather forecasting tools Pest and disease prediction models Irrigation scheduling algorithms Crop rotation and planning tools	Mobile apps for farmers to access data and insights Remote monitoring and control of farm equipment Notifications and alerts for critical events
Drones and remote sensing	*Blockchain for supply chain transparency*
UAVs (unmanned aerial vehicles) for aerial imagery Satellite data for monitoring large agricultural areas Remote sensing technologies for crop health assessment	Blockchain technology to track the origin and journey of agricultural products Smart contracts for automated payments and compliance

- Soil moisture sensors optimize irrigation to meet crop needs, save waste, and boost yields [19].
- Weather stations assist farmers plan planting and harvesting dates and reduce weather risks using real-time weather data.
- GPS-enabled tractors and machines automate agricultural work and reduce fuel use.
- Tracking livestock health and behavior improves animal welfare, while crop health sensors like NDVI cameras help spot illnesses and manage nutrients.
- Water quality sensors promote sustainability by ensuring the purity of irrigation and drinking water for both plants and animals.

These technologies provide farmers with data and control to optimize resource allocation and productivity in modern agriculture.

4.2 Communication

Cloud computing is seamlessly integrated with data collection and connectivity to revolutionize modern smart agricultural practices.

- Wireless networks including Wi-Fi, LoRaWAN, and NB-IoT connect agricultural sensors and devices for data transfer [20].
- Edge computing devices enable real-time analysis and decision-making on the farm, lowering latency and enabling quick reactions to changing conditions.
- Intermediaries like data aggregators and gateways collect and prepare data for transmission.
- Cloud-connected IoT platforms deliver this important data to cloud servers for storage, analysis, and actionable insights.

This networked ecosystem allows farmers to access real-time data, optimize resource allocation [21], and make informed decisions to improve crop yields, decrease resource waste, and promote sustainable agriculture.

4.3 Cloud Infrastructure

Smart farming relies on cloud computing infrastructure to use data to improve agricultural practices. The enormous processing capabilities of the cloud enable the aggregation, cleansing, and transformation of diverse data sources into actionable information. Cloud computing services play a significant role in enabling smart farming.

4.4 Infrastructure as a Service (IaaS)

IaaS can offer the fundamental cloud infrastructure required for a range of agricultural applications in smart farming. This covers the supply of networking, storage, and virtual machines. IaaS can be used by farmers to set up and maintain edge computing devices, sensor networks, and data storage options. These tools are adaptable to seasonal variations and data expansion, allowing for adjustments as needed. In addition, IaaS can facilitate the hosting of vital elements such as data processing clusters and IoT gateways, which are necessary for gathering and processing data from sensors and other agricultural equipment [22].

Amazon Web Services (AWS) is a well-known infrastructure as a service (IaaS) platform that can be utilized in smart farming. AWS offers a wide range of cloud computing services and resources that are appropriate for different agricultural applications.

4.5 Platform as a Service (PaaS)

Custom agriculture apps and analytics tools can be developed and deployed more quickly with the help of PaaS services. Applications that process agricultural data can be created and hosted using PaaS by farmers and agricultural software developers. Examples include data analytics, predictive modeling, and decision support. PaaS solutions also improve stakeholder collaboration by providing a common platform for the development and sharing of farm-related software applications.

Google Cloud Platform (GCP) is one of the top platform as a service (PaaS) systems that can be utilized in smart farming. GCP provides a wide range of tools, services, and development platforms that can help simplify agricultural application development, deployment, and management [23].

4.6 Software as a Service (SaaS)

Farmers, agricultural consultants, and other stakeholders can easily manage farm operations and obtain vital data with the use of SaaS applications. Specialized farm management software, tools for weather forecasting, models for predicting pests and diseases, and more are available through a number of cloud-based SaaS providers. These SaaS applications frequently have user-friendly interfaces that allow farmers to access data, generate reports, and make decisions without extensive technical knowledge.

One of the well-known software as a service (SaaS) platforms in the field of smart farming is "FarmLogs." FarmLogs offers a cloud-based SaaS solution designed to assist farmers in managing their agricultural operations efficiently [24].

4.7 Data Analytics

Agricultural practices have undergone a radical transformation due to the convergence of Big Data analytics, machine learning, AI algorithms, and data visualization tools, which are enabled by cloud computing.

- Cloud-enabled smart farming utilizes the vast capabilities of Big Data analytics tools to analyses large datasets, hence facilitating the development of machine learning models [25].
- Cloud-based machine learning models predict crop yields, insect infestations, and optimal planting times by utilizing historical and real-time data. These forecasts are crucial for planning and resource allocation.
- Cloud-based AI algorithms are capable of analyzing images captured by sensors or drones to identify indicators of crop diseases. This early detection enables producers to safeguard their crops and maximize yields through proactive measures.

- This convergence, along with data visualization tools, empowers farmers and stakeholders to make educated decisions, visualize agricultural patterns, and improve modern farming practices, increasing productivity and sustainability [26].

4.8 Decision Support System

Cloud technology in smart farming has revolutionized agriculture by providing decision support systems (DSS) like crop management software, weather forecasting tools, pest and disease prediction models, irrigation scheduling algorithms, and crop rotation and planning tools. These cloud-based DSS give farmers real-time knowledge to optimize planting plans, manage resources, and respond quickly to weather issues. Farmers can improve crop yields, reduce pests and illnesses, and encourage sustainable practices using data-driven decisions, creating a more resilient and productive agricultural [27].

4.9 Mobile Application

Smart farming in the cloud relies heavily on the incorporation of mobile applications to provide producers with real-time connectivity and control. These mobile applications enable producers to make informed decisions on the go by facilitating seamless access to vital data and insights [28]. In addition, they enable remote monitoring and control of agricultural machinery, thereby enhancing operational efficiency and decreasing the necessity for physical presence in the field. In addition, these applications provide punctual alerts and notifications for critical events, enabling farmers to promptly adapt to shifting circumstances, thereby promoting improved resource efficiency, resilience, and productivity in contemporary agricultural practices.

4.10 Drones and Remote Sensing

The agricultural sector can utilize drones for a variety of purposes, including data collection regarding crop health, weather report, etc., crop mapping, soil analysis, irrigation, and insect management [29]. The use of drones in agriculture offers the following main advantages:

- Increased efficiency: Drones can quickly cover large amounts of land, helping farmers gather data and monitor crops. Early detection allows for faster and more effective interventions.

- Increased crop yields: The use of drones to collect data on crop health enables producers to pinpoint areas that need attention. By effectively tackling these concerns, agricultural practitioners have the potential to enhance their crop productivity and augment their financial gains.
- Cost reduction: By identifying problem areas on the farm, drones can reduce the consumption of pesticides and other chemicals, as well as the need for manual labor.
- Enhanced accuracy: Drones can take high-resolution photos and data of crops for farmers. This can identify problem areas and guarantee targeted and effective solutions.

4.11 Blockchain for Supply Chain Transparency

The combination of cloud and smart farming with blockchain and smart contracts transforms agricultural supply chain management. Blockchain methodically tracks agricultural products from farm to consumer, ensuring transparency and traceability. Cloud-based smart contracts automate compliance and payments, thereby expediting transactions and reducing administrative burden. This innovative combination fosters a new era of secure and transparent agricultural commerce by enhancing efficiency, reducing fraud, and enabling real-time visibility into product provenance and quality, thereby bolstering confidence in the supply chain [30].

4.12 Benefits of Cloud Computing in Smart Farming

- The cloud enables farmers to dynamically increase or decrease their access to computational and storage resources.
- Farmers can derive valuable insights from their data by utilizing cloud-based data analytics applications.
- Cloud-based solutions enable producers to oversee their farms and equipment remotely.
- Cloud-based applications offer a centralized platform that enables seamless collaboration among multiple users from diverse organizations, irrespective of their technical proficiencies. As a consequence, collaboration is enhanced, and decision-making is expedited.
- Real-time data helps farmers and agronomists make quick changes [31].
- Data and applications hosted in the cloud are accessible from a wide range of devices, such as smartphones and tablets.
- Cloud-based AI and machine learning can create crop management forecasting models [32].

5 Key Challenges

Even though cloud computing presents smart farming with a number of advantages, it also presents a number of challenges.

Expenses

- Expenses constitute a significant challenge for farmers. This requires the purchase of technological devices like sensors, drones, and IoT devices. In addition to the initial investment, routine maintenance may be necessary, which can be costly.
- Overestimating resources can increase expenses, while underestimating can lower performance.

Solutions

- Calculate the ROI by comparing technology procurement, installation, and maintenance expenses to predicted gains.
- Sharing the expense of drones, sensors, and irrigation systems with other farmers.
- Investigate grants, programs, and subsidies from the government that are intended to foster smart agricultural initiatives.
- Effectively manage costs by closely monitoring resource utilization, implementing cloud cost management tools, and adopting a pay-as-you-go model.

Infrastructure

- The implementation of smart farming in rural locations necessitates a high-performance infrastructure, given its reliance on the Internet. Internet access in remote farms may be unreliable.

Solution

- Reduce reliance on continual internet access by utilizing satellite Internet, long-range wireless technologies, or local edge computing solutions.

Training

- Farmers in agriculture may lack the abilities to employ cloud-based technologies. Smart agriculture produces vast quantities of data. Farmers therefore require the capacity to efficiently gather and analyze data.

Solution

- Training and support farm staff to close the skill gap and maximize cloud-based technologies and applications.

Data Security and Privacy

- Preventing unauthorized access to sensitive farm information and assuring adherence to data privacy regulations. Smart farming generates vast quantities of data from a variety of sources, including sensors, devices, and the environment. Those data are vulnerable. Leaking agriculture anti-jamming devices information can let an attacker evade these security measures, while leaking soil, crop, and agriculture purchasing information might cause farms serious economic losses if utilized by competitors or hostile parties [33]. Therefore, ensuring data security and privacy is a fundamental necessity and a key goal in establishing a dependable functioning smart agricultural ecosystem.

Solution

- Strong encryption, access limits, and authentication. Select cloud service providers that offer robust security measures and guarantee adherence to data protection regulations.

6 Cloud-Enabled Platforms for Smart Farming

- FarmLogs is a cloud-based farm management application that helps farmers make data-driven decisions for precision agriculture. It helps farmers optimize operations and boost productivity using field mapping, crop health monitoring, and financial tracking [34].
- The Bushel Farm app is a multifunctional and intuitive mobile application designed specifically for contemporary farmers who employ smart farming practices. This application enhances the profitability and efficacy of agricultural operations by equipping users with critical tools for grain management, pricing, and marketing [35].
- Agrivi, a cloud-based farm management platform, supplies farmers with crop management, inventory monitoring, financial analysis, and decision assistance tools. Agrivi uses real-time data and analytics to help farmers choose planting, harvesting, resource allocation, and pest control [36].
- A powerful cloud-based technology, John Deere Operations Center, helps farmers use data analytics and precision agriculture. It has many features, including data collecting and analysis, field and equipment management, and connection with connected John Deere equipment. Farmers are able to assess the performance of their machinery, monitor real-time field conditions, and make data-driven decisions regarding planting, harvesting, and resource management using the platform [37, 38].
- The IBM Watson Decision Platform for Agriculture provides producers with weather forecasts, soil data, and crop insights through the use of artificial intelligence (AI) and cloud computing. Farmers can plant, irrigate, and harvest more wisely [39].

7 Related Research Works

Using Internet of Things (IoT) and long range (LoRa) technologies, Saban et al. [40] created a customized smart farming system that makes advantage of a low-cost, low-power, and wide-range wireless sensor network. Their system incorporated a recently developed web-based monitoring application that is hosted on a cloud server. This application enables remote visualization and control of all connected devices by processing data collected from the agricultural environment. Moysiadis et al. [41] proposed a smart farming application built on cloud computing. Agronomists and farmers can utilize this information to enhance their decision-making when using measurements from ground sensors and photos taken by UAVs or ground cameras. Docker containers are used as the virtualization technology in an implementation based on the microservices architecture. Rajak et al. [42] proposed a plant health monitoring system to solve several issues pertaining to both farmers and plants. These monitoring tools help the farmer make sure the plant stays healthy in a particular environment and conserves water.

8 Future Trends and Research Opportunities

The agriculture industry is experiencing huge changes and facing lot of problems like climate change, growing population, more demand, food quality, and availability of agro items. In order to overcome all these problems, a smart farm with smart application for the agriculture industry is needed. The integration of digital technologies into agriculture needs to be opened for new opportunities. This brings revolution in the farming to manage crops and resources. Smart farming and precision agriculture continue to evolve, offering numerous opportunities for research and innovation.

- Research AI-driven techniques for early detection of diseases, pests, and nutrient deficiencies in crops using image analysis and sensor data.
- Investigate the use of IoT and wireless sensor networks to enhance data collection, transmission, and integration on farms.
- Research drought-resistant crop varieties and climate-smart farming techniques.
- Explore AI-driven pest control methods that use data analytics to identify pest outbreaks and deploy targeted interventions.

9 Conclusion

Cloud computing has emerged as a transformative force in smart farming, revolutionizing traditional agricultural practices by providing scalable, data-driven solutions that enhance productivity, sustainability, and efficiency. From data storage and analytics to precision resource management and remote monitoring, the cloud empowers farmers with real-time insights, enabling them to make informed decisions and optimize their operations. This technology not only improves crop yields and resource utilization but also fosters collaboration, traceability, and environmental stewardship, heralding a promising future for agriculture in the digital age.

References

1. Web resource: https://smartertechnologies.com/guides/the-complete-guide-to-smart-agriculture-farming/#chapter-1. Last access: 2/10/2023
2. Web resource: https://www.iotforall.com/smart-farming-future-of-agriculture. Last access: 2/10/2023
3. V. Mekala, A survey: Smart agriculture IoT with cloud computing, 1–7 (2017). https://doi.org/10.1109/ICMDCS.2017.8211551
4. A. Akshay, Kabir, Applications of IoT for smart agriculture or farming. Int. J. Res. Anal. Rev **6**(2), 537 (2019)
5. R. Onkar, S. Aditi, Sushma, Smart farming. Int. J. Res. Appl. Sci. Eng. Technol **10**(4) (2022). https://doi.org/10.22214/ijraset.2022.43691
6. N. Prasad, D. Naveen, Dharani, Cloud computing in agriculture - an affordable way to achieve smart farming. Int. J. Comput. Sci. Commun **12**(2), 52 (2021)
7. U. Rekha, The impact of cloud computing & ICT in agricultural productivity. J. Pharma. Negat. Results (2022). https://doi.org/10.47750/pnr.2022.13.S07.680
8. K. Ahamed, P. Niranjanaa, Rekha, Cloud computing in smart farming. Int. Res. J. Eng. Technol **7**(8), 200 (2020)
9. S. Vaishnavi, A. Sumithra, Potential of precision farming technologies for eco-friendly agriculture. Int. J. Plant Soil **35**(19) (2023). https://doi.org/10.9734/IJPSS/2023/v35i193528
10. Web resource: https://www.techtarget.com/searchcloudcomputing/definition/cloud-computing. Last access: 2/10/2023
11. T. Peter, *The NIST Definition of Cloud Computing* (Special Publication, 2011), pp. 800–145
12. U. Deepali, G. Anita, PaaS providers and their offerings. Int. J. Sci. Technol. Res. **9**(02), 4009 (2020)
13. M. Kondraju, Software as a Service for Efficient Cloud Computing. Int. J. Res. Eng. Technol **03**, 178–181 (2014). https://doi.org/10.15623/ijret.2014.0301028
14. V. Lewis, X. Qianwen, Ben, Adoption of cloud computing as innovation in the organization. Int. J. Eng. Bus. Manag **14**. https://doi.org/10.1177/18479790221093992
15. J. Kenji, John, Diffusing the cloud: Cloud computing and implications for public policy. J. Ind. Compet. Trade **11**, 209–237. SpringerLink (2011). https://doi.org/10.1007/s10842-011-0106-5
16. Lola, A study literature of critical success factors of cloud computing. Proc. Comput. Sci **59** (2015). https://doi.org/10.1016/j.procs.2015.07.548
17. S. Rajani, IoT based smart farming. Int. J. Res. Appl. Sci. Eng. Technol **11**(4) (2023). https://doi.org/10.22214/ijraset.2023.54057
18. R. Kalpesh, R. Roshan, R. Rajvardhan, Rushikesh, Smart farming using IOT. Int. J. Res. Appl. Sci. Eng. Technol **11**(4). https://doi.org/10.22214/ijraset.2023.53363
19. Nura, Toward adoption of drip irrigation and soil-moisture sensors by small scale farmers. FUDMA J. Sci **6**(3), 259–270 (2022). https://doi.org/10.33003/fjs-2022-0603-961
20. J. Mohamed, M. Marcos, N. Gonzalo, Rodrigo, LoRa based IoT platform for remote monitoring of large-scale agriculture farms in Chile. Sensors **22**(8), 2824 (2022). https://doi.org/10.3390/s22082824
21. G. Fiedrich, Rickers, Optimized resource allocation for emergency response after earthquake disasters. Safety Sci. Elsevier. ScienceDirect **35**(1–3) (2000). https://doi.org/10.1016/S0925-7535(00)00021-7
22. S. Tamoghna, Raghuwanshi, Sensing-cloud: Leveraging the benefits for agricultural applications. Comput. Electron. Agric. **135**(1), 96–107 (2017). https://doi.org/10.1016/j.compag.2017.01.026
23. D. Bharath, Exploration of research on Internet of Things enabled smart agriculture. Sci. Direct **80**(3), 1936–1939 (2023). https://doi.org/10.1016/j.matpr.2021.05.652
24. M. George, C. Florin, C. Mariana, Aniello, Advanced services for efficient management. J. Parallel Distrib. Comput **116**, 3–17 (2018). https://doi.org/10.1016/j.jpdc.2017.10.017

25. M. Ashay, R. Praveen, Turki, Intelligent data analytics framework for precision farming using IoT and regressor machine learning algorithms. Appl. Sci **12**, 9992 (2022). https://doi.org/10.3390/app12199992
26. M. Mohammad, M. Samira, Rozita, A platform approach to smart farm information processing. Agriculture **12**, 838 (2022). https://doi.org/10.3390/agriculture12060838
27. T. Girma, H. David, N. Benjamin, A. Gherardo, T. Vitalijs, Abozar, A concept for application of integrated digital technologies to enhance future smart agricultural systems. Smart Agric. Technol **5**, 100255 (2023). https://doi.org/10.1016/j.atech.2023.100255
28. T. Khongdet, Mohammad, Applicability of Internet of Things in smart farming. J. Food Qual. **2022** (2022). https://doi.org/10.1155/2022/7692922
29. E. Kenji, S. Yasuhiro, Yoichi, Plant functional remote sensing and smart farming applications. Int. J. Agric. Biol. Eng **15**(4) (2022). https://doi.org/10.25165/j.ijabe.20221504.7375
30. V. Rajasekhar, T. Venkata, Vinayakumar, Blockchain-based cloud-enabled security monitoring using Internet of Things in smart agriculture. Future Internet **14**, 250 (2022). https://doi.org/10.3390/fi14090250
31. S. Kenny, A. Puja, V. Siddhi, S. Bhaskar, Viable smart sensors and their application in data driven agriculture. Comput. Electron. Agric. **198**, 107096 (2022). https://doi.org/10.1016/j.compag.2022.107096
32. Z. Elbasi, A. Topcu, C. Zreikat, S. Shdefat, Crop prediction model using machine learning algorithms. Appl. Sci **13**, 9288 (2023). https://doi.org/10.3390/app13169288
33. A.A. Jens, K. Sofia, C. Håkan, I. Karin, U. Jonathan, Future threats to agricultural food production posed by environmental degradation, climate change, and animal and plant diseases – A risk analysis in three economic and climate settings. Food Secur **6**, 201–215, Springer (2014). https://doi.org/10.1007/s12571-014-0331-y
34. Web resource: https://u.osu.edu/agsoftwarelibrary/2018/03/22/farmlogs/. Last Access 12/10/2023
35. Web resource: https://bushelfarm.com/. Last Access 12/10/2023
36. https://www.agrivi.com/. Last Access 12/10/2023
37. https://www.deere.com/en/technology-products/precision-ag-technology/data-management/operations-center/. Last Access 12/10/2023
38. Web resource: https://www.deere.com/en/technology-products/precision-ag-technology/operations-center/. Last Access 12/10/2023
39. Web resource: https://www.ibm.com/downloads/cas/PRJL6ZW4. Last Access 12/10/2023
40. B. Saban, G. Amdaouch, C. Ahmed, R. Alzola, Aghzout, A smart agricultural system based on PLC and a cloud computing web application using LoRa and LoRaWan. Sensors **23**, 2725 (2023). https://doi.org/10.3390/s23052725
41. T. Moysiadis, P. Sarigiannidis, Boursianis, A cloud computing web-based application for smart farming based on microservices architecture. 11th International Conference on Modern Circuits and Systems Technologies (MOCAST) (2022). https://doi.org/10.1109/MOCAST54814.2022.9837727
42. Rajak, Emerging technological methods for effective farming by cloud computing and IoT. Emerg. Sci. J **6**(5), 1017–1031 (2022)

Index

B. Sundaravadivazhagan et al. (eds.), *Intelligent Robots and Drones for Precision Agriculture*, Signals and Communication Technology, https://doi.org/10.1007/978-3-031-51195-0

The manufacturer's authorised representative in the EU is Springer
Nature Customer Service Centre GmbH, Europaplatz 3, 69115 Heidelberg,
Germany. If you have any concerns regarding our products, please
contact ProductSafety@springernature.com

Printed and bound by CPI Group (UK) Ltd, Croydon, CR0 4YY
30/09/2026
02225573-0015